Springer Series in Materials Science

Volume 355

The Springer Series in Materials Science covers the complete spectrum of materials research and technology, including fundamental principles, physical properties, materials theory and design. Recognizing the increasing importance of materials science in future device technologies, the book titles in this series reflect the state-of-the-art in understanding and controlling the structure and properties of all important classes of materials.

Mehran Roohnia

Properties and Treatment of Tonewood for String Instrument Construction

A Professional Guide

Springer

Mehran Roohnia
Turkish Handcraft Arts Association
Konya, Türkiye

ISSN 0933-033X ISSN 2196-2812 (electronic)
Springer Series in Materials Science
ISBN 978-3-032-00823-7 ISBN 978-3-032-00824-4 (eBook)
https://doi.org/10.1007/978-3-032-00824-4

This Springer imprint is published by the registered company Springer Nature Switzerland AG
The registered company address is: Gewerbestrasse 11, 6330 Cham, Switzerland

To my kind wife Negin and my cute daughter Rojina

Foreword

As three professional luthiers with diverse specializations—historical Persian instruments, classical and flamenco guitars, and violins—we reflect the broader community of artisans who have benefited from the author's instruction. For generations, craftsmen like ourselves have transformed wood into instruments that resonate with musicality. This journey has consistently presented challenges related to the selection, treatment, tuning, and preservation of suitable timber. For those of us whose expertise is rooted more deeply in workshop experience than in academic curricula, access to practical, scientifically grounded knowledge has been consistently difficult. This context underscores the exceptional importance of the present volume.

Dr. Mehran Roohnia's Properties and Treatment of Tonewood for String Instrument Construction—A Professional Guide is more than a technical reference for researchers; it is a foundational bridge between academic science and the applied art of instrument construction. This book effectively translates complex scientific principles into an accessible and practical guide for luthiers, without sacrificing scientific rigor.

We have each had the distinct privilege of receiving direct instruction, both theoretical and practical, from Dr. Roohnia. His research and teaching have fundamentally influenced our methodologies for wood selection and preparation. From the precise acoustic characterization of wood samples using sophisticated instrumentation—including tools we learned to construct ourselves—to the understanding of orthotropic wood properties, spectral analysis via FFT, and the precise tuning of resonances, these concepts have advanced our practices from purely empirical approaches to optimized, scientifically informed methods. A significant proportion of our current technical decisions regarding wood selection are directly informed by the principles taught by Dr. Roohnia, principles that, until now, were primarily accessible to us within the Middle Eastern instrument-making community and are now, through this publication, available to luthiers worldwide.

The distinction of this book lies not only in its scientific content but also in its remarkable clarity and directness. Many academic texts in this domain are often couched in language that is challenging for empirically trained luthiers. In contrast, this work maintains its scientific precision while adopting a language accessible and relevant to all instrument makers, irrespective of their background. It is a rare achievement to present such profound subject matter with such clarity, engaging both intellectual understanding and practical application. Furthermore, the author's consistent responsiveness to questions from luthiers via email, LinkedIn, and other platforms is truly exceptional.

- My professional life, as Siamak Ghalami (the main author of this foreword), has been dedicated to the hereditary tradition of reviving and constructing ancient Persian instruments such as the Ghaychak and Robab. The insights within this book have enabled me to integrate rigorous scientific methodology with established ancient traditions, thereby advancing the international recognition of these historical instruments.
- Ramin Fallah, the second contributor to this foreword, is renowned for crafting professional classical and flamenco guitars that have achieved significant acclaim among musicians in our region and across Europe and North America. The exceptional acoustic precision of his instruments is largely attributable to the precise criteria for wood selection, treatment, and adjustment detailed in this publication.
- Mikaeil Ghazal, the youngest co-author of this foreword, has, even in his third decade of life, established himself as a respected figure in violin making. This rapid ascent is due to his early and consistent integration of these scientific teachings with practical workshop methodologies.

Based on our collective experience, we unreservedly recommend this book to all luthiers globally, whether their focus is violins, guitars, or other traditional string instruments. If you have previously perceived a disconnection between scientific principles and practical application in the selection of wood for musical instruments, this book effectively bridges that gap. For those already incorporating acoustic or engineering principles, this volume will undoubtedly augment your existing analytical toolkit.

We consider it an honor to have studied under the esteemed author and are proud to contribute this foreword in recognition of a work that is poised to significantly shape the future of wood science and the art of instrument making.

Tehran, Iran

Siamak Ghalami
Luthier—Specialist in Historical
Persian Instruments (Ghaychak,
Robab)

Istanbul, Türkiye

Ramin Fallah
Classical and Flamenco Guitar Maker
info@raminguitars.com

Mahabad, Kurdistan Region, Iran

Mikaeil Ghazal
Violin Maker—Acoustic Optimization
Specialist

Preface

Wood has resonated with humanity for millennia. Long before the age of steel, silicon, and synthetic polymers, humans shaped nature's most organic offering—wood—into tools of expression. Among these tools, musical instruments have held a particularly sacred place. The warmth of a bowed string, the gentle resonance of a plucked tone, and the clarity of a hammered note—all owe their character not merely to design, but to the innate properties of the wood from which they are made. The story of tonewood is, in essence, the story of music itself.

This book, *Properties and Treatment of Tonewood for String Instrument Construction—A Professional Guide*, is the conclusion of nearly three decades of my interdisciplinary research at the crossroads of wood science, acoustics, luthiery, and ethnomusicology. Its genesis lies in both professional necessity and intellectual curiosity. As a researcher of material sciences who has worked closely with luthiers and musicians, I observed a growing gap between traditional knowledge and modern scientific understanding. In an age where digital technologies can simulate sounds and AI can compose symphonies, it is more critical than ever to preserve, analyze, and advance our knowledge of the organic materials that continue to define the soul of acoustic music.

Tonewood, as a term, has often been used with a mixture of reverence and ambiguity. Instrument makers instinctively understand its value; scientists aim to quantify it. Yet, it remains one of the few materials in modern craft whose evaluation often depends on a knock, a feel, or a listening ear. This book aims to bridge the art and the science. It is designed not just for researchers or academics in wood science, but also for practical instrument makers, conservators, restoration specialists, musicians, and even policymakers concerned with the sustainable use of natural resources in cultural industries.

In the course of this work, special attention has been given to the orthotropic nature of wood—a key to its acoustic behavior. I have attempted to clarify concepts such as stiffness anisotropy, damping, longitudinal modulus, and internal friction, not through oversimplification, but through careful explanation, supported by diagrams, case studies, and reference data. Just as importantly, the book ventures into the treatment methods that have historically enhanced tonewood properties—from natural seasoning and thermal modification to chemical applications and varnish techniques used by master luthiers like Stradivari and Guarneri.

This book is divided into thematic chapters that trace both the scientific underpinnings and historical applications of wood in string instrument construction. The initial chapters provide a historical overview, highlighting the centrality of wood in traditional instrument making from various cultures. From Persian barbats and Chinese pipas to Baroque violins and African koras. It is critical to remember that tonewood is not solely the domain of European classical instruments—it is a global phenomenon, shaped by local ecosystems and cultural traditions.

Subsequent chapters go into wood anatomy and identification, providing technical classifications of hardwoods and softwoods, cell structure, pore distribution, and the mechanical properties that determine tone. Later sections investigate acoustic properties and how they vary across species and processing methods. These are mainly informed by my own laboratory findings and fieldwork, and cross-referenced with recent peer-reviewed literature and international standards.

Beyond material properties, the book includes chapters on wood treatment methods. This encompasses both traditional empirical practices and modern engineering techniques. Methods such as thermal, non-chemical, and chemical treatments, along with the application of natural additives and varnishes, are described with attention to their scientific basis and acoustic outcomes. The reader will find ideas about the resurgence of neglected regional wood species that, with proper treatment and understanding, may substitute for endangered imports like rosewood or ebony.

Although the language of this book is primarily scientific and its structure methodical, it does not shy away from cultural narratives. Much of what we know and treasure about the acoustic character of wood comes from artisans, oral traditions, and regional philosophies of sound. For this reason, significant effort has been made to include ethnographic details about traditional instruments from Persia, India, Central Asia, Africa, the Middle East, Europe, and Indigenous American cultures. These traditions provide invaluable insights into how different societies interpreted and optimized their local woods for musical purposes.

This book would not have been possible without the collective wisdom of many colleagues. I am deeply indebted to master luthiers in Iran, Europe, and Türkiye who generously shared their knowledge, experience, and methods, teaching me to "hear" the wood as much as I could measure it.

I extend my gratitude to:

- **Professional Oud and/or Guitar Makers:** Mr. Ramin Fallah, Mr. Soroush Javadi Ashrafi, Mr. Mohammad Motiefar, Mr. Giuliano Nicoletti and Mr. Sina Asadi.
- **Experienced Luthiers of Persian Setar and Tar or Other Persian Instruments:** Mr. Mahdi Karimi Haghighi, Mr. Hamid Mahmoodi, Mr. Saeed Peymani, Mr. Hamidreza Samadi, Mr. Siamak Ghalami, Mr. Ehsan Ghassemi, and Mr. Abbas Salehi.

- **Renowned Violin Makers:** Mr. Mikaeil Ghazal, Mr. Ali Chavoshi, Mr. Masoud Ghafouri, Mr. Uluer Vardaloğlu, and Mr. Mohammad Norouzi, who have tested and continue to test my academic theories on their tonewood and instruments.
- **Tonewood supplier:** Mr. Sina Faramarzi

I also wish to thank my academic teachers in forestry and materials science, especially those who initially introduced me to the concept of acoustics of wood. Particular thanks goes to my French Ph.D. advisor, Dr. Joseph Gril, Research Professor of CNRS at the LMGC, University of Montpellier II, and my sabbatical host at the CIRAD Research Institute in Montpellier, France, Dr. Loïc Brancheriau, from whom I received invaluable training and assistance in working with wood acoustics. I am also grateful to my Ph.D. supervisor, Pr. Kazem Doost Hosseini from the University of Tehran, Iran, as well as the Turkish academic researcher and master luthier, Dr. Emir Değirmenli from Ankara Hacı Bayram Veli University, Turkish Music State Conservatory, whose support on double-top subject has been a great benefit.

I owe the techniques of instrumentation and the construction of measurement and evaluation tools for wood to the Hungarian professor, Dr. Ferenc Divos. His extensive teaching in the workshops of FAKOPP enterprise and the University of West Hungary, Sopron, along with numerous consulting sessions, has been and continues to be a profound inspiration throughout my journey. I would also like to express my gratitude to my esteemed academic colleague, Mr. Behzad Behnam, and his electronics team for their invaluable contribution in designing and constructing the laboratory equipment essential for my research.

I particularly thank the musicians who have shared their passion and taught me music, especially Mr. Behzad Mirkhani, Mr. Vaheed Roohnia, and Mr. Alireza Tafaghodi, masters of the classical guitar in Iran. They consistently reminded me that the ultimate goal of all these analyses is not theoretical perfection, but emotional connection.

I am grateful to AI services and their providers for their assistance in editing the grammar, spelling, and fluency of my written texts. Furthermore, AI has been instrumental in enhancing the quality of provided images and transforming my hand-drawn sketches into more aesthetically pleasing and expressive versions. I acknowledge this valuable contribution here and will specifically remind AI services in the caption of each relevant image.

Finally, my sincerest gratitude goes to the Turkish Handcraft Arts Association, which supports my scientific endeavors as a newly immigrated individual in Türkiye.

To the listeners, the audience, music students, and craftspeople who strive to achieve the perfect tone from musical instruments: this work is for you.

This book is not intended as a static manual, but as a living document of evolving practices. Readers are encouraged to approach it critically, to test its recommendations, and to expand upon its findings. Whether you are analyzing the cellular microstructure of spruce under a microscope, or selecting a well-seasoned mulberry blank by feel and instinct, your work matters to me. Every luthier and every researcher adds to the collective knowledge of our craft.

We live in a time where tradition meets innovation at every turn. Preserving the authenticity of tonewood does not mean resisting progress—it means understanding the essence of what gives music its breath and resonance. If this book contributes even a little to that understanding, it will have fulfilled its purpose.

Konya, Türkiye Mehran Roohnia

Competing Interests The author has no competing interests to declare that are relevant to the content of this manuscript.

Contents

About the Author

Mehran Roohnia was born in September 1971 in Tehran, into an Azeri Turkish family. He has an older brother who's a physician and a younger sister who's an accountant. His wife is a forestry engineer and his lonely daughter is still a middle-school student. The world of wood captivated him from a very young age. His maternal grandfather was a hereditary carpenter, and during summer school breaks, he would immerse himself in his workshop with incredible passion. He learned the invaluable crafts of sawing, gluing, joinery techniques, and the names of various wood species from him in those early years.

Academic Pursuits and Research

After graduating from Alborz High School in Tehran, a beacon of educational excellence in Iran at the time, he entered the University of Tehran in 1990 to study Wood Science. In 1995, he pursued his master's degree in Wood Science at the Gorgan University of Agricultural Sciences and Natural Resources. His M.Sc. thesis, focused on the anatomy of wood and agricultural wood waste, was completed under the guidance of esteemed Professors, Prof. Seyed Zia-al-Din Hosseini and Prof. Ahmad Jahan Latibari.

Following a two-year hiatus for mandatory military service, he was accepted into the Ph.D. program in Wood Science at Islamic Azad University, Science and Research Branch, in 2000. His doctoral proposal, centered on the acoustic properties and damping capacity of mulberry and cypress wood, was approved with the guidance of Dr. Bernard Thibaut from the University of Montpellier 2, France, and Prof. Kazem Doosthosseini from the University of Tehran. Upon Dr. Thibaut's retirement, Prof. Kazem Doosthosseini took over as his thesis supervisor, with Dr. Joseph Gril, a research professor at CNRS France, serving as his advisor. He also spent time as a visiting Ph.D. student at the University of Montpellier 2, France, where he not only received thesis consultation but also utilized the university's advanced equipment for his acoustic and vibration tests.

Career Path and Scientific Endeavors

Starting in February 2003, he began working as a scholarship instructor at Islamic Azad University. After defending his doctoral dissertation in May 2005 and a review of his educational and research portfolio, he was officially employed as an assistant professor at the university. He completed his 10-month sabbatical leave, during 2012 and 2013, focusing on wood vibration and acoustic assessments, under the supervision of Dr. Loïc Brancheriau at the CIRAD research institute in Montpellier, France. Over the years, through numerous research projects and the publication of their findings in prestigious reputable journals, he was promoted to an associate professor in 2011 and then to full professor in 2015.

Entrepreneurship and Relocation

In 2016, he founded Darkoob NDT (instagram: @darkoob.ndt), a research company specializing in the acoustic assessment of wood and standing trees, within the Science and Technology Park of Islamic Azad University. The company was recognized as a knowledge-based company by the Vice-Presidency for Science and Technology of the Iranian Presidency in the same year.

Since 2021, he has been on extended leave from his university and relocated with his family to Türkiye. There, the Turkish Handcraft Arts Association has hosted and supported his scientific endeavors. This environment, free from conventional work concerns, allows him to calmly dedicate himself to writing and authorship. Many of the research projects he has conducted over the years have not yet been converted into articles, and he is currently focused on analyzing their data. Although he has transferred the responsibilities of Darkoob NDT to a partner company in Iran, he continues to provide online, remote consulting services.

Chapter 1
The Importance and History of Wood for Musical Instruments

Abstract This chapter explores wood as the primary material in the construction of musical instruments, emphasizing its enduring relevance through history. Wood's distinct acoustic properties, workability, and aesthetic appeal have made it irreplaceable in various instruments for centuries. The chapter highlights wood's orthotropic nature, which contributes to its superior acoustic qualities, and the versatility offered by different species of wood, enabling instrument makers to create specific tonal effects. The historical overview traces the evolution of wooden instruments, from the primitive tools of early civilizations to the intricate instruments of the Baroque and Classical periods. It discusses how luthiers like Antonio Stradivari revolutionized instrument making through innovative wood treatments and craftsmanship, further enhanced by environmental factors such as the Maunder Minimum, which influenced wood properties. The chapter also describes wood's significance in global musical traditions, from the lutes of Central Asia and the barbat or oud in the Persian-Turk-Arab world to the stringed instruments of Africa and the Americas. Despite modern technological advancements, wood remains the preferred material for many acoustic instruments today, attesting to its unmatched acoustic and aesthetic value. The chapter concludes by recognizing wood's continued relevance in modern music, while exploring the potential for alternative materials in the future.

1.1 The Importance of Wood in Musical Instruments Industry and the Reasons for Choosing It as the Main Material

Wood has been a primary material in musical instrument construction for centuries. This preference is rooted in its distinctive acoustic properties, workability, aesthetic appeal, and accessibility.

- **Acoustic properties** Wood exhibits exceptional acoustic characteristics. In this book, we will discuss this aspect in detail. Wood's unique combination of physical and mechanical properties makes it an irreplaceable material in various industries. Unlike most artificial materials, wood is orthotropic, exhibiting different

M. Roohnia, *Properties and Treatment of Tonewood for String Instrument Construction*, Springer Series in Materials Science 355, https://doi.org/10.1007/978-3-032-00824-4_1

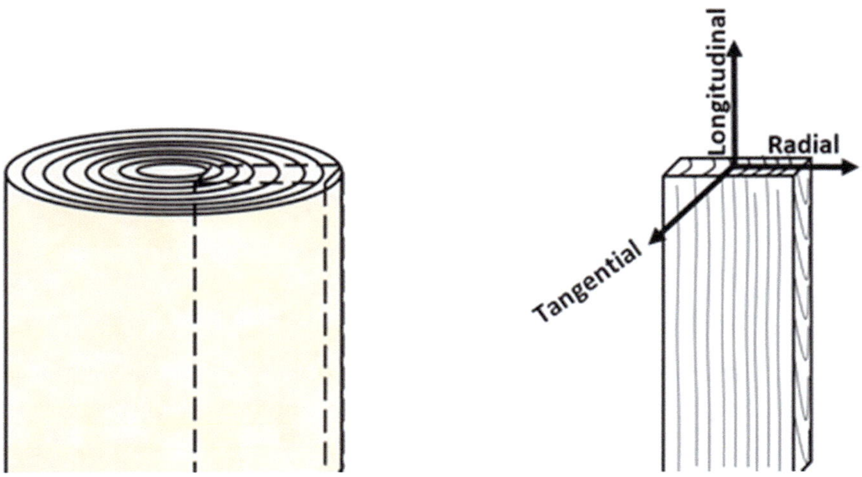

Fig. 1.1 The three orthotropic directions of wood. The left diagram shows a cross-section of a cylindrical log, revealing concentric annual rings. Dashed lines indicate how a radial section (extending from the pith outwards) and a tangential section (following the curve of the growth rings) would be cut. The right diagram shows obtained radial sawn board with arrows indicating the three orthotropic directions: longitudinal (parallel to the grain), radial (perpendicular to the growth rings), and tangential (tangent to the growth rings)

properties along its three primary axes due to its cellular structure (Fig. 1.1). This orthotropic nature, determined by cell composition and arrangement, contributes significantly to wood's overall acoustic performance [45].

The diversity of wood species offers a wide range of acoustic properties, enabling instrument makers to achieve specific tonal qualities [5, 12, 34].

- **Workability** Wood is a highly workable material, allowing for intricate shaping and carving [5]. Its relatively low density facilitates the portability of instruments.
- **Aesthetic appeal** The visual beauty of wood, characterized by diverse grain patterns and colors, contributes significantly to the aesthetic appeal of musical instruments [46]. Moreover, wood's natural affinity with the human body enhances the musician's connection to the instrument [25].
- **Availability** Except for a number of exotic wood species, the widespread availability of wood across various geographic regions renders it a cost-effective material compared to many alternatives.

These combined attributes have solidified wood's enduring status as the preferred material for a vast array of musical instruments.

1.2 Historical Overview of the Use of Wood in Musical Instruments Industry

The relationship between wood and musical instruments is as old as human civilization itself. Wood, with its unique acoustic properties, workability, and aesthetic appeal, has been the primary material for crafting musical instruments for millennia. This historical overview explores the evolution of wood use in instrument making, from prehistoric times to the present day.

1.2.1 Early Instruments and Wood

Early musical instruments, dating back to prehistoric times, relied heavily on natural materials, especially wood. While instruments made from bone or ivory, like prehistoric flutes, have survived due to their durability, wooden instruments have largely decomposed. This scarcity of wooden artifacts underscores the challenges of studying the material's early use in musical instrument construction. Despite limited material options in pre-industrial times, wood became the predominant choice for a variety of instruments, including stringed (Chordophones), wind (Aerophones), self-sounding (Idiophones) and percussion (Membranophones) types [38, 43, 46]. As Bucur [8] has already described, the origins of bowed stringed instruments can be traced back to India and the Far East. These early instruments, precursors to the rebec and vielle families, were introduced to Europe through various cultural exchanges.

Rebecs, characterized by a pear-shaped body, three strings tuned in fifths, and a lack of nut at the pegbox, were prominent among these early bowed instruments. The lyre, a notable rebec variant with a rectangular sound hole, bore resemblance to ancient Greek and Roman lyres.

During the twelfth to sixteenth centuries, ensembles of four to six bowed instruments emerged in Europe, mirroring the vocal ranges of soprano, alto, tenor, and bass. The lira da braccio, a smaller instrument held in the arms, and the lira da gamba, a larger instrument played between the legs, both appeared in Italy during the fifteenth century. These fretless instruments, often depicted in Renaissance artwork, laid the foundation for the development of the modern violin family [8]. The violin with today's appearance and acoustic development also appeared in the sixteenth century, which was made by Andrea Amati (1505–1578) in Cremona, Italy. Bucur accurately highlights the significant influence of the violin family on Western music, which emerged during the sixteenth century, shaping both musical practices and aesthetics. These examples of the earliest instruments all crafted primarily from wood. Since these ancient times, luthiers have sought ways to treat the wood used in musical instrument making.

1.2.2 The Baroque Era

The Baroque period, flourishing between approximately 1600 and 1750, marks a zenith in the evolution of musical instruments, particularly those of the string family. The Baroque period marked a pivotal shift in musical composition. While vocal music persisted, it increasingly relied on instrumental accompaniment, eclipsing the dominance of vocal works. Instrumental music flourished independently, developing unique styles tailored to specific instruments or ensembles. Even when combined with vocal elements, composers demonstrated a growing sophistication in orchestration [16].

The Cremona region in Italy emerged as a focal point for violin making, with the Amati family laying the groundwork for subsequent masters. Their legacy was amplified by the extraordinary talents of Andrea Guarneri, Giuseppe Guarneri Del Gesù, and most notably, Antonio Stradivari. Stradivari's meticulous craftsmanship and innovative designs set a standard of excellence that has influenced violin making for centuries. The culmination of this era witnessed the violin family attaining its quintessential form, a paradigm that has proven remarkably resistant to significant alteration [8, 16].

While the Baroque era encompassed a rich set of instruments, including woodwinds, brass, and percussion, the violin, viola, cello, double bass, Baroque guitar, and harpsichord reigned supreme as stringed instruments. These iconic instruments, meticulously crafted primarily from wood [8], have endured as cornerstones of classical music. Studies show that many attempts have been made to replace wood used for musical instruments with non-conventional wood species. For example, there have been reports of attempts to use spruce or Douglas fir for the top of the resonance boxes of violin family old instruments [15].

Historical evidence suggests that since these ancient times, when humans began crafting musical instruments professionally, they experimented with various treatments for wood. These included applying substances like Roman hydraulic cement and volcanic ash, often applied with or separate from varnish, to violin wood [4]. Interestingly, these techniques are traceable today. The philosophy behind them, including acoustic development and durability treatments against pests, warrants further study. Recently, a hypothesis has been raised and partially confirmed through dendrochronological investigations. Stradivari crafted his instruments during a period of unusually low solar activity known as the Maunder Minimum. The colder temperatures and specific environmental conditions of the north European spruce forests where his wood originated contributed to unique timber properties and exceptional sound quality. These particular climatic and ecological circumstances have not been replicated anymore since Stradivari's time [9].

Now, speaking of Cremona and the Baroque violin industry, it's worth noting that nearby Brescia was also a significant center. This smaller town introduced renowned luthiers like Gasparo da Salo (1540–1609) and his student Maggini (1580–1632). However, Cremona's violin-making school eventually surpassed Brescia in fame, after the seventeenth century.

The violin was undeniably the star of Baroque music. Its primary material, wood, was transformed into masterpieces through the artistry and craftsmanship of the era's greatest luthiers [8].

1.2.3 The Classical Era and Beyond

The Classical period, flourishing roughly between 1750 and 1820, produced iconic composers like Joseph Haydn (1732–1809), Wolfgang Amadeus Mozart (1756–1791) and Ludwig van Beethoven (1770–1827). This era is celebrated for its clear, melodic music, often featuring simple chordal accompaniments. The Classical period witnessed a sophisticated development in instrument making, with wood playing a central role.

The Classical period marked the beginning of widespread international recognition for violin making beyond Italy. While renowned luthiers such as Jacob Steiner (1621–1683) in Austria, who crafted instruments for musical giants like Johann Sebastian Bach (1750–1685), and the Klotz brothers (Matthias I Klotz (1656–1743) and Aegidius Klotz (1805–1733) in Germany) and Nicolas Médart (1628–1672) in France preceded this era, their full impact was realized during and after the Classical period. For instance, Mozart is believed to have owned a violin made by Aegidius Klotz. The legacy of Nicolas Médart (1628–1672) was further developed by later French masters like Nicolas Lupot (1758–1824) and Jean Baptiste Vuillaume (1798–1875), who studied the work of Stradivarius and collaborated with scientists such as Felix Savart (1791–1841).

In other European countries, such as England, Barak Norman (1678–1740) crafted violins using traditional methods. After 1800, the art of traditional violin making declined as the aristocracy favored modern English instruments. These instruments featured a higher bridge, increased string tension, a modified fingerboard angle, a longer neck and fingerboard, and reinforced bass bar and sound post. Some consider these innovations an irreparable catastrophe for the violin industry. In 1820, Louis Spoor (1784–1859) invented the chin rest to aid left-hand finger movement. This innovation has been standard on violins and violas ever since [8].

Since the mid-twentieth century, instrument makers, collaborating with university scientists, have endeavored to understand the Cremona violins of the Baroque era. Their goal has been to develop new engineering tests, modified woods, and alternative materials to replicate the sound and style of Stradivari instruments. This research continues, leading to innovative designs such as the violin octet [6], guitar family innovations [10] and guitar-violin hybrids resulting in its own acoustic signature [29].

So, despite advancements in materials science, wood still maintained its prominence in numerous acoustic applications in musical instruments today. While modern techniques and materials have been introduced, the core of many instruments still relies on wood. For example, acoustic guitars, violins, and pianos continue to be crafted primarily from wood. Even in the realm of electronic instruments, wood is often used for aesthetic and tonal reasons. The history of wood in musical instrument making is a testament to its enduring importance. From the earliest human attempts at sound production to the complexities of modern instrument design, wood has consistently been the material of choice. Its acoustic properties, workability, and aesthetic appeal make it an irreplaceable component in the world of music.

1.2.4 Wood and Musical Instruments in Asia

Stringed instruments from Turkey, Caucasian, Volga-Ural region, Afghanistan, Persia and the Persian Gulf states including Arab world countries, Central Asia, and Japan all within the Middle East and other Asian regions, predominantly utilize wood and especially mulberry wood as their primary material [3, 21, 23, 33, 38, 39, 42, 49].

As previously mentioned, the rebec and vielle, originating from India and Central Asia, have significantly impacted European and Western stringed musical instruments.

Noteworthy is the close affinity between the musical instruments of India and Central Asia with those of Afghanistan, including the Robab, tambour, dotar, panjtar, dambura, and qhichak (Ghaychak) (Figs. 1.2 and 1.3).

These instruments have mutually influenced each other in the region and subsequently shaped Western instruments. Sakurai [38] contends that this impact relationship that is considered for Asian instruments, is absent for Persian instruments. However, his idea is contrasting with the undeniable role of Ziryab (789–857), the Persian and mythical musician. While Ziryab's ethnicity remains debated in various literature, his background aligns with Persian origins [24, 30]. As a Persian speaker, I confirm that the name "Ziryab" derives from the Persian "Zaryāb," meaning "Gold Finder." Tar also is a Persian word that means "string" which has come as a suffix in Persian setar, tar, dotar, panjtar and also appeared in guitar.

Ziryab was instrumental in transforming the Persian barbat (barbad) or Arabic oud into the precursor of the guitar, which evolved through the middle ages, Baroque, and Classical periods to become the modern acoustic guitar (Fig. 1.4). Ziryab has become synonymous with the musical cultures of Muslim Spain [26]. So, this is an important document that shows at least one important influence of Persia on western music knowledge and art.

Not much information is available about Persian music before the seventh century. Because in three different periods, Persia's written works were attacked by book burnings, and books and libraries of Persia were comprehensively burned. The first was the attack (330 BC) of Alexander the Great (356–323 BC) during the Achaemenid (559–330 BC) dynasty, during which the written works on cowhide and clay historical tablets of Persian civilization and its prehistoric documents were destroyed [48].

The second was the Arab attack on Persia in the seventh century, when the second period of book burning and Persian historical documents took place.

Fortunately, after the conquest of Persia, the Arab Muslims quickly realized the importance of some scientific centers of Persia, including the ancient "Gundi-Shapur"

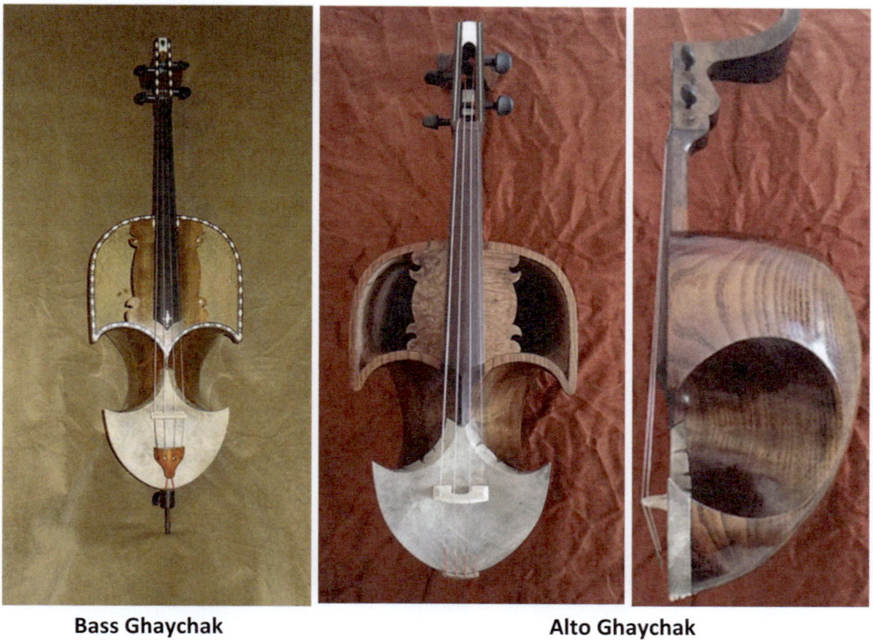

Fig. 1.2 Reconstructed Ghaychak instruments by luthier Siamak Ghalami. The left shows a Bass Ghaychak, the right an Alto Ghaychak. These reconstructions represent the cultural heritage of ancient Central Asian musical instruments in the Persian region

(Jundishapur) University, and sought to preserve their former glory. Therefore, a number of historical documents from the libraries of these centers, having survived destruction, have been used as literary and scientific references by Arabs and Muslims for years [41].

Considering that the territorial capital of Persia during the occupation by the Arabs, namely the city of Ctesiphon, with a history of being the capital for many hundreds of years [28], was located near Baghdad that is now recognized as one of the Arabic lands. The location of Gundi-Shapur the ancient university that now remains in Persian territories, is not far from Ctesiphon. Ctesiphon, the ancient capital of the Persian kingdom, now situated within Arab territories near Baghdad. Due to the forced Arabicization of occupied Persia for two hundred years, many of Persia's achievements were recorded in world history in the name of the Arabs.

After that, the 3rd, was the attack of the Mongols (thirteenth century) which also caused serious damage to the written works of Persia [27].

Following the conquest of Persia by Arab rulers, Arab and Persian musicians engaged in significant cultural exchange. Persian musicians claim that their music profoundly influenced Arabian music due to their prominence in Arab royal courts,

Fig. 1.3 Reconstructed Robab instruments by luthier Siamak Ghalami, showing the back (left), side (middle), and top (right). These reconstructions represent the cultural heritage of ancient Central Asian musical instruments in the Persian region

the introduction of Persian instruments, and the adoption of Persian musical terminology. Conversely, Arab musicians assert that their own music served as the foundation for Persian music, citing the numerous Arabic words used in Persian musical terminology [52]. In any case, this means the unification of Persian and Arab knowledge of music and instrumentation during the Arab domination of Persia.

Therefore, based on the remaining evidence of these attacks within Persia or in the literary works of neighboring countries, the history of Persian music will be presented here, with sources cited.

The most final form of the Persian barbat (Fig. 1.5) was revived by Ebrahim Ghanbarimehr, one of Persia's most professional and renowned luthiers [19], using imagery from the Sassanid civilization (the last Persian dynasty before the Arab Muslim conquests).

Fig. 1.4 The image displays six stringed instruments. Left is an oud, right is a modern classical guitars. The instruments demonstrate the evolution from the oud to the modern classical guitar, highlighting visual similarities in their development. Both instruments are made by Luthier Sina Asadi

Footprints similar to barbat lute have been seen for the first time in Persia [36]. In the clay tablets obtained from the eighth century BC, a lute-like instrument with a short neck can be identified, but its strings and appendages cannot be recognized. A similar instrument also documented to be existed in Chinese Han dynasty (3rd BC to 3rd AD). The clear influence of Persian musical instruments and their descendants is evident on this Chinese instrument and also in the musical instruments of other Asian regions (Fig. 1.6). The Chinese pipa instrument is a descendant of the Persian

Fig. 1.5 Ebrahim Ghanbarimehr, a renowned Persian luthier, holding a revived Persian barbat. The instrument, with a light-colored, pear-shaped body and short neck, reflects imagery from the Sassanid civilization. Ghanbarimehr is credited with bringing this final form of the barbat back into practice [19]. Photo from IRNA, news agency free resources: https://irna.ir/xj6GW4

short-necked lute, both of which have ancient histories in their respective regions [18, 47, 50, 51].

From the Chinese pipa, we can trace the lineage of the Japanese biwa, the Korean bipa, and Vietnam's dan ti ba, all of which are undeniably related to the lute family, Persian barbat, Arabic oud or Turkish baglama. Initially constructed primarily from mulberry and chestnut, the pipa's design evolved with the incorporation of acoustic wood science. The substitution of paulownia and rosewood for the top plate and backs, respectively, led to improved acoustics and widespread popularity [44]. However, in the short-necked lute family's native regions of Persia and Khorasan, a strong preference for traditional mulberry wood persists in instrument making [42]. Beyond the pipa and biwa, numerous Far Eastern musical instruments, even those claimed to have indigenous origins [31], exhibit striking similarities in shape and pattern to Persian instruments. This suggests a potential cross-cultural influence during periods of innovation.

For instance, the Japanese and Korean koto, and the Chinese kin, share characteristics with the Middle Eastern qanun and santur (Fig. 1.7). Similarly, the shamisen resembles the Persian setar and Turkish baglama, while fiddles like the keikin, kokin, and teikin closely resemble the kemanche. Interestingly, wood remains the primary material for the resonance boxes of all these instruments, with little to no alternative in use within these regions.

Fig. 1.6 A traditional Chinese pipa. The pear-shaped, wooden-bodied instrument with four strings and raised frets shows clear design continuity with its ancestor, the Persian barbat, sharing a similar lute-like structure and elegant craftsmanship rooted in cross-cultural musical heritage. Photograph is Ms. Zhao Zhuoqun (Chinese musical instrument scientist and player)'s art, gifted to this book

Fig. 1.7 The image displays a Persian Santur, a hammered dulcimer, from an elevated perspective. The instrument has a wooden, trapezoidal body with numerous strings stretched across its surface. Two sets of small, dark, cylindrical bridges are visible, supporting the strings. The instrument is an artwork from Luthier Ehsan Ghassemi

Another Persian historical document goes back to the period of the kingdom of the Medes (sixth and fifth centuries BC), in which the performance of shakh nafir (horn trumpets) and kous (big drum) is mentioned in the celebration of the reign of the first king of the Medes [37].

Roman and Greek historians recorded the existence of music programs in Persia long before the seventh century. These programs included military music, the singing of martial poems and hymns, and simple songs accompanied by various instruments. These instruments included barbat (oud), harp, reed, organ, chaghana, and hand instruments such as trumpets (kermil, marnai), karna and surna (Fig. 1.8), as well as kous, damameh, dohol, gavdam and jam [37, 52].

Traces of Persian influence are evident in present-day Indian music too. There is a vast correspondence in terminology; for example, some Indian names of instruments tambura, sitar, dohol, nuy, surnai, nafari, naqara, and table and names of forms and styles gazal, tarana, durbari—appear to have a Persian origin [52].

Instruments similar to the Persian surna are found all over North Africa and Asia [52]. This oboe originated in the Middle East, according to Sachs [36], and was carried to the Far East during the Islamic conquests [36, 52].

After those introduced above, another trace of the Persian lute family is not found until the seventh century AD. About the Arabicization of Persian musical instruments, especially the barbat, and its impact on Western music, Zonis writes, relying on other references, that before the advent of Islam, Persia had an important instrument called the barbat, which was played by prominent musicians such as Barbad before Arabicization of Persia in seventh century. Barbad had entered the court of the kingdom of Persia as a musician [1]. The barbat, i.e., Barbad's instrument, was made of wood and had a pear-shaped resonance box and a graduated wooden neck, unlike the Arabic oud, whose neck is neither graduated nor fretted (Fig. 1.9).

After the Arab attack on Persia in the seventh century AD, a number of Persian musicians were captured and enslaved, and the Arabs, who were not familiar with the instrument, gave it the Arabic name "oud", which means a flexible stick, simply because it was made of wood. The oud was a popular instrument among the Arabs, and became popular as an Arabic version due to the changes that were made on it, especially by Mansour Zalzal (a Persian musician of the same era as Ziryab). In the years following the Arab invasion of Spain, between the second half of the eighth century and the first half of the ninth century AD, this instrument was moved to Spain by the Muslims. This occurred during the Arabicization of the capital of Persia and other Persian territories occupied by the Arabs, which led to the mixing of Persian and Arab populations. From Spain, the instrument then reached Europe. This narrative is historically compatible with the previous narrative and Ziryab's influence on Spanish music and his mastery of playing the instrument, and has been validated in this way [52].

A large part of Turkey has been part of the territory of Persia or the Roman Empire for many years. Therefore, about the history of its music before the establishment of the Ottoman Empire (fourteenth to twentieth century), we should refer to the history of Persian music and Europe.

Documents from regions of Europe outside the south and west reveal an instrument resembling the oud and barbat but bearing a different name. Despite its unfamiliar nomenclature, the instrument's Persian origins are evident. This instrument, referred to as a "Kopuz" in historical sources, likely reached Eastern Europe via the Anatolian Plateau and the Byzantine Empire, as suggested by its Turkish etymology. The name Kopuz for this instrument can be seen in German poems of the fourteenth century and in Greek sources of the eighth and ninth centuries. While the Spanish Christians called it the Moorish (Spanish Muslim) guitar in the fourteenth century [36]. In all of these instruments, such as the Kopuz, lute, and oud, wood is mentioned as the

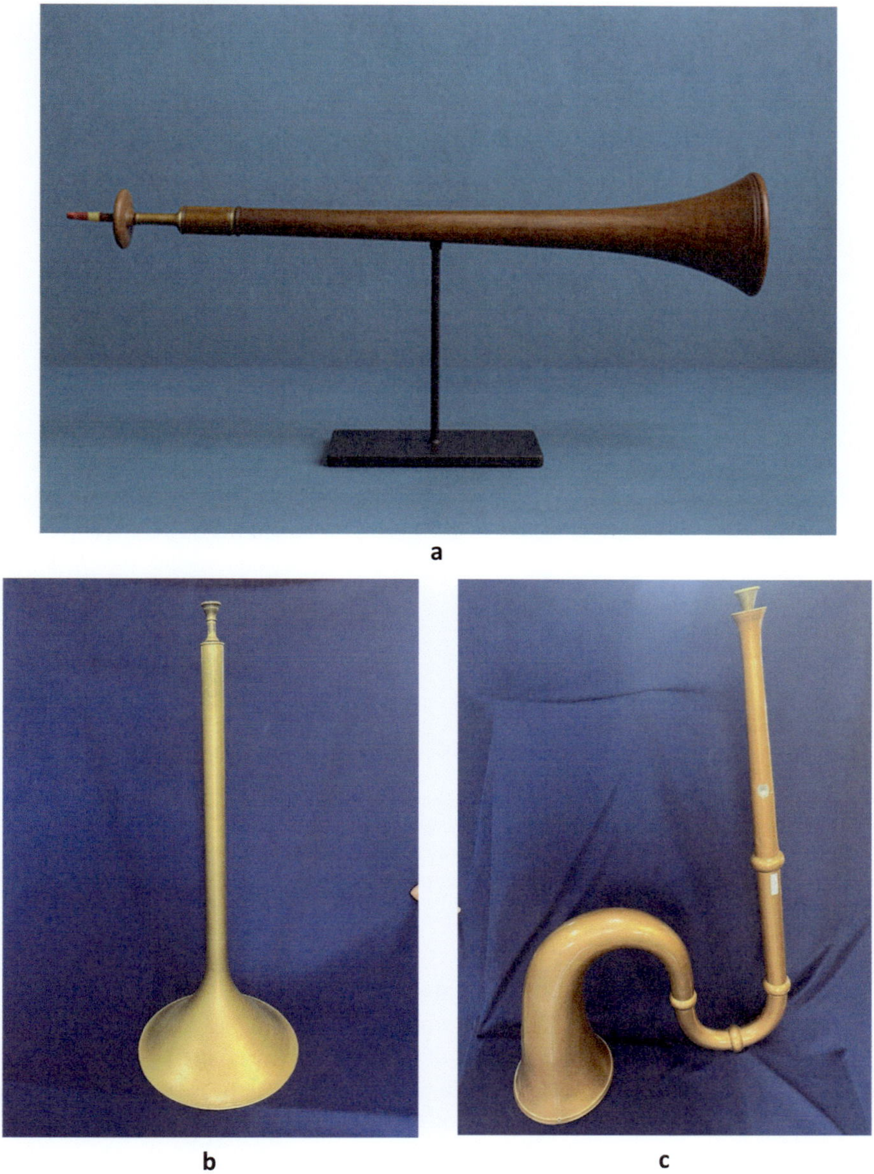

Fig. 1.8 Three wind instruments. **a** A recent wooden surna on a stand. **b** A brass kermil with a wide bell. **c** A brass marnai with a curved body. These revivals of ancient Persian instruments, similar to their ancestors, were crafted by Luthier Ebrahim Ghanbarimehr in 1971 for the 2500-year celebration of the Persian Empire [19]. Photographs are Mr. Luthier Siamak Ghalami's art, gifted to this book

Fig. 1.9 A stone relief carving within Sangshekanan cave in Jahrom—Iran, depicts Barbad holding an Ancient Persian Barbat. The figure, with flowing hair and a beard, is carved into the cave wall. The photograph is by Amir Sadeghian and is freely available under the Creative Commons Attribution license. https://www.tasnimnews.com/fa/. https://commons.wikimedia.org/wiki/File:Barbod_relief.jpg

main material. However, some sources also mention that parts of the soundboard were made of animal skin in addition to wood.

However, following the establishment of the Ottoman government, which marked a clear division between Persia and the Ottoman-controlled territories after the shared rule of the Seljuk dynasty over both regions, certain traditional instruments became associated with Turkey and the Caucasus. Western soldiers had faced the Ottoman army many times, encountering the intoxicating music of cymbals, oboes, triangles, and drums. After the eighteenth century, when the Ottoman Empire's power declined, victorious Europeans imitated and learned Ottoman music. This influence is evident in Mozart's Turkish march, which showcases elements of Turkish music [13, 36].

In addition to military instruments, the Ottoman Empire period saw a rich exchange of musical influences between East and West. This exchange led to the adoption of instruments such as the violin, kemanche, oud, and tanbur, all crafted primarily from wood. The Ottoman kemanche differs from its Persian counterpart, likely due to Western influence.

Dendrochronology studies revealed that Turkish luthiers have been skilled in selecting different wood pieces based on their characteristics since the seventeenth and eighteenth centuries, a period of peak violin craftsmanship in Europe [13]. For instruments like the oud and violin, they often used fine European wood species, while traditional instruments like the kemanche were typically made with wood species sourced from Anatolia, the Caucasus, or regions around the Black Sea. Violin production in Turkey appears to have begun after the 1950s, despite the instrument's introduction (the ancestors of the violin) in the sixteenth century. A dendrochronological analysis of one of the oldest Turkish violins, housed at historical Chapa Science College in Istanbul, revealed that its spruce wood originated from the Swiss Alps in the early twentieth century [2].

Unlike most Islamic governments, which often held negative views of music, the Ottoman government considered it a crucial cultural component. As a result, Turkish musical instruments flourished compared to those of neighboring Muslim states. While our knowledge of Ottoman musical instruments primarily focuses on the sixteenth, seventeenth, and later centuries, numerous accounts from tourists and travelers of that era confirm music's significance in Ottoman culture.

Beyond the above mentioned references, historical records indicate that the Ottoman Empire's musical heritage included instruments such as santur (zither), cheng (harp), chogur (short-necked baglama), panflute, Robab, daire, finger cymbals, bozuk, sine, mazhar or bendir, nefir (horn), tanbur, qanun, ney, kemanche, oud, def, and kopuz. While certain instruments have been lost to history, others have endured and continue to be utilized today. Furthermore, the Ottoman government's conquests often fostered the growth of music in the conquered territories. For example, Sultan Murad IV's capture of Baghdad from the Safavid dynasty in the seventeenth century contributed to the development and elevation of music in that region [17]. Prior to the Safavid period, Shia Islamic culture in Persia generally discouraged public musical expression. As a result, written works on musical instrumentation from this era are limited. To supplement this lack of domestic sources, researchers have relied on European travelogues for insights into local and foreign instruments used in Persia during the Safavid period, providing valuable historical and cultural context [32].

Among traditional Turkish musical instruments, the baglama, a descendant of the Asian kopuz, Persian barbat, or Arabic oud, has undergone significant development [11]. Influenced by modern technology and the innovative approaches of Turkish instrument makers, the baglama has evolved beyond its traditional use of mulberry wood. Today, it incorporates imported species commonly used in guitar making, such as high quality European spruces for the soundboards and maple, mahogany and juniper wood for the sound box [11]. This openness to new materials demonstrates a scientific and unbiased approach within the Turkish music industry.

In contrast, many neighboring regions, including Persia, still consider the substitution of any wood other than mulberry in instruments such as Persian setar, dotar, and tanbur to be a reprehensible act. While many superstitions about mulberry wood in traditional musical instruments may seem humorous in twenty-first-century, some musicians in these regions still adhere to this belief [42]. This adherence to tradition can hinder the instrument's evolution and integrity. While the baglama has benefited from innovation, it's important to note that it still lacks standardized production guidelines and has been subject to changes influenced by the preferences of musicians and instrument makers.

This book will explore the need for alternative woods in musical instrument construction, comparing the tonal qualities of spruce and Rosewood to Mulberry for either the sound boards or the sound boxes. It also will provide methods for replicating the sound color of mulberry using other wood species, satisfying both enthusiasts and those with mulberry preferences.

1.2.5 Wood and Musical Instruments in America, Africa and Oceania

Oceania appears to have a smaller amount of documented traditional music compared to other regions. Their indigenous musical instruments, primarily wind and percussion types, rely heavily on wood but also incorporate other natural materials. Stringed instruments (chordophones) were scarce in Oceania before European contact [22].

African music has greatly influenced western music such as blues. But globalization has destroyed traditional African music, and older forms are at risk of extinction [7].

The history of African music has been studied through oral sources, analysis of ancient artifacts, and the examination of geographical and ethnographic data, and tools such as ancient writings and artworks from Egypt and Ethiopia have been helpful [7].

Stringed instruments have played a vital role in African music for millennia, with evidence pointing to their presence as early as Ancient Egypt. The lyre and harp, found in Egyptian tomb paintings dating back to around 3000 BC, are among the earliest known stringed instruments in Africa [7, 35]. These instruments were likely introduced to Nubia and other regions of the Nile Valley during early trade and cultural exchanges [35].

In West Africa, the kora, a wooden 21-string harp-lute, emerged as a defining instrument in the Mande culture. The kora's construction, with a large gourd resonator, dates back several centuries, with its origins linked to the Mali Empire (1235–1600 AD). It remains central to traditional music in modern countries like Mali, Senegal, and Guinea [20].

The ngoni, a small lute from West Africa, has been in use. Its design bears striking similarities to the oud and other North African and Middle Eastern lutes, which may have arrived via the trans-Saharan trade routes [20]. The kamalengoni, a more recent variant, was developed in the twentieth century as a lighter and smaller version, becoming popular in modern Malian music [20].

In East Africa, the krar, a bowl-shaped lyre, has a history in Ethiopia and Eritrea. The krar is a descendant of the ancient lyres found in Egypt and has remained a key instrument in Abyssinian music [40].

Thus, African stringed instruments reflect a complex history of local innovation and external influences, with roots stretching back to Ancient Egypt and continuing to evolve in different cultural contexts across the continent.

In ancient America, American Indian wooden musical instruments, sound the depths of their civilization. American Indian wooden musical instruments are deeply rooted in the rich and ancient culture of these people and are known as one of the most important elements of their identity and connection with nature. These instruments are not only musical tools but also symbols of American Indian beliefs, legends, and traditions.

It should be noted that Central American Indians were relatively underdeveloped in instrumental music. In contrast, while the Indians of South America were culturally and literarily less advanced than their Central American counterparts, they excelled in areas such as social organization, art, architecture, and music [36].

The exact date of American Indian instruments' origins remains unknown due to a lack of accurate written records, but archeological evidence shows that American Indians used wooden musical instruments many of years ago. These instruments played an important role in religious ceremonies, celebrations, hunting, and even in wars. American Indian musical culture, renowned for its rich diversity, is reflected in its vast array of musical instruments. While these instruments can be broadly categorized into percussion and wind instruments, the sheer variety within each category is astounding. While stringed instruments are less prevalent in traditional American Indian music, their influence has grown significantly since the colonial era [14].

Wooden flutes are one of the most well-known types of American Indian wooden musical instruments. These flutes were usually made of different native wood species and different notes were produced by making holes in their bodies.

Wooden drums were also among the most used musical instruments of the American Indians. These drums were made in various sizes and shapes and were usually made from hollow tree trunks.

Rattles were a type of percussion instrument that produces sound by shaking wooden parts together. This instrument was often used in religious ceremonies and to communicate with supernatural forces.

Two traditional stringed instruments are found in the Indian world, only north of Mexico, though their use is uncommon. The first, often referred to as the "Apache fiddle," is a distinctive instrument that likely emerged after European contact. While the concept of a string stretched over a sound box may have European origins, the instrument's design and the music played on it are undeniably indigenous.

The second stringed instrument, found in California, is a mouth bow. Essentially a hunter's bow, it is played similarly to a mouth harp or Jew's harp. Held straight out with one end in the player's mouth, the string is tapped with the index finger. Different tones are produced by adjusting the bow's tension or the shape of the player's mouth. It is believed that this quiet instrument was primarily used by lovers [14].

American Indians used raw materials available in nature to make wooden musical instruments. Wood, animal skin, plant fibers and bone were some of the raw materials used in making these instruments. The construction of these instruments was usually done manually with simple tools and each instrument had its own unique characteristics.

1.3 Final Words

The chapter on the importance and history of wood in musical instrument construction underscores wood's unmatched role as a material of choice across cultures and centuries. Its unique acoustic properties, workability, and aesthetic appeal have solidified its position as a cornerstone in the creation of instruments, from ancient to modern times. Wood's orthotropic nature, varying along different grain orientations, allows it to produce distinctive tonal qualities that synthetic materials have yet to replicate fully. Instrument makers, from early civilizations to contemporary luthiers, have relied on its versatility, using different wood species to achieve specific sound characteristics.

Historically, wood has shaped the development of many iconic instruments, from stringed instruments in Europe to wind and percussion instruments across Asia, Africa, and the Americas. Despite advancements in materials science and engineering, no alternative has supplanted wood in terms of its tonal richness, resonance, and natural warmth. Even in the age of electronic instruments, wood continues to be a vital material, providing both acoustic and aesthetic qualities that modern musicians appreciate.

The chapter concludes by reflecting on the continued significance of wood in musical instrument making, despite modern innovations. As scientific research seeks to replicate the qualities of legendary instruments like the Stradivarius, wood remains irreplaceable. However, ongoing experimentation with alternative materials suggests that the future may hold new possibilities, offering both traditionalists and innovators options to explore while maintaining wood's legacy in music.

References

1. N. Ahmadian, Affective Hal and the aesthetic education of sorrow in performing Persian classical music. Ph.D. thesis, Department of Music, University of Alberta, 2023
2. Ü. Akkemik, M.N. Sargın, N. Köse, Dating of a violin from Turkey. Arkeoloji Sanat Derg. **150**, 133–138 (2015)
3. U. Aslan, S. Karahasanoğlu, Sound ethnobiology of musical instruments: a sound view of nature in manufacturing kemençe. Musicologist **5**(2), 240–263 (2021). https://doi.org/10.33906/musicologist.988011
4. C.Y. Barlow, P.P. Edwards, G.R. Millward, R.A. Raphael, D.J. Rubio, Wood treatment used in Cremonese instruments. Nature **332**, 313 (1988). https://doi.org/10.1038/332313a0

5. B.C. Bennett, The sound of trees: wood selection in guitars and other chordophones. Econ. Bot. **70**(1), 49–63 (2016). https://doi.org/10.1007/s12231-016-9336-0
6. G. Bissinger, The Hutchins–Schelleng violin octet after 50 years, in *The Science of String Instruments*, ed. by T. Rossing (Springer, New York, NY, 2010). https://doi.org/10.1007/978-1-4419-7110-4_18
7. R. Blench, Reconstructing African music history, in *Presented at the SAFA Conference*, Tucson, 17–21 May 2002. Retrieved from https://www.rogerblench.info/Ethnomusicology/Papers/Africa/General/Reconstructing%20African%20music%20history%20web.pdf
8. V. Bucur, Description of string instruments for classical music, in *Handbook of Materials for String Musical Instruments* (Springer, Cham, 2016), pp. 5–91. https://doi.org/10.1007/978-3-319-32080-9_2
9. L. Burckle, H.D. Grissino-Mayer, Stradivari, violins, tree rings, and the Maunder Minimum: a hypothesis. Dendrochronologia **21**(1), 41–45 (2003). https://doi.org/10.1078/1125-7865-00033
10. G. Caldersmith, Designing a guitar family. Appl. Acoust. **46**(1), 3–17 (1995). https://doi.org/10.1016/0003-682X(95)93949-I
11. H.K. Celik, S. Gok, N. Caglayan, A.E.W. Rennie, A numerical method-based analysis of the structural deformation behaviour of a Turkish string instrument (Cura Baglama) under varying string tensions. Appl. Sci. **13**(17), 9682 (2023). https://doi.org/10.3390/app13179682
12. F. Dinulică, M.D. Stanciu, A. Savin, Correlation between anatomical grading and acoustic–elastic properties of resonant spruce wood used for musical instruments. Forests **12**(8), 1122 (2021). https://doi.org/10.3390/f12081122
13. Y. Doğan, Ü. Akkemik, N. Köse, K. Sakowska, M. Bernabei, Tree-ring dating of Turkish musical instruments. Tree-Ring Res. **80**(1), 1–12 (2024). https://doi.org/10.3959/TRR2022-11
14. G.S. Fields, P.P. Schools, *American Indian Music Traditions and Contributions* (1993)
15. M. Fioravanti, G. Di Giulio, G. Signorini, A non-invasive approach to identifying wood species in historical musical instruments. J. Cult. Herit. **27**(Supplement), S70–S77 (2017). https://doi.org/10.1016/j.culher.2016.05.012
16. K. Geiringer, The Baroque era (1600–1750), in *Instruments in the History of Western Music*, 1st edn. (Routledge, 1978), p. 44. ISBN: 9781003543404
17. D. Göktürk Cary, Ottoman music in travel books: a path to restructure the forgotten Ottoman musical instruments. Tarih İncelemeleri Derg. **32**(1), 73–92 (2017). https://doi.org/10.18513/egetid.327734
18. R. Harris, Applied experiments in collaboration along the Silk Road. World Music **7**(1+2), 37–59 (2018). https://eprints.soas.ac.uk/26018/
19. H. Jalilvand, Sāzšenāsī-ye farhangī-ye mūsīqī-ye dastgāhī-ye Īrān (barrasī-ye taġyīrāt-e farhangī-ye ān az 1304 tā emrūz) [Cultural organology of Iranian classical music (a study of its cultural changes from 1925 to today)]. Master's thesis, University of Tehran, 2002. http://thesis2.ut.ac.ir/thesisinfo/ThesisPdf15PagesFiles/2013-10/261447-15Page-.pdf
20. L. Kaba, B. Charry, Mamaya: renewal and tradition in Maninnka music of Kankan, Guinea (1935–45), in *The African Diaspora: A Musical Perspective*, ed. by I. Monson, 1st edn. (Routledge, 2000). https://doi.org/10.4324/9781315805344
21. E. Karami, K. Pourtahmasi, M. Shahverdi, Wood anatomical structure of *Morus alba* L. and *Morus nigra* L., native to Iran. Not. Sci. Biol. **2**(4), 129–132 (2010). https://doi.org/10.15835/nsb244985
22. R.F. Kennedy, Musics of Oceania. Music. Educ. J. **59**(2), 59–64 (1972). https://doi.org/10.2307/3394143
23. G. Khayrullina, Musical instruments of the Turkic peoples of the Volga region and Urals. Liet. Istor. Stud. 315–323 (2023). https://doi.org/10.15388/Totoriai-Lietuvos-istorijoje.2023.22
24. S.A.R. Khezri, Ziryab and his role in transferring Persian music and customs to the Islamic Spain (Andalusia). Honar-Ha-ye-Ziba (22), 107–112 (2005). Retrieved from https://sid.ir/paper/5818/en
25. V. Kotradyova, E. Vavrinsky, B. Kalinakova, D. Petro, K. Jansakova, M. Boles, H. Svobodova, Wood and its impact on humans and environment quality in health care facilities. Int. J. Environ. Res. Public Health **16**(18), 3496 (2019). https://doi.org/10.3390/ijerph16183496

26. M. Machin-Autenrieth, Ziryab and us: tradition and collaboration in the interpretation of an Arab-Andalusian musical myth. J. Intercult. Stud. **42**(4), 494–514 (2021). https://doi.org/10.1080/07256868.2021.1939273
27. J.D. McDonald, M. Levine-Clark (Eds.), *Encyclopedia of Library and Information Sciences*, 4th edn. (CRC Press, 2017). https://doi.org/10.1081/E-ELIS4
28. A. Mousavi, T. Daryaee, The Sasanian Empire: an archaeological survey, c. 220–AD 640, in *A Companion to the Archaeology of the Ancient Near East*, ed. by D.T. Potts (2012), pp. 1076–1094. https://doi.org/10.1002/9781444360790.ch57
29. T.J. Nania, Archtop guitar dynamics. Violin Soc. Am. Pap. **29**(1), 1–13 (2021)
30. J.F. O'Callaghan, *A History of Medieval Spain* (Cornell University Press, 2013), 735 pp. ISBN: 080146871X, 9780801468711
31. F.T. Piggott, *The Music and Musical Instruments of Japan* (Kelly & Walsh, Yokohama; B.T. Batsford, London, 1909), 196 pp
32. I. Rahbar, H. Asadi, R. Gaja, Musical instruments of the Safavid period according to Kaempfer and Chardin. J. Fine Arts Perform. Arts Music **2**(41), 23–32 (2011)
33. Z.Z. Rashitovich, National musical instruments. J. Soc. Sci. Humanit. Res. Fundam. **4**(03), 51–54 (2024). https://doi.org/10.55640/jsshrf-04-03-11
34. M. Roohnia, Wood: vibration and acoustic properties, in *Reference Module in Materials Science and Materials Engineering*, ed. by S. Hashmi (Elsevier, Oxford, 2019), pp. 1–13. ISBN: 978-0-12-803581-8
35. D. Rowan, The universal lyre: three perspectives. Am. Harp J. 2 pp. 55+. Gale Academic OneFile. link.gale.com/apps/doc/A343156517/AONE?u=anon~9c3f7f78&sid=googleScholar&xid=4f0290bf. Accessed 21 Sept 2024 (2013)
36. C. Sachs, *The History of Musical Instruments* (W W Norton & Co Inc., 1940)
37. H. Safakish, M. Seyed, Historical process of Persian music from ancient era until emergence of Safawids. Sci. J. Hist. Res. **32**(8), 1–26 (2013)
38. T. Sakurai, *The Musical Instruments of Afghanistan, Senri Ethnological Studies (SES)* (National Museum of Ethnology, 1980), pp. 47–74
39. A. Se Golpayegani, I. Brémaud, J. Gril et al., Effect of extractions on dynamic mechanical properties of white mulberry (*Morus alba*). J. Wood Sci. **58**, 153–162 (2012). https://doi.org/10.1007/s10086-011-1225-7
40. G.A. Sinshaw, The analysis of Ethiopian traditional music instrument through indigenous knowledge (kirar, masinko, begena, kebero and washint/flute). Int. J. Sci. Res. Manag. (IJSRM) **8**(01), 591–596 (2020). https://doi.org/10.18535/ijsrm/v8i01.sh02
41. M.M. Söylemez, The Jundishapur school: its history, structure, and functions. Am. J. Islam Soc. **22**(2), 1–27 (2005). https://doi.org/10.35632/ajis.v22i2.455
42. F. Vahabzadeh, The music of the mulberry: wood science, know-how and symbolism in instrument-making in Khorassan (Iran) and Central Asia, in *Wooden Musical Instruments—Different Forms of Knowledge: Book of End of WoodMusICK COST Action FP1302*, ed. by M.A. Pérez, E. Marconi (2018). ISBN: 979-10-94642-35-1. hal-02086598
43. C. Waltham, S. Yoshikawa, Construction of wooden musical instruments, in *Springer Handbook of Systematic Musicology*, ed. by R. Bader. *Springer Handbooks* (Springer, 2018), pp. 63–79. https://doi.org/10.1007/978-3-662-55004-5_4
44. C. Waltham, E. Koster, A. Kotlicki, J. Simard, N. Wolfe, Acoustic radiation from the pipa and yueqin. Proc. Meet. Acoust. [Online] **15**(1), 035004 (2012). https://doi.org/10.1121/1.4801410
45. U.G.K. Wegst, Wood for sound. Am. J. Bot. **93**(10), 1439–1448 (2006). https://doi.org/10.3732/ajb.93.10.1439
46. U.G.K. Wegst, Bamboo and wood in musical instruments. Annu. Rev. Mater. Res. **38**(1), 323–349 (2008). https://doi.org/10.1146/annurev.matsci.38.060407.132459
47. D. Wei, Historical development of the pipa. Thesis, University of Tasmania, 1989. https://doi.org/10.25959/23242583.v1
48. Y. Yamanaka, From evil destroyer to Islamic hero: the transformation of Alexander the Great's image in Iran. Ann. Jpn. Assoc. Middle East Stud. **8**, 55–87 (1993). https://doi.org/10.24498/ajames.8.0_55

49. S. Yoshikawa, Acoustical classification of woods for string instruments. J. Acoust. Soc. Am. **122**(1), 568–573 (2007). https://doi.org/10.1121/1.2743162
50. N. Zaker Jafari, The place of oud and its various types in the Islamic eras of Persian music history. J. Hist. Sci. **11**(2), 191–206 (2013)
51. N. Zaker Jafari, M. Farahani, Mutual relations between Iranian and Chinese musical instruments. J. Drama. Arts Music **6**(11), 113–126 (2016). https://doi.org/10.30480/dam.2016.295
52. E. Zonis, *Classical Persian Music: An Introduction* (Harvard University Press, Cambridge, MA, 1973)

Chapter 2
The Basics of Wood Science and Wood Anatomy for Musical Instruments

Abstract Wood science and anatomy are fundamental to understanding wood's unique characteristics, which make it suitable for diverse applications, from construction to crafting musical instruments. This chapter introduces key concepts in wood science, emphasizing the structural, chemical, and mechanical properties that distinguish different types of wood. Wood has been integral to human life for centuries, evolving from a basic material for tools and shelters to a highly valued resource with specialized uses. Musical instruments, in particular, require precise knowledge of wood properties to ensure tonal quality, durability, and aesthetic appeal. The study of wood science explores the factors that influence wood properties, including species, growth conditions, and environmental factors. A comprehensive understanding of these aspects enables sustainable forest management, improved processing techniques, and better preservation methods. Additionally, it aids archaeology by revealing insights into ancient wood usage and ecological conditions. This chapter discusses key topics within wood science, including wood classification, physical and mechanical properties, and cellular structure, all of which affect wood's functionality. A major focus is on anatomical distinctions between hardwoods and softwoods, which are critical for selecting wood for specific applications. Hardwoods, typically from deciduous trees, have a complex cellular structure and are often chosen for their strength and durability. Softwoods, from coniferous trees, have simpler structures and are valued for their workability and lighter weight. Understanding these distinctions allows optimization of wood selection for purposes ranging from construction to musical instrument components. The chapter also covers the physical and mechanical properties of wood, such as density, moisture content, grain, strength, and hardness, all of which impact its performance and suitability for various uses. For example, density and moisture content are crucial for stability and workability, while grain patterns and color add aesthetic value. The orthotropic nature of wood—where properties vary along different grain directions—is especially relevant for applications needing precise load-bearing or acoustic qualities, as in instrument making. Wood anatomy examines the cellular structure and organization within wood, focusing on features like annual rings, wood rays, and microstructure. These elements influence wood's appearance and strength, with annual rings indicating growth history and wood rays aiding nutrient transport. The microscopic

M. Roohnia, *Properties and Treatment of Tonewood for String Instrument Construction*, Springer Series in Materials Science 355, https://doi.org/10.1007/978-3-032-00824-4_2

structure, including cell wall composition, significantly impacts wood's mechanical and acoustic performance, essential for instrument crafting. This chapter provides a foundational understanding of wood science and anatomy, enabling informed decisions when working with wood. From its structural complexity to its physical and aesthetic qualities, wood remains a remarkable natural material with broad applications, especially in fields where performance, durability, and beauty are paramount. Consider the information in this chapter as a foundational building block for your understanding of wood in musical instruments industry. While the practical applications and measurement techniques will be explored in later chapters, this chapter provides the essential theoretical knowledge you'll need to successfully apply those techniques. Therefore, it's crucial to thoroughly understand the concepts presented here.

2.1 Wood Science: General Knowledge

Wood science is a branch of knowledge that studies the structure, properties, uses and processes related to wood. Wood, as one of the oldest and most widely used natural materials, has played an important role in human life for a long time. From making tools and weapons in the early days to making houses, furniture, musical instruments and wooden structures in the past to the present ages, wood has always been of interest.

2.1.1 The Importance of Wood Science

- **Wood industry**: Accurate knowledge of wood properties is necessary to improve the production processes of wood products, increase their quality and durability.
- **Forest conservation**: Understanding the principles of wood science helps in sustainable management of forests and optimal utilization of wood resources.
- **Archaeology**: The study of ancient wood provides valuable information about past human history, culture and environment.
- **Biological sciences**: Wood, as a biological material, is the subject of many researches in various fields of biological sciences such as plant physiology and cell biology.

2.1.2 The Main Topics of Wood Science

- **Types of wood and their classification**: Classification of wood based on plant species, wood formation structure, properties and application.

- **Physical and mechanical properties of wood**: Examining properties such as density, strength, hardness, color, grain, texture and resistance to environmental factors.
- **Wood anatomy**: The study of wood components such as cells, tissues and vessels.
- **Wood chemistry**: Is the study of the chemical composition and properties of wood. It involves the analysis of the various organic compounds, such as cellulose, hemicellulose, lignin, and extractives, that make up wood's structure and contribute to its physical and mechanical properties.
- **Wood applications**: Examining the diverse ways wood is utilized in industries ranging from furniture production and construction to paper manufacturing and chemical processes.
- **Wood changes over time**: Study of wood degradation processes such as decay, insectivory and changes caused by environmental factors.
- **Dendrochronology**: Is the scientific method of dating tree rings to the exact year they were formed. By analyzing these rings, scientists can reconstruct past climates, track historical events, and even date archaeological artifacts.

2.1.3 Definition of Wood and Its Formation Process in the Tree

- **What is wood?**

  Wood is a solid, porous and fibrous material that consists of dead cells of vascular plants. Most of the cells are dead at maturity even in the sapwood [55]. This substance is mainly found in the trunk, roots and branches of trees. Wood is actually a part of the plant's vascular system, which is responsible for transporting water and minerals from the roots to the leaves [33].
- **Wood formation process**

  Wood is formed by the activity of the cambium, a thin layer of living cells located between the wood and the bark (Fig. 2.1). The cambium continuously produces new cells, with those produced towards the inside developing into wood cells and those produced towards the outside forming the bark (skin) cells [20, 21].
- **Stages of wood formation**

  - **Cell division**: Cambium cells divide to produce new cells, both inward and outward. The newly formed cells on the inside become wood (xylem cells, which transport water and minerals). The cells formed on the outside become inner bark (phloem cells, which transport nutrients and organic compounds).
  - **Changing the shape of cells**: New cells gradually grow and change their shape and size.
  - **Cell wall change**: The cell wall of wood cells thickens and lignin is deposited in it. Lignin is a substance that gives wood hardness and strength.

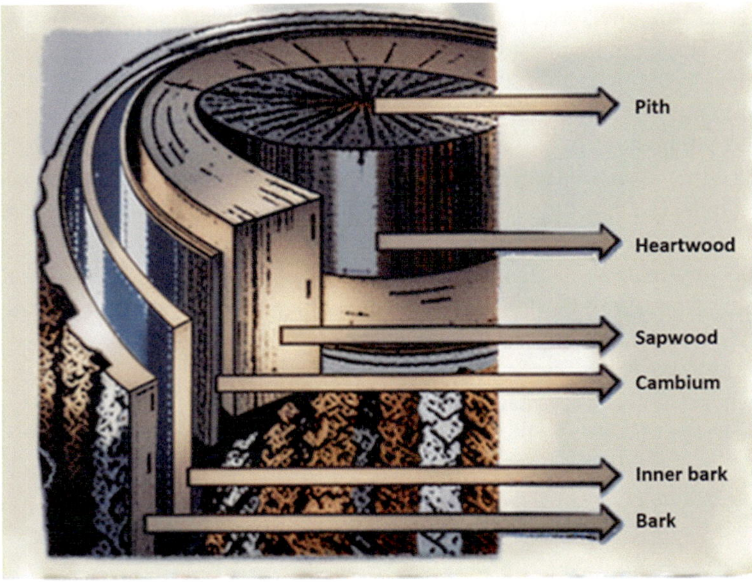

Fig. 2.1 A cross-sectional diagram illustrating the internal structure of a tree trunk. It shows the heartwood in the center, surrounded by concentric rings of sapwood, cambium, inner bark, and the outermost bark. Arrows point to each layer with corresponding labels. Hand art image with an AI enhancement

- **Cell death**: After reaching maturity, wood cells die and their contents are emptied [21].

- **Factors affecting wood properties**
  - **Plant species**: Each plant species produces wood with unique properties.
  - **Tree growth conditions**: Climate, soil and sunlight conditions affect the growth of the tree and consequently the properties of the wood.
  - **Age of the tree**: The wood of young trees, is usually softer and lighter than the wood of older trees. It is due to the presence of a high proportion of juvenile wood [59], which will be explained in the next paragraphs.
  - **Tree part**: The properties of wood are different in different parts of the tree (inner or outer layers of the trunk, branches) which will be explained later in the next paragraphs.

2.1.4 Types of Wood in Terms of Growth

- **Early wood**: Also known as springwood, is the first type of wood formed by a tree in a growing season. Wood cells produced in spring usually have thinner cell walls, that allows for more efficient water and nutrient transport and have a larger

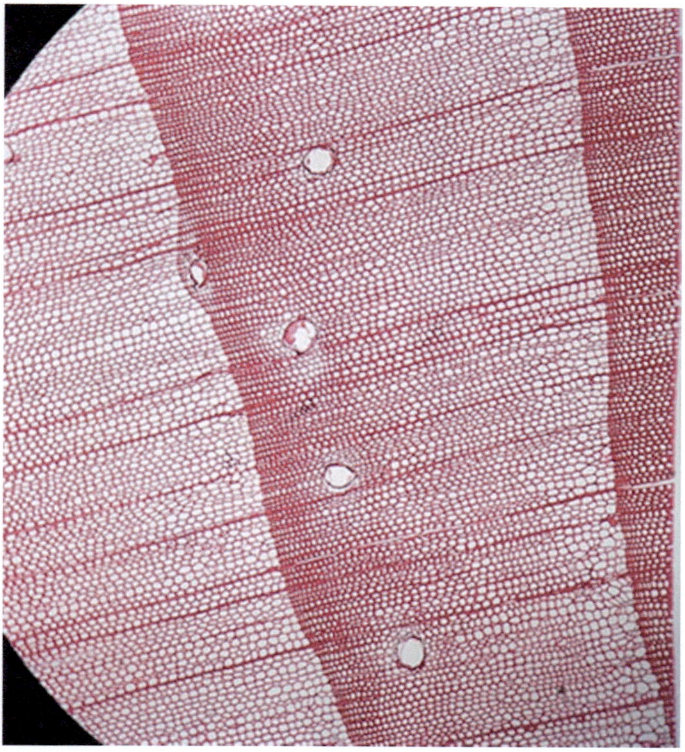

Fig. 2.2 A cross-section of pine wood showing the growth rings. Each ring has a distinct lighter-colored earlywood section and a darker-colored latewood section. Large holes are resin canals. Image from Wikimedia commons free resources (under the Creative Commons Attribution license): https://commons.wikimedia.org/wiki/File:Resin_canal.jpg

diameter that contributes to the tree's rapid growth during the spring. Early wood is typically lighter in color than latewood (Fig. 2.2).

- **Latewood**: Is the wood formed by trees during the latter part of the growing season, typically in the summer or fall [13].

  Compared to earlywood, which is formed in the spring, latewood has thicker cell walls and a smaller diameter. Latewood's thinner capillaries do indeed play a role in helping trees conserve water during dry summer months [51]. These narrower capillaries can more effectively draw water from the soil, even when moisture levels are low. This ability is particularly beneficial in regions with arid climates or during periods of drought.
- **Heartwood and sapwood**: Are two distinct regions within the trunk of a tree (Fig. 2.1). Heartwood is the central, older portion of the trunk that has lost its living cells and functions primarily as a structural support. It is often darker in color due to the presence of extractives, such as resins and tannins.

Sapwood is the outer layer of the trunk that surrounds the heartwood. It contains living cells that are responsible for transporting water and nutrients throughout the tree. Sapwood is generally lighter in color than heartwood [3].

- **Juvenile, transition, and mature wood**: These terms refer to different stages of wood development within a tree's growth rings [39].
  - **Juvenile wood**: This is the wood formed in the early years of a tree's or a tree branch's life. It is characterized by its relatively low density, weak mechanical properties, and distinct anatomical features. Juvenile wood is often more susceptible to decay and insect attack.
  - **Transition wood**: This is the intermediate stage between juvenile and mature wood. As the tree ages, its growth characteristics begin to change, and the wood produced during this transition period exhibits properties that are between those of juvenile and mature wood.
  - **Mature wood**: This is the wood formed in the later stages of a tree's or tree branch's life. It is generally denser, stronger, and more resistant to decay and insects compared to juvenile and transition wood. Mature wood is often preferred for structural and other applications due to its superior mechanical properties (Fig. 2.3).

When mature wood mixes with sapwood in the outer layers, it can have several implications. Sapwood, being the outer layer of wood, is generally equal in mechanical properties of normal heart wood. When mature wood is reached and mixed with sapwood, it can reduce the overall durability of the wood product [35]. Sapwood is

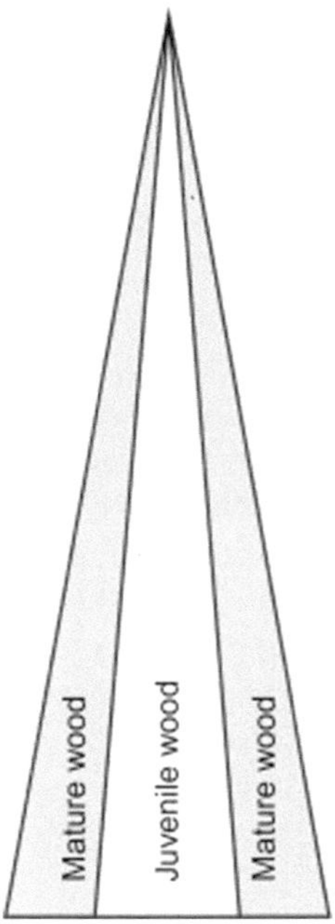

Fig. 2.3 A stylized diagram of a tree trunk showing the transition from juvenile wood to mature wood. The tree is depicted as an exaggerated cone shape with the narrow end representing the top and the wide end representing the base. The juvenile wood percentage is increased from the base to the top, transitioning to mature wood towards the outer growth rings

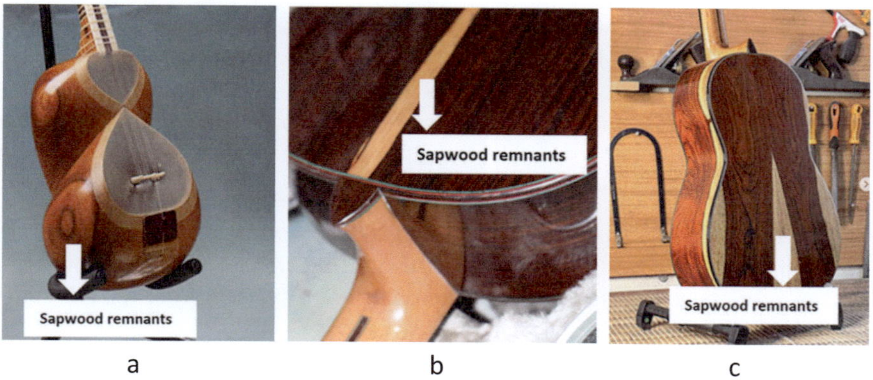

Fig. 2.4 A photo of **a** a Persian Tar by Luthier H. Mahmoodi, a long-necked lute with a pear-shaped body. **b** A Spanish Guitar by Luthier S. Ashrafi and **c** a Spanish Guitar by Luthier R. Fallah. The instruments are made of wood, with visible remnants of sapwood

more susceptible to decay and insect attack due to its higher moisture content and presence of nutrients.

When sapwood is mixed with mature wood, it can create a less uniform appearance, which may be desirable or undesirable in certain applications. To avoid these issues, it is important to carefully select and process wood to minimize the presence of sapwood, especially in applications like musical instruments industry, where strength, durability, and appearance are critical. Although it has been seen many times that thıs color difference has been favored as an aesthetic aspect in musical instruments industry (Fig. 2.4).

2.1.5 Types of Wood in Terms of Deciduous and Coniferous

Trees are divided into two categories: deciduous and coniferous. Each of these categories produces wood with different characteristics [59].

- **Wood of deciduous trees**: Also called hardwoods and also known as angiosperms, are characterized by their complex wood structure and flowering reproductive system. They typically have broad, flat leaves and can be either deciduous or evergreen. While the term "hardwood" suggests a high degree of hardness, this isn't always the case; some hardwoods e.g. paulownia and poplar can be relatively softer than the typical softwoods e.g. Douglas-fir and long-leaf pine [4, 56].
  - **Wood structure**: Hardwoods possess a more intricate wood structure compared to softwoods. They contain vessel elements, fibers, longitudinal parenchyma, and rays with varying cell types, that will be more illustrated later.

  - **Fiber length**: Hardwood fibers are generally shorter than softwood tracheids, ranging from 1.5 to less than 1 mm.
  - **Vessel production**: Hardwoods can be classified as ring-porous or diffuse-porous based on the pattern of vessel production throughout the growing season. This pattern significantly influences their suitability for different applications, that will be more illustrated later.

- **Coniferous wood**: Also called softwoods. Softwoods, or gymnosperms, are primarily characterized by their simple wood structure and lack of true flowers. They often have needle-like leaves and are typically evergreen, though some species e.g. Genkgo biloba have broad leaves.

  - **Wood structure**: Softwood primarily consists of vertical tracheids, horizontal ray cells, and specialized resin canals. They generally lack fibers.
  - **Tracheid length**: Softwood tracheids are significantly longer than hardwood fibers, averaging around 3.0 mm.
  - **Growth pattern**: Softwoods are typically evergreens, but a few species exhibit deciduous characteristics [4, 56].

  While the terms "hardwood" and "softwood" are widely used in the forestry and woodworking industries, it's important to remember that they are not strictly botanical terms. For example, as it was written before, some hardwoods can be softer than certain softwoods, and vice versa. So, these terms are primarily used for convenience and to align with common industry practices.

- **The pore distribution in hardwoods and softwoods**: The structure of hardwoods, as seen in cross-section, is often characterized by the arrangement of its pores. These pores, or vessels, are channels that transport water and nutrients throughout the tree. The pattern of pore distribution can vary significantly among different species and has a significant impact on the wood's properties, such as its grain, texture, and strength.

  The porous categorization of ring-porous, diffuse-porous, and semi-diffuse-porous is primarily used for hardwoods. Softwoods, which are primarily gymnosperms (cone-bearing trees), generally have a simpler structure and do not exhibit the same distinct pore arrangements as hardwoods. Softwoods typically have small, uniform pores that are distributed more evenly throughout the growth rings [10]. This is due to their different anatomical structure and growth characteristics compared to hardwoods.

  While there may be some variation in pore size and distribution among different softwood species, the overall pattern is generally less complex than that found in hardwoods (Fig. 2.5).

  - **Ring-porous hardwood**: Ring-porous woods are characterized by the presence of large, distinct pores arranged in earlywood bands. These bands are typically followed by narrower bands of latewood, which contain smaller pores. This alternating pattern creates a visible ring-like structure in the wood, hence the

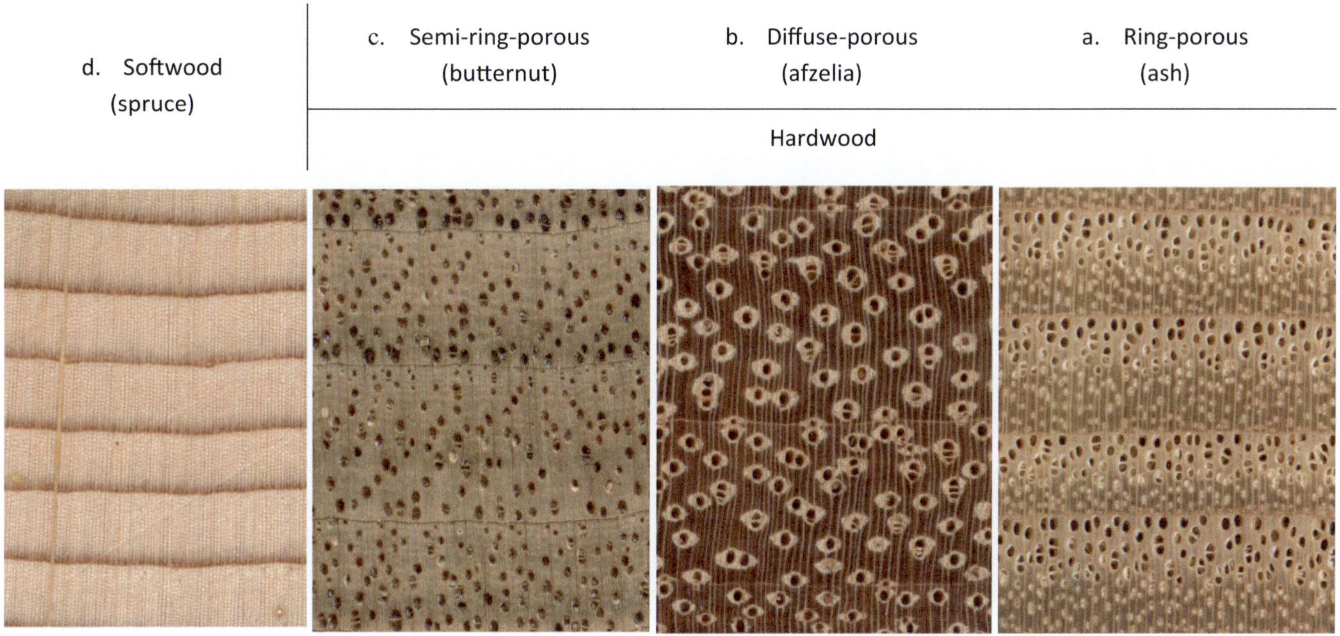

Fig. 2.5 A close-up comparison of four different wood examples, showing variations in pore distribution. The image includes: **a** ring-porous hardwood (ash): large pores are clustered in earlywood bands. **b** Diffuse-porous hardwood (afzelia): pores are evenly distributed throughout the growth ring. **c** Semi-ring-porous hardwood (butternut): pores are larger and more numerous in earlywood, with smaller pores in latewood. **d** Softwood (spruce): no visible pores, with a uniform texture. The images are used with permission from: https://www.wood-database.com

name. Examples of ring-porous woods include mulberry, European oak, European ash, elm, and hickory. These species are commonly used for different industries due to their distinctive grain patterns and strength.

 - **Diffuse-porous wood**: In diffuse-porous woods, the pores are distributed relatively evenly throughout the growth rings. There is no clear distinction between earlywood and latewood. This type of pore arrangement results in a more uniform grain pattern. Examples of diffuse-porous woods include maple, birch, and sapele. These species are often used for furniture, cabinetry, and musical instruments due to their fine grain and even texture.
 - **Semi-ring-porous wood**: Or semi-diffuse-porous woods exhibit a combination of ring-porous and diffuse-porous characteristics. The pores are arranged in a scattered pattern, with some larger pores clustered together and others distributed more evenly. This type of pore arrangement can result in a variety of grain patterns, depending on the specific species. Examples of semi-ring-porous woods include walnut and teak. These species are highly valued for their beauty, durability, and versatility, making them popular choices for furniture, cabinetry, and other high-end applications [8].

The pore distribution in wood has significant implications for its properties and uses. For example:

- **Grain pattern**: The arrangement of pores affects the grain pattern of the wood, which can influence its appearance and aesthetic value.
- **Strength and durability**: The size and distribution of pores can affect the wood's density, stiffness, strength and durability. Woods with larger pores may be more susceptible to decay and insect attack.
- **Drying characteristics**: The rate at which wood dries can be influenced by its pore structure [40]. Pore spaces are negatively related to wood density and specific gravity. Large pore volumes facilitate rapid drying. So, woods with larger pores may dry more faster and be more prone to warping or cracking.

Understanding the different types of pore distribution in wood is essential for selecting the appropriate species for various applications. By considering factors such as grain pattern, density, strength, durability, and workability, woodworkers and designers can choose the most suitable wood for their specific needs.

- **Reaction wood**: Reaction wood is a specialized type of hardwood or softwood that forms in response to stress or lean in trees. It is produced in an attempt to counteract the effects of gravity or a permanent wind load and maintain the tree's upright position. Reaction wood can have significant impacts on the growth and development of trees. For example, the formation of reaction wood can alter the shape and orientation of a tree's trunk or branches. It can also affect the tree's mechanical properties and its susceptibility to damage from wind, snow, or other factors. In forestry and wood science, understanding reaction wood is important for several reasons. It can help in the selection of trees for specific uses, such as timber for construction or pulp for paper production. It can also inform decisions about tree management practices, such as pruning or thinning, to minimize the formation of reaction wood and maintain the health and productivity of trees for musical instruments industry.

There are two main types of reaction wood: tension wood and compression wood (Figs. 2.6 and 2.7).

Fig. 2.6 A diagram comparing wood growth patterns in hardwoods and softwoods on a slope. • **Hardwood**: the cross-section shows wider growth rings on the uphill side, indicating that hardwoods primarily add wood to their tensile (upper) side. • **Softwood**: the cross-section shows wider growth rings on the downhill side, indicating that softwoods primarily add wood to their compression (lower) side. Arrows point to the growth rings, highlights the year the tree began to incline. A derivative hand art image created by author, with an AI enhancement, inspired from Degraff and Agard [18]

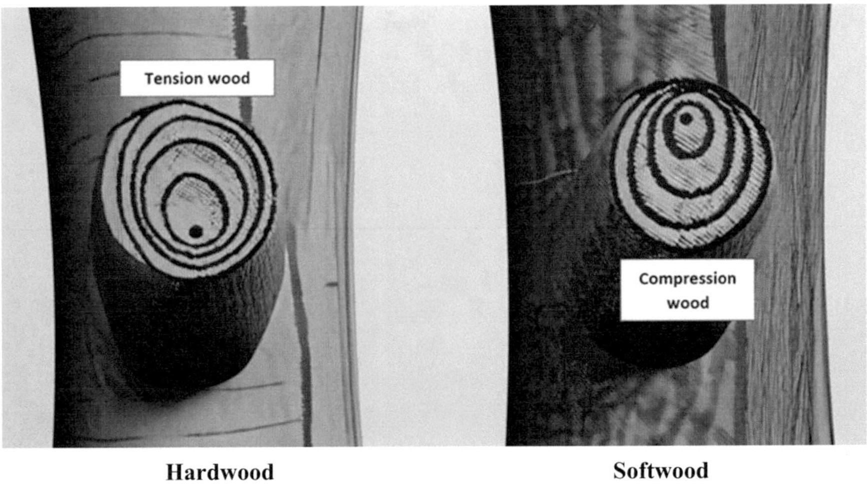

Fig. 2.7 Two cross-sections of tree branches, right is a softwood and left, a hardwood, demonstrating the formation of reaction wood in response to stress. The softwood trunk shows compression wood on the lower side (facing down-compression side), while the hardwood trunk shows tension wood on the upper side (facing up-tensile side)

- **Tension wood** is formed on the upper side of leaning stems in hardwood trees. It is characterized by its high cellulose content and the presence of gelatinous fibers in its cell walls. These gelatinous fibers have a unique structure that allows them to shrink under tension, helping to straighten the leaning stem [9]. Various studies have contrasted tension wood with normal wood, overlooking the intensity of tension wood formation. While tension wood generally displays a higher density, its structural heterogeneity and intermittent occurrence, in contrast to compression wood, impede the identification of unequivocal density trends [15].
- **Compression wood** is formed on the lower side of leaning stems in softwood trees. It is characterized by its high lignin content, which gives it a denser structure [15]. Compression wood is produced to counteract the compressive forces acting on the lower side of the leaning stem. While it is stronger in compression than normal wood, compression wood can be weaker in tension. Compression wood exhibits significant physical and mechanical property changes compared to normal wood. It has higher density, leading to increased compressive strength and hardness, but lower elastic stiffness, reduced radial, tangential, and volumetric shrinkage, and significantly higher longitudinal shrinkage [15].

2.2 Physical and Mechanical Properties of Wood

Wood, a natural material derived from trees, exhibits a unique combination of physical and mechanical properties that make it a versatile resource for various applications. This initial chapter provides a foundational overview of key physical and mechanical wood properties that are essential for understanding tonewood identification and control in string musical instruments industry. While we'll touch on several of these properties here, we'll go deeper into the most critical ones for specific practical instructions in subsequent chapters.

Our goal is to familiarize luthiers with these fundamental wood characteristics, laying the groundwork for more in-depth discussions throughout the book.

2.2.1 *Density and Specific Gravity*

Density is a crucial physical property that significantly impacts its weight, strength, and workability. It is defined as the mass per unit volume of wood. In simpler terms, it measures how heavy a piece of wood is for its size. Its unit is kilograms per cubic meter (kg/m^3).

Specific Gravity is the ratio of the density of a wood to the density of a reference substance and becomes a unitless parameter. For wood, the reference substance is usually water at 4 °C [59]. Specific gravity and density are closely related concepts, often used interchangeably when discussing wood. However, there is a subtle difference between them. In other words (Eq. 2.1):

$$\begin{aligned} Specific\_Gravity &= \frac{Density\_of\_wood}{Density\_of\_reference\_subs.} \\ &= \frac{Dry\_Weight\_of\_Wood}{Weight\_of\_an\_equal\_volume\_of\_water} \end{aligned} \tag{2.1}$$

Wood density may be determined at oven-dry, partially dry, or green condition. The term "Basic (Conventional) density" is used when the mass of the test piece is determined at absolutely dry condition i.e., oven-dry condition, but its volume is at a moisture content greater than or equal to the fiber saturation point (over 30% moisture contents for most wood species). The determination of the mass of the test piece is by weighing and its volume is by measurement of its dimensions, or by other methods. The volume of irregularly-shaped specimens can be determined by immersion in water, in mercury or in other liquids where the mass of displaced volume equals the volume of the test piece.

The density is calculated with varying moisture contents of wood. Therefore, the conditions under which density is determined shall be specified. For partially dry (or air-dry) wood, 12% moisture content is frequently used as a reference. The apparent density, "*ad*" of each test piece at the moisture content "*mc*" at the time of the test is given, in kilograms per cubic meter (or grams per cubic centimeter), by the formula [26] (Eq. 2.2):

$$ad = \frac{Weight\_at\_the\_mc}{Volume\_at\_the\_mc} \tag{2.2}$$

When required, the apparent density shall be adjusted to a 12% moisture content by the formula valid for moisture content from 7 to 17% (Eq. 2.3):

$$ad_{12} = ad_{mc}\left(1 - \frac{(1-K)(mc-12)}{100}\right) \tag{2.3}$$

where K is the coefficient of volumetric shrinkage for a change in moisture content of 1%. The value of K shall be determined according to National or International standards as will be described later in shrinkage and swelling section of this chapter. For approximate calculations, the value of K can be taken as equal to $ad \times 0.85 \times 10^{-3}$ when the density is expressed in kilograms per cubic meter (kg/m^3).

Specific gravity is directly related to density. A wood with a higher density will also have a higher specific gravity. In practice, specific gravity is often used in the wood industry because it allows for easy comparison of different wood species. In fact, while density measures the mass per unit volume of wood, specific gravity compares the density of wood to water, providing a standardized way to assess its relative heaviness. For a workshop operator, both terms have the same meaning and will not need to be sensitive to these alternative academic definitions. Because both parameters will always show the same value as long as the unit of gram per cubic centimeter is expressed for density, and in cases where density is expressed in kilograms per cubic meter, simply divide by 1000 to convert it to specific gravity.

- **Factors affecting wood density**

  The true situation is that all factors interact, and it becomes quite impossible to find a distinct causal effect for each. Density is not a simple characteristic but is determined by several characteristics of wood such as cell size and wall thickness, the ratio of earlywood to latewood, the amount of ray cells, the size and amount of vessel elements, and other factors.

  - **Species**: Different wood species naturally have varying densities. For instance, hardwoods like oak and maple are generally denser than softwoods like pine and spruce.
  - **Growth conditions—growth patterns**: Factors such as site quality, fertilization, stocking, growth rate, and provenance can all be of importance in determining wood density under certain conditions. Soil quality, climate, and growing conditions can influence wood density. Considering some exceptions, trees grown in harsh conditions may develop denser wood as a protective mechanism. The thickness of any cellwall within the late and early wood and also the percentage of latewood affects the wood density.
  - **Moisture content**: The amount of water present in wood affects its density.

  - **Defects**: Knots, stains and deteriorations, reaction wood and other defects can also influence density.
  - **Chemical deposits**: In addition to cell characteristics, factors such as chemical deposits within and between the cells can drastically affect specific gravity. When such deposits are present, wood specific gravity must be categorized as "extracted" or "unextracted".

- **Significance of wood density**

  - **Weight**: Higher density wood is heavier for a given volume. This is important in applications where weight is a concern, such as musical instruments, furniture, construction, and shipping.
  - **Strength**: Generally (not always), denser woods are stronger and more durable. They can better withstand loads and stresses.
  - **Workability**: Denser woods can be more challenging to work with, as they require more effort to cut, shape, and finish.
  - **Insulation**: Wood density can influence its insulating properties. Denser woods may provide better conductivity of heat and cold.
  - **Appearance**: Wood density can affect its appearance. Denser woods often have a denser grain and a more solid feel.

Wood density is a fundamental property that affects various aspects of wood's performance and suitability for different applications. By understanding wood density, engineers, architects, and woodworkers can make informed decisions about selecting the appropriate species and grades for their projects.

2.2.2 *Moisture Content*

The amount of water contained within wood, affecting its dimensions, weight, and mechanical properties.

Wood moisture content (*mc*) is the amount of water present in wood, expressed as a percentage of the dry weight of the wood (Eq. 2.4).

$$mc = \frac{m_h - m_o}{m_o} \times 100 \tag{2.4}$$

in which, m_h and m_o are the weights of non-dry and dry wood, respectively.

As you may already know, the relative humidity (RH) is the amount of water vapor in the air compared to the maximum amount of water vapor the air can hold at a given temperature.

- **The relationship between wood moisture content and relative humidity**

  - **Equilibrium moisture content (EMC)**: Wood will naturally absorb or lose moisture from the surrounding air until it reaches an equilibrium moisture content (EMC). This EMC is nominally estimated by the relative humidity and temperature of the environment (Table 2.1).

- **Hygroscopic nature**: Wood is hygroscopic meaning it can absorb and release moisture from the air. When the RH is high, wood tends to absorb moisture, increasing its mc. Conversely, when the RH is low, wood tends to lose moisture, decreasing its mc.
- **Hysteresis**: The relationship between wood mc and RH is not always linear. Wood often exhibits hysteresis, meaning it takes a different path to reach equilibrium when the RH is increasing compared to when it is decreasing. Wood reaches a higher moisture content through desorption than absorption under the same conditions.

  Figure 2.8 shows the relationship between moisture content and relative humidity for wood. It illustrates three curves: adsorption, desorption, and scanning curves. The adsorption curve represents the moisture uptake of dry wood as relative humidity increases. The desorption curve shows the moisture loss of wet wood as relative humidity decreases. The scanning curve is a combination of both, demonstrating hysteresis, where the desorption curve lies above the adsorption curve [32].
- **Fiber saturation point (FSP)**: Is the moisture content at which all the cell walls of wood are completely saturated with water. Beyond this point, any additional

Table 2.1 The equilibrium moisture content (EMC) of wood at different relative humidity (RH) and temperature conditions

		Temperature						
		10	15	20	25	30	35	40
Relative humidity (%)	90	21.1	21	21	20.8	20	19.8	19.3
	85	18.1	18	18	17.9	17.5	17.1	16.9
	80	16.2	16	16	15.8	15.5	15.1	14.9
	75	14.7	14.5	14.3	14	13.9	13.5	13.2
	70	13.2	13.1	13	12.8	12.4	12.1	11.8
	65	12	12	11.8	11.5	11.2	11	10.7
	60	11	10.9	10.8	10.5	10.3	10	9.7
	55	10.1	10	9.9	9.7	9.4	9.1	8.8
	50	9.4	9.2	9	8.9	8.6	8.4	8
	45	8.6	8.4	8.3	8.1	7.9	7.5	7.1
	40	7.8	7.7	7.5	7.3	7	6.6	6.3
	35	7	6.9	6.7	6.4	6.2	5.8	5.5
	30	6.2	6.1	5.9	5.6	5.3	5	4.7
	25	5.4	5.3	5	4.8	4.5	4.2	3.8

X-axis: relative humidity (%)
Y-axis: equilibrium moisture content (%)
Includes entries for RH from 25 to 90% and temperature from 10 to 40 °C

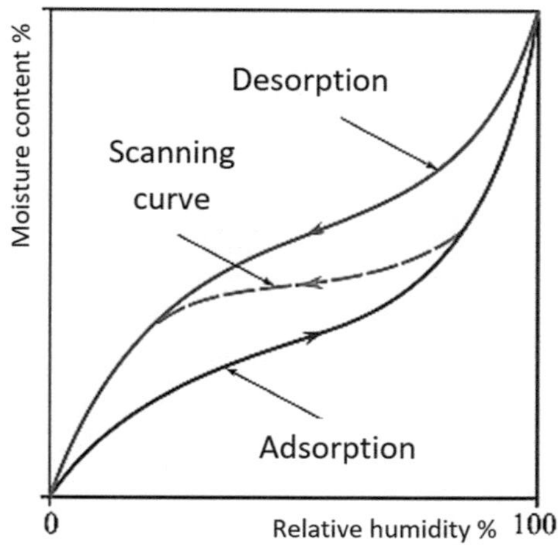

Fig. 2.8 The adsorption, desorption, and scanning curves depict the progression of equilibrium states for wood moisture content as a function of relative humidity. The directional arrows signify the progression of moisture adsorption or desorption processes

moisture is held in the cell cavities. FSP is typically around 25–30% for most wood species. It's a critical concept in understanding wood's behavior. When wood's moisture content exceeds FSP, it becomes more susceptible to decay, mold, and insect attack. Additionally, changes in moisture content below FSP can lead to dimensional instability, such as swelling and shrinking [6].

- **Factors affecting the EMC**
  - **Wood species**: Different wood species have varying EMCs. Some species, like Determa wood, are more resistant to moisture absorption than others [36].
  - **Temperature**: Higher temperatures increase the amount of water vapor the air can hold, affecting the EMC.
  - **Wood defects**: Knots, splits, deterioration and other defects can affect the EMC of wood.
  - **Extractives content**: İs another significant factor that can influence the relationship between wood moisture content and relative humidity. Extractives are naturally occurring substances found in wood, such as resins, tannins, and oils. Many extractives, especially resins and oils, have hydrophobic properties, meaning they repel water. This hydrophobic behavior can reduce the wood's ability to absorb moisture, especially from the surface. As a result, wood with a high extractives content may have a lower equilibrium moisture content (EMC) compared to wood with a lower extractives content.

  Because various factors influence wood's equilibrium moisture content (EMC), EMC tables similar to Table 2.1, typically provide only approximate values based on temperature and relative humidity.

- **Dimensional stability**: Changes in wood moisture content can cause dimensional changes, leading to shrinking, or swelling. Wood is a hygroscopic material, meaning it can absorb and release moisture from the surrounding environment.

Changes in wood moisture content can lead to dimensional changes, which can cause shrinking, or swelling.

- **Shrinking** is the reduction in the dimensions of wood as it loses moisture. When wood dries, it shrinks in all directions, but the shrinkage is typically greater along the grain than across the grain. Shrinking can cause gaps to appear between boards or joints, and it can also affect the fit of wood components.
- **Swelling** is the increase in the dimensions of wood as it absorbs moisture. When wood becomes wet, it swells in all directions, but the swelling is typically greater across the grain than along the grain. Swelling can cause wood to expand, leading to tight joints or even splitting.

Different wood species have varying degrees of dimensional stability. For example, some species, like redwood [36], are more resistant to dimensional changes than others. Or, tropical hardwoods that are stronger than temperate zone hardwoods of the same basic density, show less shrinkage [1]. The nominal value of coefficient of sweling or shrinkage for a wood sample shall be determined according to National or International standards [2, 27, 28]. The initial moisture content of the wood and the relative humidity of the environment significantly influence dimensional stability. Also the direction of the wood grain can affect how much the wood shrinks or swells (Eq. 2.5).

$$\begin{aligned} \beta_l &= \frac{l_{fsp} - l_0}{l_{fsp}} \times 100 \\ \beta_r &= \frac{r_{fsp} - r_0}{r_{fsp}} \times 100 \\ \beta_t &= \frac{t_{fsp} - t_0}{t_{fsp}} \times 100 \\ \beta_V &= \frac{r_{fsp} \cdot t_{fsp} - r_0 t_0}{r_{fsp} t_{fsp}} \times 100 \end{aligned} \tag{2.5}$$

where β is the total shrinkage/swelling in longitudinal (l), radial (r), tangential (t) and volumic domails. The indexes "*fsp*" and "*0*" are the are the dimensions of the green or fully saturated test piece and at absolutely dry (oven-dry) condition, respectively, measured in the longitudinal, radial and tangential directions. Longitudinal shrinkage in normal wood is so insignificant that it is safely ignored in standard volumetric shrinkage calculations.

Now:

$$K_i = \frac{\beta_i}{fsp} \tag{2.6}$$

In which, K_i is the coefficient of shrinkage in i = *longitudinal* or *radial* or *tangential* or *volumetric* domain. *fsp* is the fiber saturation point as introduced earlier.

Knots, splits, and other defects can weaken the wood's structural integrity and make it more susceptible to dimensional changes. To minimize dimensional changes in wood, it is important to select the appropriate wood species that are known for their dimensional stability or maintain a relatively constant moisture content in the wood by using appropriate finishes and storing wood in a controlled environment. The average values of radial and tangential shrinkages for lots of species as well as the ratio of tangential to radia shrinkage of them are available in some references i.e. Ruffinatto and Crivellaro [45]. Especially the ratio of T to R, which plays a key role in the disadvantages and damages caused by wood drying must be taken into an account. To minimize dimensional changes in wood, it's important to also consider and ensure that wood is thoroughly dried before use to reduce its moisture content and employ joinery methods that can accommodate some dimensional changes.

- **Durability**: Wood durability refers to its resistance to degradation and decay over time. It is influenced by a combination of factors, including the wood's inherent properties, environmental conditions, and any treatments or finishes applied.

  Different wood species exhibit varying levels of durability. Some species, like redwood and cedar, are naturally resistant to decay, while others, like pine, are more susceptible. The conditions under which a tree grows can influence its durability. Wood Defects: Knots, splits, and other defects can weaken wood and make it more vulnerable to decay.

  High moisture content can promote the growth of decay-causing fungi and bacteria. Factors like temperature, humidity, and exposure to sunlight can affect wood decay. Naturally occurring chemicals in wood, such as tannins and resins, can provide some level of protection against decay. Preservatives and finishes can enhance wood durability by protecting it from moisture, insects, and other harmful factors. Ensuring proper ventilation can help reduce moisture levels and prevent decay. Direct contact with the ground can increase the risk of decay. Regularly inspecting wood structures and addressing any signs of damage can help prevent further deterioration.

  Types of Wood Decay are:

  - **Brown rot** This type of decay causes wood to become brittle and crumbly. It is typically caused by fungi that break down cellulose [49].
  - **White rot** This type of decay causes wood to become soft and spongy. It is caused by fungi that break down both cellulose and lignin [11].
  - **Soft rot** This type of decay is characterized by the formation of small pits or cavities in the wood. It is often caused by fungi that attack the cell walls [11].
  - **Stain** Discoloration caused by mold, mildew, or fungi. Wood stain is indeed a type of weak deterioration that can affect the appearance of wood. It occurs when certain substances, such as mold, mildew, or fungi, discolor the wood's surface [11]. These organisms can thrive in humid environments and can cause unsightly stains or discoloration.

Ensuring proper ventilation can help reduce moisture levels and prevent decay. The relationship between wood moisture content and relative humidity is crucial to understanding how wood behaves in different environments. By controlling the moisture content of wood, it is possible to maintain its dimensional stability, durability, and appearance.

2.2.3 Color

The color of wood arises from a complex interaction of its chemical components, extractives, environmental factors, and external treatments during the drying process. Sandoval-Torres et al. [46] details several key factors and mechanisms that contribute to wood color, which can change due to these interactions.

- **Chemical composition of wood**
  Wood is primarily made of cellulose, hemicelluloses, and lignin, with additional extractives (such as tannins, phenolic compounds, and pigments) that are species-dependent. These chemical components are the primary contributors to the natural color of wood. Two critical chemical groups related to color are:
  - **Chromophores** These are atomic groups with unsaturated chemical bonds, like –C=C–, which can easily transition to an excited state and absorb light. They give wood its inherent colors by selectively absorbing certain wavelengths of light.
  - **Auxochromes** These are atomic groups with isolated electron pairs, like –OH or –COOH, that modify the behavior of chromophores, intensifying or shifting the wood's coloration.

Softwoods and hardwoods differ in their lignin and hemicellulose structures, which leads to species-specific coloration.

The yellow tones of wood are primarily governed by the photochemistry of wood, particularly in lignin. Extractives (such as polyphenols, tannins, and quinonoid pigments) play a significant role in color by giving red hues to heartwood. Phenolic Compounds are responsible for much of the dark color in heartwood. The oxidation of phenolic compounds, particularly in species like beech, birch, and oak, leads to browning and reddish hues.

Tannins and Flavonoids, often present in hardwoods, react with environmental factors such as light, heat, and moisture, changing the wood's color. For instance, when oxidized, tannins produce dark browns and reds.

When wood is exposed to heat, especially during kiln drying, its color darkens due to the thermal degradation of hemicelluloses and lignin. At temperatures as low as 65 °C, heat can cause hydrolysis and oxidation of wood components, shifting the color to darker tones. Higher temperatures also promote the migration of extractives to the wood surface, which can cause surface staining.

Light Exposure and UV light triggers photochemical reactions, especially in lignin, breaking it down into smaller compounds such as quinones and carbonyl groups, which contribute to the yellowing and darkening of wood. Prolonged light exposure can also bleach certain pigments, changing the appearance over time. For example, heartwoods containing quinonoid pigments may lighten, while other woods like pine, with more tannins, may redden.

Oxygen exposure can cause oxidation reactions in the extractives, especially under weakly alkaline or neutral conditions. For example, species like red alder or Sugi (*Cryptomeria japonica*) may develop reddish or orange hues due to oxidative reactions between the extractives and atmospheric oxygen.

Certain fungi (e.g., blue stain fungi) may cause visible discoloration, typically blue or gray, though they don't affect the wood's strength. Microbial activity often targets sapwood, where the nutrients are concentrated. While these fungi don't decay the wood, they create unsightly discolorations that can reduce its value for decorative uses.

Microorganisms can also interact with the chemical extractives, enhancing the oxidation of phenolic compounds or catalyzing other biochemical reactions that lead to color changes.

Similar to the browning of fruit, enzymatic reactions in living wood cells (like those catalyzed by polyphenoloxidases) create quinones, which then polymerize into brown pigments. These reactions are particularly significant in species like maple or birch, which are prone to internal darkening when exposed to heat and oxygen during drying.

The drying process, whether air-drying, kiln-drying, or vacuum drying, can have a profound impact on wood color. During drying, moisture and volatile organic compounds migrate to the surface, along with phenolic extractives. These extractives can undergo oxidation or hydrolysis, leading to discoloration. High drying temperatures accelerate these reactions, especially in woods with high tannin or sugar content. Some species, like European oak, exhibit noticeable darkening under high temperatures due to changes in the heartwood's tannin and ellagitannin contents [46].

2.2.4 Grain and Texture of Wood

Grain and texture are two essential terms used to describe the visual appearance and feel of wood. They are influenced by the arrangement of wood fibers and other characteristics. The terms "grain" and "texture" are often used interchangeably when discussing wood, but they have distinct meanings. While "grain" typically refers to the direction and pattern of the wood fibers, "texture" describes the appearance and feel of the wood surface. Grain can be further classified as fine or coarse, straight or curly, and open or closed, depending on the arrangement of fibers, annual rings, and pores. Texture, on the other hand, is influenced by the size and uniformity of wood cells within growth rings. A coarse texture often results from large vessels, while a fine texture is characterized by small, even-sized cells. When using these terms, it is important to specify the intended meaning to avoid confusion [48].

- **Grain**

  The grain refers to the direction and pattern of the wood fibers. It is visible as the lines or markings on the surface of the wood. The grain can affect the wood's strength, workability, and appearance. Types of grain of wood are:

  - **Straight grain**: The fibers run parallel to the length of the board, creating a uniform appearance. Straight-grained wood is generally longitudinally stronger and easier to work with than other grain types.
  - **Interlocked grain**: The fibers are interwoven, creating a zigzag or wavy pattern. Interlocking grain is a common feature in tropical hardwoods e.g. Mahogany and sapele that produces a ribbon like beautiful wavy pattern (Fig. 2.9), but it's rare in temperate hardwoods and almost nonexistent in softwoods.

Fig. 2.9 A longitudinal section of a Nigerian satinwood (*Distemonanthus benthamianus*), showcasing a distinctive interlocked grain pattern. The interlocked grain appears as a ribbon-like or wavy pattern across the wood surface. This image with permission is from: https://www.wood-database.com/wp-content/uploads/satin-1.jpg

- **Curly or spiral grain**: The fibers are tightly curled or twisted, resulting in a unique and decorative appearance (Fig. 2.10).
- **Figured grain**: The grain exhibits distinctive patterns or figures, such as burl, birdseye, or fiddleback (Fig. 2.11).

- **Texture**

  Texture refers to the feel and appearance of the wood surface, determined by the arrangement of fibers, pores, and other characteristics. Types of Texture of wood are:

  - **Coarse texture**: The fibers are widely spaced, creating a rough or grainy feel.

Fig. 2.10 A close-up image of curly maple wood grain (radial sawn), showcasing its distinctive swirling pattern. This image is courtesy of Violin Luthier Mikaeil Ghazal, who gifted it for inclusion in this book

Fig. 2.11 A close-up image of a mulberry wood board showcasing a beautiful burl grain pattern. The wood has a rich, golden-brown color with intricate swirling patterns. This image is courtesy of Mr. Sina Faramarzi, tonewood supplier, who gifted it for inclusion in this book

- **Fine texture**: The fibers are closely spaced, resulting in a smooth and even texture.
- **Uniform texture**: The texture is consistent throughout the wood.
- **Variable texture**: The texture varies within different areas of the wood.

Texture can affect the wood's overall appearance and its suitability for different applications. For example, a fine-textured wood may be more suitable for musical instruments, furniture or cabinetry, while a coarse-textured wood may be better suited for flooring or structural applications.

Grain and texture are interrelated properties that contribute to the unique appearance and feel of wood. Understanding these characteristics is essential for selecting the right wood species and grades for various applications.

2.2.5 *Strength*

Wood strength refers to its ability to resist deformation or failure when subjected to various types of loads. This property is crucial in determining the suitability of wood for different applications.

- **Factors affecting wood strength**
  - **Species**: Different wood species exhibit varying levels of strength due to their genetic makeup and growth conditions. Hardwoods, such as oak and maple, are generally stronger than softwoods like pine and spruce for fracture.
  - **Growth conditions**: Factors like soil quality, climate, and growing conditions can influence wood's growth in terms of the early and latewood composition and percentage resulting in its varying strengths.
  - **Grain orientation**: The direction of the wood fibers significantly impacts its strength. Wood with straight grain is typically stronger in longitudinal axis than wood with interlocked or curly grain.
  - **Moisture content**: The amount of water in wood affects its strength. As wood dries, it becomes stronger. However, excessive drying can lead to brittleness.
  - **Defects**: Knots, splits, deterioration and other defects can significantly reduce the strength of wood.
  - **Load type**: The type of load applied to the wood (e.g., bending, tension, compression) influences the specific strength properties required.
- **Types of wood strength**
  - **Bending strength (modulus of rupture)**: The ability of wood to resist bending or deflection under a load.
  - **Tensile strength**: The ability of wood to resist pulling or stretching forces.
  - **Compressive strength**: The ability of wood to resist crushing or compression.
  - **Shear strength**: The ability of wood to resist forces that cause it to slide or shear.
  - **Torsional strength**: The ability of wood to resist twisting or torsion.
- **Importance of wood strength**
  - **Structural applications**: Wood strength is crucial for the structural integrity of buildings, bridges, musical instruments, and other structures, particularly those subjected to complex loads.
  - **Product design**: Designers consider wood strength when crafting instruments, structures, and other wood products to ensure they can withstand expected loads.
  - **Safety**: Understanding wood strength is crucial for ensuring the safety and durability of wood-based structures and products.

Wood strength is a complex property influenced by various factors. By understanding the different types of wood strength and the factors that affect it, engineers, architects, and woodworkers can select appropriate wood species and grades for their specific applications.

2.2.6 Stiffness

Stiffness is a mechanical property of wood that measures its resistance to a reversible elastic deformation under load. It is a measure of how much a piece of wood will bend or deform when a force is applied to it. Stiffness is typically quantified using the modulus of elasticity (MOE). The MOE is a material property that relates stress (force per unit area) to strain (change in length per unit length).

Stress (σ) is the force (F) applied to the wood divided by its cross-sectional area (A) and strain (ε) is the change in length of the wood (δ), divided by its original length (L) (Eqs. 2.7–2.10).

$$\sigma = \frac{F}{A} \tag{2.7}$$

$$\varepsilon = \frac{\delta}{L} \tag{2.8}$$

$$E = \frac{\sigma}{\varepsilon} = \frac{FL}{A\delta} = K\frac{L}{A} \tag{2.9}$$

$$K = \frac{F}{\delta} \tag{2.10}$$

where, E and K both are the indicators for the elastic stiffness, the first is a material property (modulus of elasticity) and the second is also a similar material property but changing with the size and shape of the object (springy constant) (Fig. 2.12).

These equations are valid for isotropic materials, where properties are the same in all directions. In an orthotropic material, such as wood, the elastic modulus varies significantly depending on the direction in which the stress is applied. This is due to the orthotropic nature of wood, meaning its properties are different in different directions. For wood, applying the equations to three different orthotropic axes results in a 6×6 stiffness matrix. But first, we have to deal with other concepts such as shear modulus (shear stiffness) and Poisson's ratio in order to show the orthotropic nature of wood in this complex 6×6 matrix.

Figure 2.13 shows a stress–strain curve illustrating the mechanical behavior of wood in its growing axis.

Yield Strength is the point on the stress–strain curve beyond which a wooden material begins to deform plastically. In simpler terms, it's the point where the wood

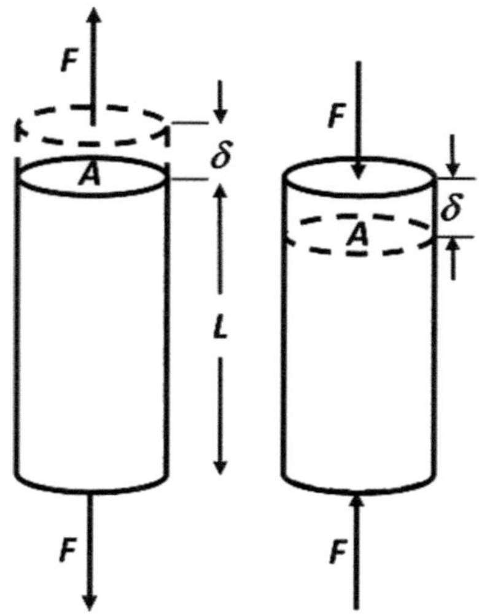

Fig. 2.12 A diagram illustrating the deformation of an object under tensile (left) and compressive (right) force (F). The object is shown in two states: initial and deformed. The initial state shows the object's original length (L) and cross-sectional area (A). The deformed state shows the object stretched or compressed by a distance (δ), indicating an increase or decrease in length. Arrows indicate the direction of the applied force and the resulting deformation

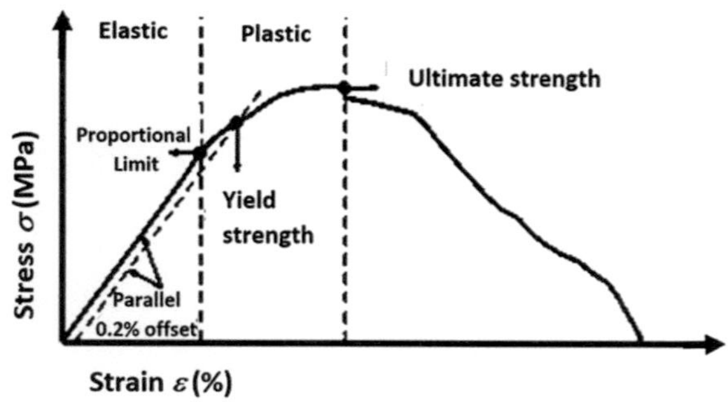

Fig. 2.13 A stress–strain curve illustrating the mechanical behavior of wood. The curve shows three distinct regions: elastic, plastic, and post-failure. Key points include the proportional limit, yield strength, and ultimate strength. The 0.2% offset method is used to determine the yield strength

starts to permanently deform rather than returning to its original shape after the stress is removed.

In the provided stress–strain curve, the yield strength is indicated by the point where the curve deviates from the linear elastic region and starts to curve upwards. The 0.2% offset method is often used to determine this point more precisely. The 0.2% offset method is a technique used to determine the yield strength of a material, especially when the yield point is not clearly defined on the stress–strain curve [38].

This method provides a consistent way to estimate the yield strength, even for materials that don't exhibit a distinct yield point.

As the diagram shows, wood uniquely enters a post-failure stage after reaching its ultimate strength, continuing to withstand some stress [7]. While this post-failure behavior is less relevant for acoustic instruments, it's crucial for earthquake-resistant construction. By delaying complete failure, it provides precious time for building occupants to evacuate during seismic events.

The longitudinal direction is parallel to the grain or the long axis of the tree. The elastic modulus in the longitudinal direction (E_L) is typically the highest of the three directions. This is because the wood fibers are aligned in this direction, providing the greatest resistance to deformation (Eq. 2.11).

$$E_L = \frac{\sigma_L}{\varepsilon_L} \tag{2.11}$$

The high longitudinal elastic modulus makes wood well-suited for applications that require stiffness, such as musical instruments resonators [42].

The tangential direction is perpendicular to the grain but within the plane of the growth rings (Eq. 2.12). The elastic modulus in the tangential direction (E_T) is generally lower than in the longitudinal direction. This is due to the less organized arrangement of wood cells in this direction.

$$E_T = \frac{\sigma_T}{\varepsilon_T} \tag{2.12}$$

The lower tangential elastic modulus can affect the wood's resistance to bending and twisting. It is important to consider this direction when designing the musical instruments or components.

The radial direction is perpendicular to both the grain and the growth rings (Eq. 2.13).

$$E_R = \frac{\sigma_R}{\varepsilon_R} \tag{2.13}$$

The elastic modulus in the radial direction is typically (without considering the radial ray cells) smaller than the longitudinal and tangential directions. It is influenced by the arrangement of axial or lateral wood cells and the presence of early and latewood different compositions in growth rings. So, in lots of cases, the elastic modulus in the radial direction becomes typically equal to or greater than tangential [58].

The radial elastic modulus can affect the wood's resistance to splitting and shrinkage. It is important to consider this direction when selecting wood for specific applications, such as musical instruments industry.

When required, the elastic modulus shall be adjusted to a 12% moisture content [12, 19, 52, 58].

About this property of wood we must know that the static elastic modulus and dynamic elastic modulus are two key mechanical properties of wood that measure its stiffness or resistance to deformation under load. While they both describe the relationship between stress and strain in wood, they differ in the type of loading and the rate at which the load is applied [7]. They and their practical measurements will be discussed in next chapters alot.

- **Static modulus of elasticity**: The static modulus of elasticity is the ratio of stress to strain in wood when the load is applied slowly and steadily, such as in bending or above illustrated compression methods. It is a measure of the wood's stiffness under static conditions. Static modulus of elasticity is typically determined by subjecting wood samples to a gradually increasing load until failure and measuring the corresponding stress and strain.
- **Dynamic modulus of elasticity**: The dynamic modulus of elasticity is again the ratio of stress to strain in wood when the load is applied rapidly and cyclically, such as in vibration or impact tests. It is a measure of the wood's stiffness under dynamic conditions. Dynamic modulus of elasticity is typically determined by subjecting wood samples to a rapidly oscillating load and measuring the resulting stress and strain. The dynamic modulus of elasticity is influenced by the same factors as the static modulus of elasticity, but it is generally higher due to the frequency dependence or the rapid loading rate [14] and some other operational errors that will be discussed in details in next chapters while the measurement methodologies and operations are discussed. To address the impact of rapid loading errors on vibration, it's necessary to recognize the viscoelastic nature of wood. This means that wood requires time to fully deform under stress. Rapid loading, however, limits this deformation time, leading to incomplete response and potential errors in vibration analysis, that would not be a concern in musical instruments industry. It is enough here to highlight that the static modulus of elasticity is measured under static conditions, while dynamic modulus of elasticity is measured under dynamic conditions. Static modulus of elasticity is used in applications where the wood is subjected to static loads, such as in structural design and construction. Dynamic modulus of elasticity is used in applications where the wood is subjected to dynamic loads, such as in sports equipment, musical instruments, and impact-resistant structures. Therefore, the dynamic modulus of elasticity, despite its differences with the static modulus, is actually the parameter that is important and needed in the musical instrument industry. Both the static and the dynamic moduli of elasticity are important mechanical properties of wood that characterize its stiffness. The choice between the two depends on the specific application and the type of loading that the wood will be subjected to.

  Now that the name of viscoelastic behavior of wood is mentioned, it is necessary to give a brief explanation about it here.

  Wood, like many natural materials, exhibits viscoelastic behavior. This means it responds to stress with both elastic and viscous properties [7].

- **Elastic behavior**: When a stress is applied to wood, it initially deforms elastically, meaning it returns to its original shape when the stress is removed. This is a rapid response and decribed above.
- **Viscous behavior**: However, wood also exhibits time-dependent deformation, known as creep. When a constant stress is applied, the wood continues to deform over time, even after the initial elastic deformation. This delayed deformation is due to the viscous nature of the wood's structure.

The behavior of wood is neither completely viscous nor completely elastic. Something between these two behaviors is called viscoelastic. This indicates that wood experiences a time lag between the application of force and the initiation of deformation. This delay can be quantified and visualized by analyzing the phase angle difference between the stress applied and the resulting deformation. Therefore the viscoelastic behavior of wood has significant implications for its use in various applications. The viscoelastic properties of wood, such as phase delay and vibration attenuation in soundboards after string excitation, significantly influence the sound quality of stringed musical instruments. We will go deeper into these properties in the following chapters.

But something that finalizes the topic, is the concept of the phase angle (δ) in the complex modulus of elasticity (CME) definition. CME is a measure of wood's stiffness and damping properties under dynamic loading conditions. It's a complex number composed of a real part (elastic modulus) and an imaginary part (loss modulus). Damping capacity is a property that will be discussed in detais in future chapters and refers to its ability to absorb and dissipate energy, typically from vibrations. Damping or the energy attenuation must be considered in complex modulus definition. Higher damping capacity means the wood can better reduce vibrations and noise. Components of the CME are:

- **Storage modulus (E')**: This represents the material's stiffness and is the ratio of stress to strain in the elastic range. It's a measure of how much a material deforms under a given load.
- **Loss modulus (E'')**: This represents the material's damping capacity, or its ability to dissipate energy as heat during vibration. It's related to the internal friction within the material.

To calculate the complex modulus, theoretically [7] (Eq. 2.14),

$$\mathrm{E}* = \mathrm{E}' + \mathrm{i}\mathrm{E}'' \tag{2.14}$$

where, $E*$ is the complex modulus and i is the imaginary unit ($i^2 = -1$).

Now that we have considered the CME as a complex number, it would be calculated in a triangle as (Fig. 2.14).

So, these relationships are considered (Eqs. 2.15 and 2.16):

$$E*^2 = E'^2 + E''^2 \tag{2.15}$$

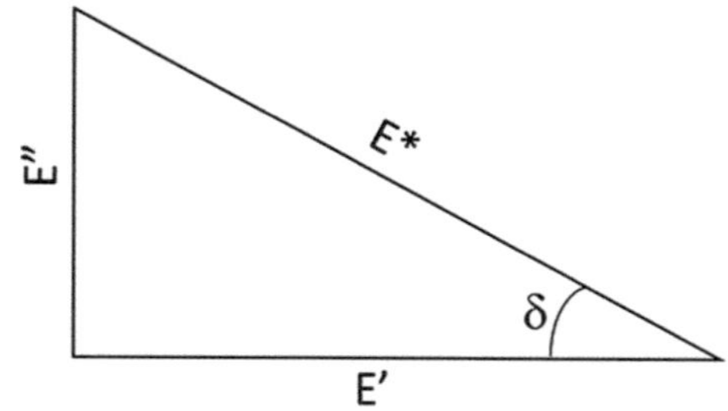

Fig. 2.14 A right triangle illustrating the complex modulus of elasticity (E*). The hypotenuse represents the magnitude of E*, while the angle δ (delta) represents the phase angle between the stress and strain. The adjacent side E′ is the storage modulus, representing the elastic component of the material's response to stress. The opposite side E″ is the loss modulus, representing the viscous or energy-dissipating component of the material's response

$$\tan\delta = \frac{E''}{E'} \tag{2.16}$$

where, tan δ represents the attenuation of the energy of vibration and named the "damping capacity".

CME is essential in understanding the dynamic behavior of wood structures, such as musical instruments. We have a lot of the practical works with this parameter through this book. It provides valuable information about the material's internal structure and properties, which can be used for quality control and product development.

- **Shear modulus**

  Shear modulus of elasticity also known as the modulus of rigidity, is a measure of a material's resistance to deformation under shear stress [7]. In simpler terms, it indicates how much force is required to twist or shear a material without causing permanent deformation (Fig. 2.15).

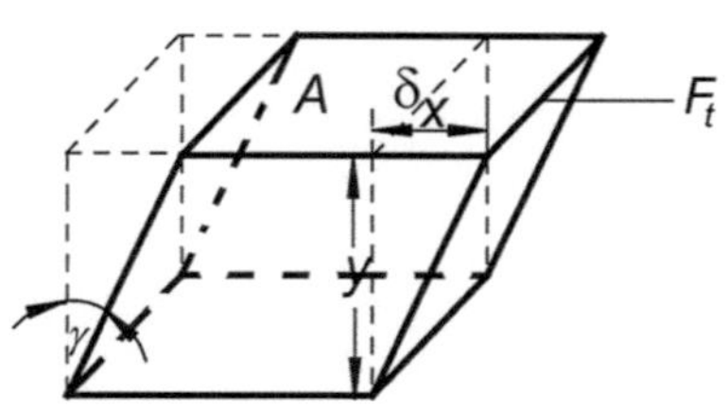

Fig. 2.15 A diagram illustrating shear stress and deformation in a rectangular block. A force (F_t) is applied parallel to one face of the block, causing it to deform. The original shape of the block is shown with dashed lines, while the deformed shape is shown with solid lines. The deformation is characterized by a change in angle (γ) and a displacement (δ_x) in the direction of the applied force. The cross-sectional area of the block is labeled as A

The shear modulus (G) can be calculated using the following formula (Eqs. 2.17–2.19):

$$\tau = \frac{F_t}{A_{||}} \tag{2.17}$$

$$\gamma^r = \tan\gamma = \frac{\delta_x}{y} \tag{2.18}$$

$$G = \frac{\tau}{\gamma} \tag{2.19}$$

where, τ is the shear stress (force, F_t, per unit area, $A_{||}$, parallel to F_t) applied to the wood. γ is the shear strain (angular deformation) resulting from the applied shear stress. For small shear angles, the tangent function is a good approximation of the angle in radians. This allows us to simplify the formula by using the angular deformation directly. δ_x is the deformation or displacement parallel to x axis.

Shear modulus is a critical property for understanding the behavior of wood under various loading conditions. It is particularly important in applications where wood is subjected to twisting or torsional forces, such as in beams, shafts, and structural components. While the shear modulus is less commonly used for experimentally assessing wood's vibrational properties in musical instrument manufacturing, understanding it is crucial for learning the concept of orthotropic wood. Therefore, it remains an essential component of this knowledge base.

In the context of wood, shear modulus varies depending on the direction of the applied force relative to the wood's grain, i.e. LR, LT, and RT Directions [37].

- LR plane (Longitudinal-Radial): The force is applied parallel to the wood's length and perpendicular to its radial direction.
- LT plane (Longitudinal-Tangential): The force is applied parallel to the wood's length and perpendicular to its tangential direction.
- RT plane (Radial-Tangential): The force is applied perpendicular to both the length and radial direction of the wood.

There are no obvious differences between the shear Modulus values in *ij* or *ji* planes, so we consider the G_{LR} same as the G_{RL}. The same can be considered for the other planes.

Generally, the shear modulus in the LR direction is higher than in the LT direction, which is, in turn, higher than in the RT direction ($G_{LR} > G_{LT} > G_{RT}$).

2.2.7 *Poisson's Ratio*

Poisson's ratio is a dimensionless material property that relates the transverse (lateral) strain to the axial strain of a material subjected to a uniaxial stress. In simpler terms, it describes how much a material expands or contracts laterally when it is stretched or compressed lengthwise [7].

To calculate Poisson's ratio for wood, you need to measure both the axial and the lateral strains. Here's the formula:

$$\nu = -\frac{\varepsilon_l}{\varepsilon_a} \tag{2.20}$$

Poisson's ratio (ν)
Lateral strain (ε_l)
Axial strain (ε_a).

For an example, to measure axial strain in a wood specimen (Fig. 2.16), in which the growth and orthogonal directions are nominally matched with each other apply a known axial load to the specimen in its longitudinal growth direction and measure the change in its length (ΔL).

Determine the longitudinal (axial) strain (ε_L), using the formula:

$$\varepsilon_L = \frac{\Delta L}{L} \tag{2.21}$$

where L is the original length of the specimen.

Measure the change in width in Radial (ΔR) or tangential (ΔT) direction of the specimen perpendicular to the direction of the applied load.

Calculate the lateral strain using the formula:

$$\varepsilon_R = \frac{\Delta R}{R} \tag{2.22}$$

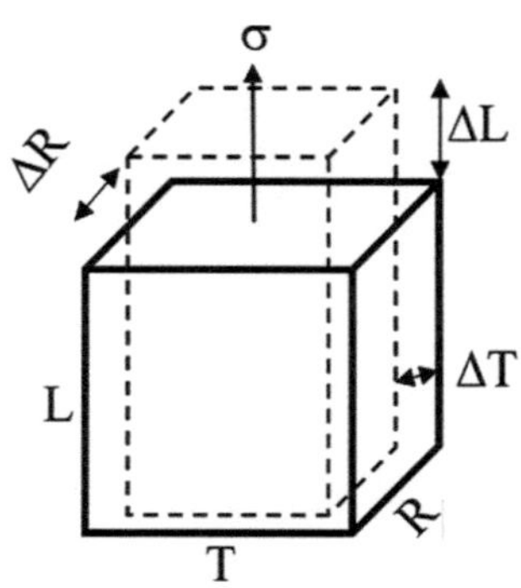

Fig. 2.16 A diagram illustrating the setup for measuring axial strain in a wood specimen. A cubic wood block is subjected to an axial stress (σ) applied in the longitudinal direction (L). The resulting change in length (ΔL) is measured to determine the axial strain. The radial (R) and tangential (T) directions are also labeled for reference

$$\varepsilon_T = \frac{\Delta T}{T} \tag{2.23}$$

where R and T are the original width or thickness of the specimen in radial or tangential derection of the specimen, respectively.

Substitute the calculated axial and lateral strains into the Poisson's ratio formula.

$$\nu_{LR} = \frac{\varepsilon_R}{\varepsilon_L} \tag{2.24}$$

$$\nu_{LT} = \frac{\varepsilon_T}{\varepsilon_L} \tag{2.25}$$

Due to its orthotropic nature, wood exhibits varying Poisson's ratios in different directions [37]:

- Longitudinal-Radial (ν_{LR})
- Radial-Longitudinal (ν_{RL})
- Radial-Tangential (ν_{RT})
- Tangential-Radial (ν_{TR})
- Longitudinal-Tangential (ν_{LT})
- Tangential-Longitudinal (ν_{TL}).

The Poisson's ratio in the LR direction would be higher than in the RL direction. When a wood specimen is compressed or stretched radially, it exhibits a longitudinal response (ν_{RL}). However, the wood is most stiff in the longitudinal direction, resulting in significantly smaller deformations compared to the radial response in LR loadind and deformation (ν_{LR}).

Likewise, the Poisson's ratio is higher in the LT ratio than in the TL, and the TR than the RT ratio. This is because longitudinal stiffness, E_L, typically exceeds tangential stiffness, E_T, while radial stiffness E_R is generally a different value.

Now, imagine that due to variations in Poisson's ratio within wood, additional moduli of elasticity besides the longitudinal, E_L, radial, E_R and tangential E_T ones could be considered. Orthotropic Poisson's ratios are asymmetrical and vary, but when combined with their related orthotropic stiffness, they form a symmetrical matrix.

$$\begin{aligned} \nu_{ij}E_i &= \nu_{ji}E_j \\ \nu_{LR}E_L &= \nu_{RL}E_R \\ \nu_{LT}E_L &= \nu_{TL}E_T \\ \nu_{RT}E_T &= \nu_{TR}E_T \end{aligned} \tag{2.26}$$

When a load is applied to wood along one axis, the Poisson effect results in different moduli of elasticity perpendicular to that axis. This concept seems to be a bit complicated, then it is simplified as:

$$\begin{aligned}
\varepsilon_L &= \frac{\sigma_L}{E_L} - \frac{\nu_{RL}\sigma_R}{E_L} - \frac{\nu_{TL}\sigma_T}{E_L} \\
\varepsilon_R &= -\frac{\nu_{LR}\sigma_L}{E_R} + \frac{\sigma_R}{E_R} - \frac{\nu_{TR}\sigma_T}{E_R} \\
\varepsilon_T &= -\frac{\nu_{LT}\sigma_L}{E_L} - \frac{\nu_{RT}\sigma_R}{E_L} + \frac{\sigma_T}{E_L} \\
\gamma_{LR} &= \frac{\tau_{LR}}{G_{LR}} \\
\gamma_{LT} &= \frac{\tau_{LT}}{G_{LT}} \\
\gamma_{RT} &= \frac{\tau_{RT}}{G_{RT}}
\end{aligned} \tag{2.27}$$

And in a matrix form:

$$\begin{bmatrix} \varepsilon_L \\ \varepsilon_R \\ \varepsilon_T \\ \gamma_{LR} \\ \gamma_{LT} \\ \gamma_{RT} \end{bmatrix} = \begin{bmatrix} \frac{1}{E_L} & \frac{\nu_{RL}}{E_R} & \frac{\nu_{TL}}{E_T} & 0 & 0 & 0 \\ \frac{\nu_{LR}}{E_L} & \frac{1}{E_R} & \frac{\nu_{TR}}{E_T} & 0 & 0 & 0 \\ \frac{\nu_{LT}}{E_L} & \frac{\nu_{RT}}{E_R} & \frac{1}{E_L} & 0 & 0 & 0 \\ 0 & 0 & 0 & \frac{1}{G_{LR}} & 0 & 0 \\ 0 & 0 & 0 & 0 & \frac{1}{G_{LT}} & 0 \\ 0 & 0 & 0 & 0 & 0 & \frac{1}{G_{RT}} \end{bmatrix} \begin{bmatrix} \sigma_L \\ \sigma_R \\ \sigma_T \\ \tau_{LR} \\ \tau_{LT} \\ \tau_{RT} \end{bmatrix} \tag{2.28}$$

The Poisson's ratio for wood for LT or LR typically ranges from 0.3 to 0.5 [37]. This means that for every 1% increase in length of a wood specimen, it will decrease in width or thickness by 0.3–0.5%.

While Poisson's ratio serves as a general indicator of mechanical deformation, it's essential to measure its specific values in various directions for accurate analysis of any wood sample.

2.2.8 Toughness

Wood toughness is a mechanical property that measures the ability of wood to absorb energy (total work) before breaking (exceeding the elastic limit). It's a combination of strength and resilience, meaning the wood can withstand impact and bending without fracturing [7].

$$W = \frac{\sigma\varepsilon}{2} = \frac{F\delta}{2V} \tag{2.29}$$

and for the proportional limit (the max point of the linear elastic slope in Fig. 2.13 reshown in Fig. 2.17):

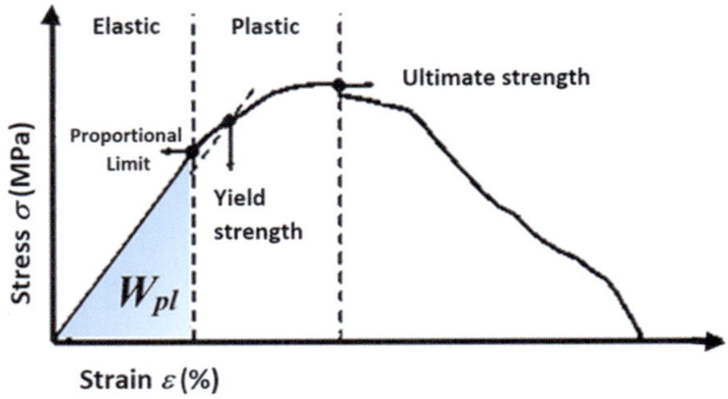

Fig. 2.17 A stress–strain curve illustrating the mechanical behavior of a material. The curve already shown in Fig. 2.13, repeated again to demonstrate that the area under the elastic region, highlighted in blue, represents the toughness or total work done in the elastic region

$$W_{pl} = \frac{\sigma_{pl}\varepsilon_{pl}}{2} = \frac{F_{pl}\delta_{pl}}{2V} \tag{2.30}$$

where, *pl* index corresponds to the proportinal limit in the elastic stress-strain or force-deformaion chart. Joules per cubic meter (J/m^3) is a common unit for measuring wood toughness.

2.2.9 *Hardness*

Wood hardness, a mechanical property measuring the resistance of wood to indentation or penetration, is a critical factor in applications ranging from tool handles to fingerboards on musical instruments. Its importance extends to vibrating string supports such as saddles and nuts.

There are several methods used to measure wood hardness, including:

- **Janka hardness test**: This is the most common method used to measure the hardness of wood. It involves driving a steel ball into the wood and measuring the force required to penetrate a specific distance. The result is expressed in Newtons (N) (Fig. 2.18) [7].
- **Brinell hardness test**: This method uses a larger steel ball and measures the indentation diameter [25].
- **Rockwell hardness test**: This test uses a diamond indenter or a steel ball and measures the depth of the indentation [47].

Generally, there is a direct correlation between wood hardness and density. Denser woods tend to be harder due to their compact structure. The closer the wood fibers are packed together, the more resistant they are to indentation. So that in non-destructive testing, wood density is often estimated by measuring its hardness and the depth to which a pin can penetrate under a given force [17, 29, 30].

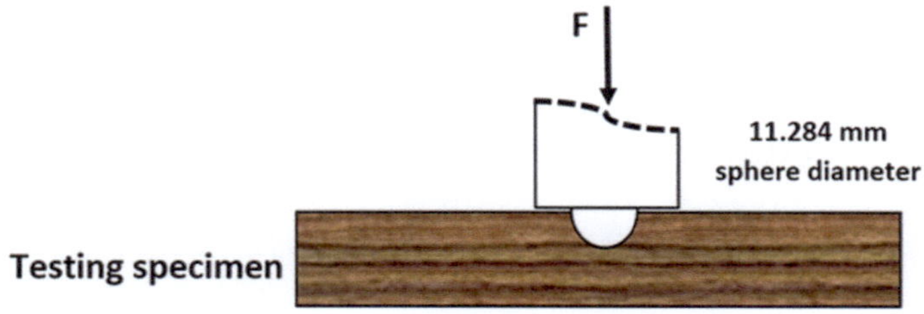

Fig. 2.18 A diagram illustrating a hardness testing device. A steel sphere with a diameter of 11.284 mm (resulting in a contact area of 1 cm^2) is pressed into a wooden specimen until it reaches the maximum depth of penetration (half the sphere's diameter, 5.642 mm). The force (F), required to achieve this penetration is used to report the hardness of the wood at that surface and direction, measured in Newtons (N)

While density is a significant factor, it's not the sole determinant of hardness. Other factors, such as surface layer properties, grain orientation and extractives content, can also influence hardness. For example, some woods with relatively low densities may exhibit high hardness due to their unique surface layer treatments or chemical composition.

2.2.10 *Workability*

Wood workability refers to the ease with which wood can be cut, shaped, and finished. It is a crucial property that influences its suitability for various applications, from woodworking projects to construction. The workability of wood depends on a large number of parameters that are caused by physical, mechanical, chemical and aesthetic properties of wood on the one hand, and parameters of the processing regime on the other hand, which can be technologically geometric [44].

- **Grain orientation**: Wood with straight grain is generally easier to work with as it cuts cleanly and evenly. Interlocked grain can be more challenging to work with due to its irregular pattern, which can cause tools to bind or tear the wood. Curly grain is also difficult to work with due to its unpredictable nature and tendency to splinter.
- **Density**: Harder woods like oak and maple can be more difficult to work with due to their higher density related to their hardness [30] and resistance to cutting. Softer woods like pine and spruce are generally easier to work with due to their lower density.
- **Moisture content**: Dry wood is easier to work with than wet wood, as it is less prone to warping, splitting, and dulling tools [57]. Wet wood can be more difficult to work with due to its increased toughness and tendency to splinter.

- **Defects**: To name a few, knots can interfere with cutting and can weaken the wood, making it more difficult to work with. Splits can make it difficult to cut or shape wood and can reduce its strength.
- **Tool selection**: Using the right tools for the job is essential for achieving good workability. For example, a sharp saw is necessary for clean cuts, while a well-maintained router is essential for precise shaping.
- **Work environment**: A clean work environment can help prevent tools from dulling and reduce the risk of accidents.
- **Lighting**: Adequate lighting is essential for accurate cutting and shaping.

Wood workability is a multifaceted property influenced by several factors. By understanding these factors and selecting the appropriate tools and techniques, woodworkers can improve their efficiency and produce high-quality results.

2.2.11 Wood Acoustic and Vibration

Is a specialized field that explores the interaction of wood with sound and mechanical vibrations. This topic is central to understanding how wood behaves as a material in various applications, particularly those involving sound transmission, absorption, and insulation [42].

Key aspects of this field include:

- **Density**: The mass per unit volume of wood, which affects its vibration properties.
- **Specific stiffness**: The ratio of stiffness to density, indicating how efficiently wood can transmit vibrations.
- **Sound velocity**: The speed at which sound travels through wood, influencing its acoustic properties.
- **Sound radiation coefficient**: The specific acoustic velocity with which wood can radiate sound into the surrounding environment.
- **Mechanical impedance**: The resistance of wood to mechanical vibrations, affecting its resonance characteristics.
- **Damping capacity**: The ability of wood to absorb and dissipate energy from vibrations, reducing noise and vibration transmission.
- **Acoustic conversion efficiency**: The efficiency with which wood can convert mechanical vibrations into sound waves. It is an overall combination of radiation and damping.

These topics, the main subjects of our goals, will be discussed in detail in the book, providing a comprehensive understanding of wood's acoustic and vibrational behavior.

2.3 Wood Anatomy

Wood, a natural material derived from trees, is a complex structure composed of various cell types. As seen, the primary components of wood are cellulose, hemicellulose, and lignin. Cellulose provides strength and rigidity, while lignin acts as a binding agent, cementing the cells together.

The basic unit of wood structure is the cell. Wood cells are elongated and arranged in layers, forming the annual rings visible in cross-sections of many tree trunks. The arrangement of cells and their properties influence the wood's physical and mechanical properties.

For example, the density, strength, and hardness of wood vary depending on the proportion of earlywood and latewood cells, as well as the thickness of cell walls. Understanding wood anatomy is crucial for various applications, including the wood for the musical instruments industry.

2.3.1 The Cellular Structure of Wood

Wood, a complex natural material, is composed of millions of tiny cells. These cells are the fundamental building blocks that give wood its unique properties. The cellular structure of wood are broadly divided into two main types of softwood and hardwood.

Softwood is primarily composed of tracheids (Fig. 2.19). These elongated cells are responsible for both water conduction and structural support. They have thick cell walls and a narrow lumen (central cavity). While softwoods primarily consist of tracheids, they may also contain small pockets of parenchyma cells. These cells, which are smaller and specialized for food storage, are absent in some softwoods like redwood [50].

Hardwoods have a more complex cellular structure than softwoods. They contain vessel elements that are specialized cells for water transport. These cells are larger and more open-ended than tracheids, allowing for efficient water flow. Hardwoods also contain fibers for structural support and parenchyma cells for storage (Fig. 2.19).

Regardless of the wood type, the cell wall of wood is primarily composed of three main components (Fig. 2.20):

- **Cellulose**: A long-chain polymer of glucose molecules, cellulose provides strength and rigidity to the cell wall.
- **Hemicellulose**: A complex carbohydrate that binds cellulose fibers together.
- **Lignin**: A complex organic polymer that provides stiffness and water resistance to the cell wall.

The arrangement and proportions of these components vary between different wood species, influencing the wood's properties.

Fig. 2.19 Cellular structure of hardwood (oak, top) and softwood (pine, bottom). This image illustrates the key differences in the cellular structure of softwood and hardwood, which can be helpful in identifying different types of wood. Image from Wikimedia commons free resources (under the Creative Commons Attribution license): https://commons.wikimedia.org/wiki/File:Hard_Soft_Wood.jpg

Understanding the cellular structure of wood is crucial for various applications. By understanding the composition and arrangement of wood cells, we can better utilize this valuable natural resource for musical instruments.

2.3.2 Annual Rings: A Closer Look

Annual rings are visible bands in the cross-section of a tree trunk. They represent one year of growth and provide valuable insights into the tree's history and properties. The formation of annual rings is influenced by seasonal changes, primarily variations in temperature and precipitation.

Each annual ring is typically composed of two distinct layers. Earlywood is formed during the early growing season when conditions are favorable. So, earlywood cells are characterized by large lumens and thin cell walls. They are responsible for rapid growth and efficient water transport. Latewood is formed later in the growing season.

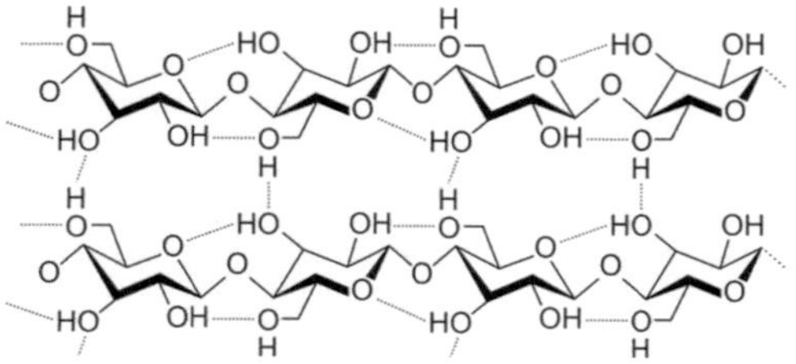

a. <u>Cellulose</u>
Image from Wikimedia commons free resources (under the Creative Commons Attribution license): https://commons.wikimedia.org/wiki/File:Cellulose_strand.svg

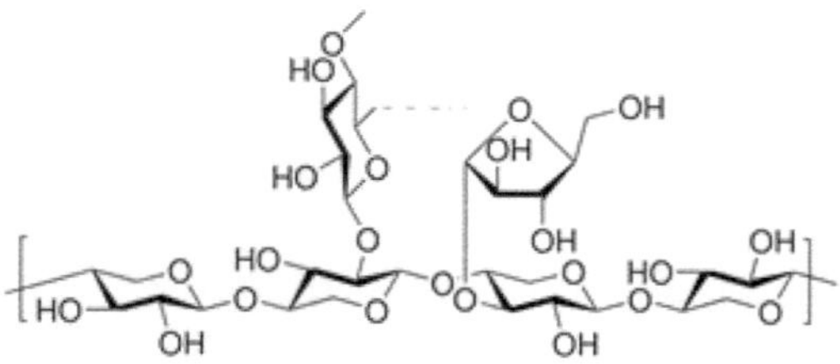

b. <u>Hemicellulose</u>
Image from Wikimedia commons free resources (under the Creative Commons Attribution license): https://commons.wikimedia.org/wiki/File:Arabinoxylan.svg

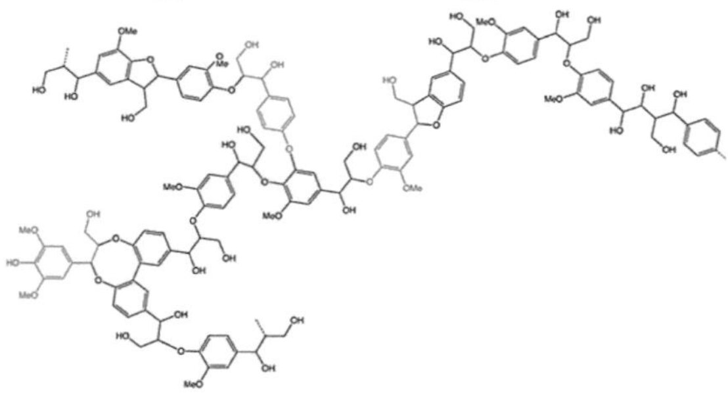

c. <u>Lignin</u>
Image from Wikimedia commons free resources (under the Creative Commons Attribution license): https://commons.wikimedia.org/wiki/File:Lignin.png

Fig. 2.20 Image of the chemical structures of **a** cellulose, **b** hemicellulose, and **c** lignin. This image shows the molecular structures of cellulose, hemicellulose, and lignin, the three main components of wood. Cellulose is a linear polymer of glucose molecules, hemicellulose is a branched polymer of various sugars, and lignin is a complex polymer of aromatic compounds. Hydrogen-type bonds link celluloses to celluloses and hemicelluloses to the cellulose polymer, while covalent-type bonds link hemicelluloses to the lignin macromolecule [41]

latewood cells have smaller lumens and thicker cell walls to enable the water intake by stronger capilary forces aided by the cellular active transmission. They also provide structural support and contribute to the wood's density and strength.

The characteristics of annual rings significantly influence the wood's properties. The proportion of latewood to earlywood affects the wood's density. More latewood results in denser wood, which is generally stronger. The thickness of cell walls and the arrangement of fibers within annual rings contribute to the wood's strength. Thicker cell walls are typically stronger. The uniformity of annual rings can affect the ease of working with wood. Wood with distinct and regular annual rings may be easier to saw, plane, and shape. By understanding the cellular structure and formation of annual rings, we can better appreciate the diversity and complexity of wood. This knowledge is essential for selecting appropriate wood species for specific musical instruments components and optimizing their use.

2.3.3 *Wood Rays: The Vertical Connectors*

In addition to the horizontal layers of cells that form annual rings, wood also contains vertical structures called wood rays (Fig. 2.19). These rays extend radially from the pith to the bark, connecting the different layers of the tree.

- **Types of wood rays and their function**
  There are two main types of wood rays [31]. Uniseriate Rays are composed of a single row of cells and Multiseriate Rays that are composed of multiple rows of cells. These wood rays serve important functions and effects:
  - **Lateral transport**: They transport water and nutrients horizontally within the tree, supplementing the vertical transport carried out by vessels and tracheids.
  - **Storage**: They store food reserves, such as starch.

Wood rays can influence the wood's appearance, workability, and durability. The presence of wood rays can create distinctive patterns on the wood's surface, such as the "silver grain" seen in some oaks and curly patterns in maples or changing dark and bright lustrous ribbon pattern of wood rays in sapele. Wood rays can affect the ease of cutting and shaping wood. Some wood rays can dull saw blades and other cutting tools. Wood rays can provide pathways for decay fungi to penetrate the wood. However, they can also contribute to the wood's resistance to decay by storing protective compounds.

The appearance of wood rays varies depending on the plane of cut, and they create distinct patterns on both radial and tangential surfaces.

On a radial surface, wood rays appear as short, linear streaks or flecks. These flecks are the cross-sectional view of the rays. The overall pattern created by these flecks is often referred to as the "ray figure." It can vary in appearance depending on the species and the specific orientation of the cut.

On a tangential surface, wood rays appear as elongated, narrow lines or stripes. These stripes represent the longitudinal view of the rays. The pattern of these stripes can also contribute to the overall appearance of the wood, especially in certain species where the rays are prominent.

The prominence of wood rays and their impact on the wood's appearance can vary significantly between different species. Some species, like beech and sycamore, have very distinct and prominent rays, while others may have less noticeable rays. By understanding the structure and function of wood rays, we can better appreciate the complexity of wood and its various properties. This knowledge is essential for selecting the right wood species for specific applications and optimizing its use.

2.3.4 *The Microstructure of Wood Cells*

Wood cells are the fundamental building blocks of wood, and their wall plays a critical role in determining the mechanical and then the acoustical properties of the wood. The cell wall of a wood cell is composed of several layers, each with specific functions and properties. Microfibrils are the fundamental building blocks of the wood cell wall. They are long, thin, and thread-like structures composed primarily of cellulose molecules. These cellulose molecules are tightly packed together, forming highly ordered crystalline structures (Fig. 2.21). The arrangement, angle and orientation of microfibrils within the cell wall play a crucial role in determining the wood's properties.

- **Primary wall**: This is the initial layer formed during cell growth. It is relatively thin and flexible, composed primarily of cellulose microfibrils embedded in a matrix of hemicellulose and pectin [24].
- **Secondary wall**: This layer is deposited after the cell has reached its final size. It is thicker than the primary wall and consists of three distinct layers:
  - **S1 layer**: The outermost layer of the secondary wall, characterized by its high cellulose content and a helical arrangement of microfibrils.
  - **S2 layer**: The thickest layer of the secondary wall, composed primarily of cellulose microfibrils arranged in a helical pattern. This layer provides the majority of the cell wall's strength and contributes significantly to the wood's tensile strength and stiffness.
  - **S3 layer**: The innermost layer of the secondary wall, thinner than the S2 layer and often composed of a more random arrangement of microfibrils.
- **Middle lamella**: This layer lies between adjacent cell walls and cements them together. It is primarily composed of lignin, hemicellulose and pectines [24].

The microstructure changes in reaction wood is interesting (Fig. 2.22). Compression wood forms on the lower side of leaning stems or branches. The cell walls of

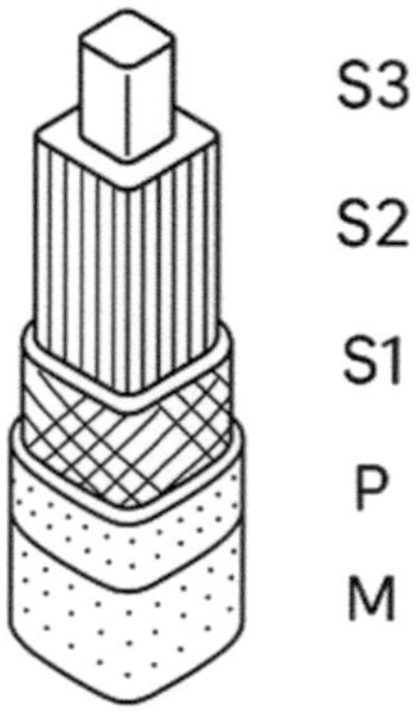

Fig. 2.21 A schematic representation of the structure of softwood at microscopic level, depicts the layered structure of the cell wall, with distinct layers labeled middle lamella (M), primary (P) and secondary (S1, S2, and S3) walls, highlights the orientation of cellulose microfibrils within each layer, which contributes to the wood's strength and stiffness

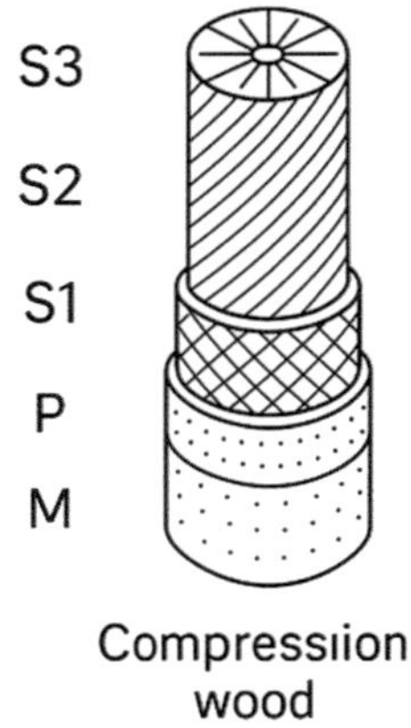

Fig. 2.22 The structure of compression wood, highlighting the thickened cell walls, reduced lumen size, and steeper microfibril angle in the S2 layer compared to normal wood. These structural differences contribute to compression wood's unique properties [5]

tracheids in compression wood are significantly thicker than those in normal wood [15]. This increased thickness provides additional mechanical support. The lumen, or central cavity, of tracheids in compression wood is smaller than that of normal wood, further increasing the cell wall thickness. The microfibrils in the S2 layer of the cell wall are oriented at a steeper angle than in normal wood. This unusual orientation gives compression wood its characteristic properties, including increased longitudinal shrinkage and decreased transverse shrinkage.

Tension wood forms on the upper side of leaning stems or branches. It is characterized by the presence of gelatinous fibers, which are specialized fibers with a distinct microstructure (Fig. 2.23). Tension wood fibers have a gelatinous layer (G-layer) that is deposited on the secondary wall. This layer is composed primarily of cellulose and has a high degree of crystallinity. The microfibrils in the G-layer are oriented at a very steep angle, often approaching 90°. This orientation gives tension wood its characteristic properties, including increased longitudinal shrinkage and decreased transverse shrinkage.

These microstructural changes in reaction wood allow trees to compensate for mechanical stress and maintain their upright growth. However, these changes can also lead to undesirable properties in wood products, such as warping and splitting, especially during drying. Tree branches, growing at an angle from the trunk, inevitably develop reaction wood. This type of wood, is unsuitable for musical instruments, even if the branch appears to be of adequate size. The musical instrument industry avoids reaction wood due to its inherent disqualifications.

As a tree matures, its wood undergoes significant microstructural changes, leading to differences in its physical and mechanical properties. Here are some key changes that occur from juvenile to mature wood [22].

Juvenile wood, typically have thinner cell walls and larger lumens. Fibers are shorter and less well-aligned, leading to lower strength and stiffness. Rays are often larger and more numerous, potentially affecting the wood's permeability and drying characteristics. Microfibrils in the cell wall are often oriented at a steeper angle, leading to lower stiffness and strength (Fig. 2.24).

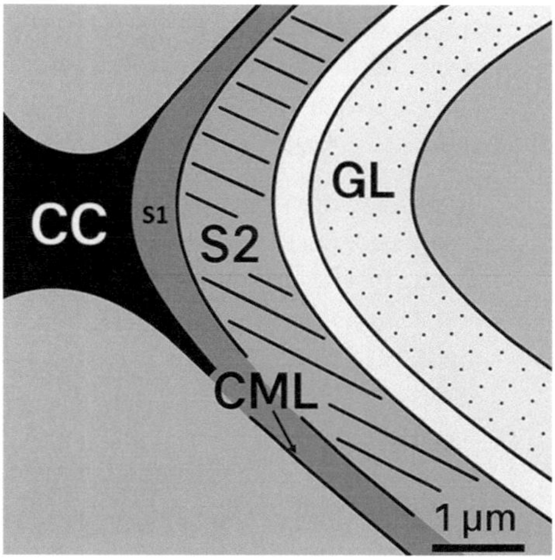

Fig. 2.23 The cross section of tension wood under electron microscope for an schematic example in *Fagus sylvatica* (beech). The images highlight the presence of a gelatinous layer (GL) in the secondary cell wall

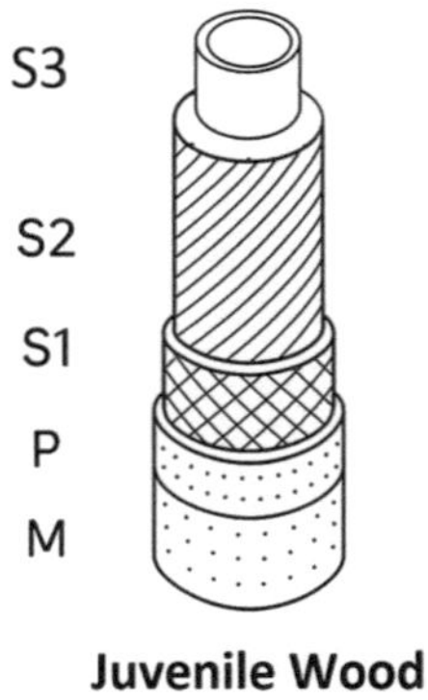

Fig. 2.24 Juvenile wood, shows the cell wall structure. Juvenile wood has thinner cell walls, larger lumens, and microfibrils in the cell wall are often oriented at a steeper angle, resulting in lower strength and stiffness [5]

Mature Wood develop thicker cell walls and smaller lumens, increasing the wood's strength and stiffness. Fibers become longer and more aligned, improving the wood's mechanical properties. Rays may become smaller and less frequent, influencing the wood's overall structure and properties. Microfibrils become more aligned with the longitudinal axis of the cell, increasing the wood's strength and stiffness.

These microstructural changes significantly impact the wood's properties, including its strength, stiffness, density, and durability. Understanding these changes is crucial for optimizing the use of wood in musical instruments industry.

- **Identifying different woods based on microstructure and anatomy**: Identifying wood species can be challenging, especially when considering the effects of stains, finishes, weathering, and aging, which can alter the wood's original appearance. While macroscopic examination of clean or freshly sanded surfaces is often sufficient, microscopic analysis is sometimes necessary for accurate identification. By comparing the microstructure and anatomy of the wood sample to reference specimens, experts can determine the species, even when the wood's original characteristics have been obscured [23].

Here are some key features that wood anatomists use to identify different wood species:

- **Growth rings**: The width, color, and distinctness of growth rings can vary between species.
- **Texture**: The feel and appearance of the wood surface, influenced by the size and arrangement of cells.
- **Color**: The natural color of the wood, which can vary between species and within a single species due to factors like genetics and environmental conditions.
- **Odor**: Some wood species have distinctive odors that can aid in identification.
- **Cell types and arrangement**: The types of cells present (e.g., tracheids, vessel elements, fibers, parenchyma) and their arrangement within the tissue can vary between species.
- **Vessel element size and distribution**: The size, shape, and arrangement of vessel elements can be characteristic of specific species.
- **Fiber length and wall thickness**: The length and thickness of fibers can vary between species and affect the wood's strength and stiffness.
- **Ray structure**: The size, shape, and arrangement of rays can be diagnostic features for certain species.
- **Tylosis**: The presence or absence of tyloses, which are ingrowths of parenchyma cells into vessel lumens, can be a useful identification feature.
- **Mineral inclusions**: The presence of mineral deposits, such as silica or calcium oxalate crystals, can be characteristic of certain species.

Using a microscope to examine thin sections of wood allows for detailed observation of cellular structure and other features. By combining these techniques and a deep understanding of wood anatomy, experts can accurately identify different wood species, even from small samples or finished products. This information is crucial for various applications, including forestry, music industry, conservation, and forensic science.

- **Examples of wood species with distinctive microscopic features**

  Here are a few examples of wood species with distinct microscopic features. These examples are merely an introduction to microscopic wood identification. Accurate species identification is a complex process requiring specialized equipment and expertise. Instrument makers should consult experts or research institutions for definitive results. Researchers and wood scientists can leverage InsideWood (https://insidewood.lib.ncsu.edu), a continually expanding database maintained by North Carolina State University, for accurate identification of unknown wood species, ensuring compliance with copyright law [54].

  To name a few, pine is characterized by tracheids with bordered pits and uniseriate rays. Earlywood and latewood are distinct, with earlywood having larger cells and thinner walls. Spruce is similar to pine, but often has smaller, more uniform tracheids and uniseriate rays (Fig. 2.25). The cell walls tend to be thicker, giving

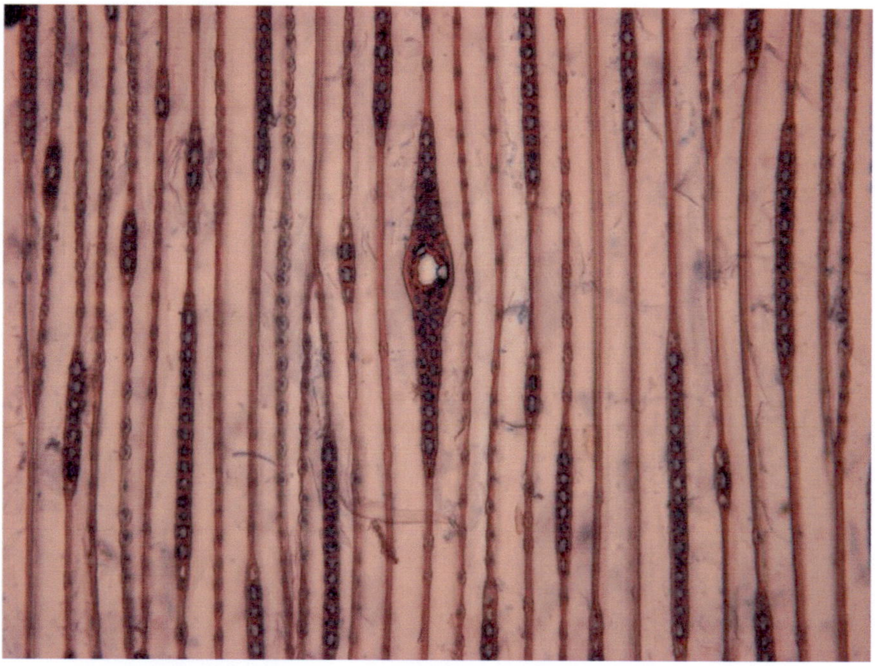

Tangential section (×10) highlighting uniseriate rays and bordered pits

Fig. 2.25 Microscopic section of spruce (*Picea abies*) wood. Photo courtesy of Dr. Turgay Birtürk, Düzce University, Türkiye, from their class teaching archive, gifted to be used in this book

spruce a denser appearance. Fir is distinctive uniseriate and biseriate rays with large, prominent ray cells. Tracheids are often similar to those of spruce.

Oak has large, open-ended vessels with tyloses. Prominent rays, often multiseriate and storied (Fig. 2.26). Also it shows distinct growth rings with a sharp boundary between earlywood and latewood. Maple has small, simple perforations in vessel elements, prominent, storied, and often multiseriate rays. It shows a uniform texture with indistinct growth rings. Walnut has large, solitary vessels with simple perforations and prominent, mostly uniseriate rays. It shows distinctive wavy grain pattern due to the arrangement of vessel elements and fibers.

- **The interplay between wood structure and acoustic properties**

  The acoustic properties of wood are intricately linked to its cellular structure, particularly the arrangement of microfibrils within the cell walls.

  Thicker cell walls and higher density generally lead to increased sound transmission and reduced sound absorption. Larger cell lumens can contribute to higher sound absorption, as they create internal spaces within the wood that can dissipate sound energy. In the following chapters, we will see that the combination of stiffness and density plays the main role in the vibration and acoustic properties of wood, which is highly dependent on the microscopic structure of wood. So,

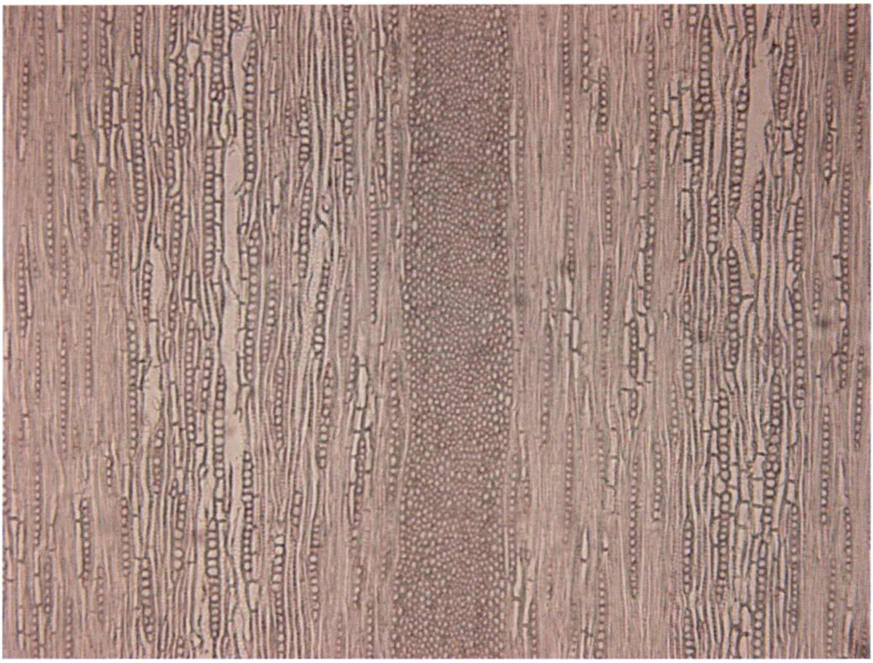

Tangential section (×10) highlighting open ended vessel elements with tyloses, Prominent rays, often multiseriate and storied

Fig. 2.26 Microscopic section of oak (*Quercus macrantera*) wood. Photo courtesy of Dr. Turgay Birtürk, Düzce University, Türkiye, from their class teaching archive, gifted to be used in this book

it would be expected that the angle at which microfibrils are oriented within the cell wall significantly impacts the wood's acoustic properties. A low MFA, where microfibrils are aligned parallel to the longitudinal axis of the cell, results in higher stiffness and sound transmission velocity. This is because the aligned microfibrils provide a more efficient pathway for sound waves to travel. A high MFA, where microfibrils are oriented at a steeper angle, leads to lower stiffness and sound transmission velocity. This can increase sound absorption and damping. This is one of the reasons that a luthier must identify the reaction wood, juvenile wood and similar disorders with microfibril angle abnormalities and avoid using them in musical instrument sound boards or other sensitive components.

White Mulberry (*Morus alba*) is a valuable wood species with distinctive microscopic features. Its vessel elements are large, solitary, with simple perforations. Fibers are short to moderately long fibers with thick walls. Rays are prominent, uniseriate and multiseriate rays, often with abundant crystalliferous parenchyma cells. Tyloses Frequently present in vessels, especially in older wood.

Cracks and microscopic defects that are hidden from the naked eye, disrupt the continuity of the wood's structure, reducing its stiffness and increasing its damping capacity. As a result, they can affect the sound transmission properties of the wood.

Small-scale cracks and fractures can scatter sound waves, leading to increased sound absorption and reduced sound transmission. Large-scale cracks and fractures can significantly reduce the wood's stiffness and strength, affecting its acoustic properties.

In conclusion, the intricate interplay between cell structure, microfibril angle, and the presence of defects significantly influences the acoustic properties of wood. By understanding these relationships, wood scientists and engineers can develop new techniques to optimize the acoustic performance of wood-based products, such as musical instruments, soundproofing materials, and architectural elements.

Luthiers often select wood for its aesthetic appeal and sounding quality, relying on visual inspection and other sensory or audible cues. However, even visually appealing wood can have hidden microscopic defects that can negatively impact the sound quality of an instrument. Improper storage and transportation conditions can exacerbate these issues, leading to moisture damage and other microscopic flaws. In the other words, while the selected wood may appear visually appealing with its fine, parallel grain and light weight, it's possible that it has suffered microscopic damage during production, storage, or transportation. This damage, such as ruptured cell walls, can negatively impact the wood's springiness and elasticity, potentially compromising the instrument's sound quality [43]. To ensure the best possible sound quality, it is essential to use accurate acoustic measuring tools to assess the wood's health.

2.4 Common Woods for Stringed Instruments

The choice of wood significantly impacts the sound, appearance, and durability of a stringed instrument. Here are some little discussions of the most common woods used in instrument making. Thus, the main philosophy behind these choices including a detailed table for acceptable ranges of the acoustic properties in each component of the instrument, will be discussed in the following chapters.

2.4.1 For Soundboards

The wood employed for soundboards in musical instruments, a pivotal component in sound production within the soundbox, must be lightweight and possess low density [42]. This allows the wood to readily overcome inertia during the back-and-forth vibrations, which involve frequent stops and changes in the direction of vibrational motion at the end of each oscillation. However, while being lightweight, this wood must not be 'weak' or 'porous', it should exhibit exceptional stiffness and elasticity.

These properties are crucial for enabling the soundboard to sustain the high-energy vibrational throws with vigor.

The selection of wood for soundboards is a delicate balance of opposing qualities. The wood must be sufficiently flexible to vibrate freely in response to the excitation from the strings or other sound-producing elements. Simultaneously, it needs to be rigid enough to resist deformation and maintain its shape under tension. The density or the specific gravity of the wood, is a critical factor. Woods with a lower specific gravity, such as spruce, are commonly used for soundboards as they offer a good combination of lightness, stiffness, and flexibility. Woods e.g. mulberry that typically do not possess the requisite qualities of stiffness and lightness must be subjected to treatments to enhance these properties prior to their integration into the resonance box. A comprehensive discussion of these treatments can be found in the chapter dedicated to the acoustic treatment of Tonewood.

The microstructure of the wood also plays a significant role in its acoustic properties. The arrangement of wood fibers, the presence of knots and other defects, and the overall grain pattern can all influence the sound produced by the instrument. For example, a straight, even grain pattern is often desirable as it can contribute to a more consistent and resonant sound. Additionally, the seasoning process, which involves drying the wood to reduce moisture content, is essential for achieving optimal stability and sound quality. By carefully selecting and preparing the wood, instrument makers can create soundboards that enhance the tonal characteristics of their instruments

- **Spruce**: Widely used for its excellent sound properties, particularly for acoustic guitars and violins. It offers a good balance of strength and flexibility. In *Picea engelmannii* from western north America, the basic specific gravity is about 0.33 and the elastic stifness in growth direction reaches over 10 GPa [34].
- **Cedar**: Often used for classical guitars, cedar provides a warmer and more mellow tone compared to spruce. In western red cedars from Pacific Northwest United States/Canada (*Thuja plicata*), the basic specific gravity is about 0.31 and the elastic stifness in growth direction reaches over 7.5 GPa [34].

2.4.2 For Backs and Sides

The back and sides of an acoustic string instrument are traditionally designed to reflect the sound produced by the vibrating top, or soundboard. This fundamental function implies that the materials used for these components must possess specific properties. If we were to disregard the subtle complexities and exceptions that contribute to the rich tonal characteristics of a string instrument, we could simplify the requirements as follows.

To effectively reflect sound, the back and sides should ideally have high inertia. This means that the material must resist changes in its state of motion. A wood with high inertia will not be easily set into vibration by the sound waves impinging upon it, but will instead act as a relatively stationary surface that reflects the sound back towards the soundboard and into the body of the instrument. Moreover, the material should be dense, providing a substantial mass that further aids in sound reflection.

However, inertia alone is insufficient. The back and sides must also possess exceptional stiffness [42]. This property is crucial for enabling the efficient transmission of vibrational energy. When the soundboard vibrates, it imparts a force to the back and sides. A stiff material can resist deformation and effectively transmit this force back to the soundboard, enhancing the overall volume and projection of the instrument. In essence, the back and sides should act as a rigid, reflective surface that can withstand and transmit the energetic vibrational throws generated by the soundboard.

- **Mahogany**: Known for its rich, warm tone and beautiful grain patterns. It's often used in guitars and other stringed instruments. In African mahogany (*Khaya senegalensis*), the basic specific gravity is about 0.65 and the elastic stifness in growth direction reaches over 11.5 GPa [34].
- **Rosewood**: Highly prized for its durability, beauty, and tonal properties. It's commonly used in high-end instruments. In Brazilian rosewood (*Dalbergia nigra*), the basic specific gravity is about 0.7 and the elastic stifness in growth direction reaches over 14 GPa [34].
- **Koa**: A Hawaiian hardwood that offers a bright, clear tone and beautiful grain patterns. In *Acacia koa*, the basic specific gravity is about 0.55 and the elastic stifness in growth direction reaches over 10 GPa. *Acacia melanoxylon* (Australian blackwood) shows a more beautiful aestetic view while benefits from a greather elastic stiffness reaching 15 GPa [34].
- **Maple**: A popular choice for its strength, beauty, and tonal qualities. It can produce a bright and resonant sound. In sugar maple (*Acer saccharum*), basic specific gravity is about 0.56 and the elastic stifness in growth direction reaches over 12.5 GPa [34]. When using maple woods other than the sugar maple (e.g. silver maple—*Acer saccharinum*, basic specific gravity: 0.44, elastic stiffness: 7.5 GPa) with lower specific gravities or stiffness, for the soundbox backs and sides, one may encounter increased activation of the back, leading to a sound that is more resonant and potentially noisier. While this extra vibration might be unwanted in classical instruments, it can enhance the signature sound of folklore instruments like flamenco guitars and Turkish bağlamas, adding a unique and expressive ringing tone.
- **Mulberry**: Mulberry wood, particularly white mulberry (basic specific gravity: 0.55, elastic stiffness: 9 GPa), is a traditional choice for stringed instruments in many regions, including the Middle East and Central Asia [53]. Its unique properties make it well-suited for backs and sides. But it is used for soundboards too. In following chapters it would be seen that there are a few mulberry pieces that marginally meet the qualifications suitable for sound board. By incorporating mulberry wood into the list, we get a more comprehensive overview of the diverse range of woods used in stringed instrument making.

2.4.3 For Fingerboards and Neck

The wood used for the fingerboard of a stringed instrument must possess exceptional hardness to withstand the constant pressure and friction exerted by the strings and the player's fingers. Over time, if the wood is too soft, it will wear down, leading to uneven fret surfaces and compromised playability.

Furthermore, the neck of the instrument and the fingerboard must maintain their shape and alignment despite fluctuations in humidity. The coefficient of expansion and contraction of the wood must be minimal, and ideally, the neck wood and fingerboard wood should have similar expansion and contraction rates. If these properties differ significantly, the fingerboard may warp or separate from the neck, rendering the instrument unplayable.

The choice of wood species is crucial for the long-term performance of a fingerboard. Reaction woods or juvenile wood with uneven grain patterns, are unsuitable for this application. These woods are more prone to warping, cracking, and other forms of dimensional instability, due to their steeper microfibril angles in cell wall's S2 layer. Additionally, the wood must be dense and have a fine, even grain structure to provide maximum durability and resistance to wear.

Commonly used woods for fingerboards include ebony, rosewood, and some times, maple, wenge and pau ferro. These woods are prized for their hardness, density, and stability. The selection of wood for a fingerboard is a critical factor in the overall quality and playability of a stringed instrument. By choosing a wood species with the appropriate physical and mechanical properties, luthiers can ensure that the instrument will remain stable and responsive for many years to come.

- **Ebony**: A dense, hard wood that is highly resistant to wear and tear. It's commonly used for fingerboards, bridges, and tailpieces. Ebony, in particular, is considered the gold standard for fingerboards due to its exceptional hardness and smooth, dense surface. In *Diospyros ebenum* from southeast asia the Janka hardness reaches 10,000 N and tangential to radial shrinkage (T/R) ratio is as low as 1.6 [34]. However, due to overharvesting, ebony is becoming increasingly rare and expensive, leading to the exploration of alternative woods such as pau ferro and wenge.
- **Rosewood**: Also used for fingerboards and other parts, it offers a smooth surface and a rich appearance. In *Dalbergia nigra* the T/R ration is 1.6 but the Janka hardness is a bit grater than the ebony reaching to 12,000 N [34].
- **Maple**: Can be used for fingerboards, especially on instruments with maple backs and sides, providing a consistent aesthetic. Hard maple's (sugar maple—*Acer saccharum*) Janka hardness is lower than the above introduced species but still over 6000 N and benefits from a tangential to radial shrinkage (T/R) ratio of 1.9 [34].

It's important to remember that the specific properties of a wood can vary depending on factors such as growth conditions, age, and drying methods. Luthiers carefully select wood for each instrument, considering factors like weight, grain pattern, and audible properties to achieve the desired sound. While traditional woods remain the preferred choice for many instruments, the availability of suitable timber is becoming increasingly limited. This has led luthiers to explore alternative woods [16] or even synthetic materials. The goal is to find substitutes that can replicate the sound and feel of traditional instruments. However, many attempts have yielded mixed results, with some successes and many failures.

2.5 Final Words

In understanding the science and anatomy of wood, we gain insights into the unique properties that make it invaluable for applications across various fields, especially in crafting musical instruments. Wood is a material that encapsulates both natural beauty and complex structure. Its formation, determined by biological processes and environmental factors, results in a range of qualities that vary among species, growth conditions, and tree parts. From the durability and strength of latewood to the lightweight, fast-growing characteristics of earlywood, each type of wood offers distinct advantages depending on its intended use. Wood's porous nature, chemical composition, and cellular structure all contribute to its density, strength, and resilience, making it a versatile material with myriad applications.

One of wood's unique characteristics is its orthotropic nature, where properties differ based on the direction of the grain. This is a critical consideration for applications requiring precise performance under stress, such as in structural beams or musical instruments. By understanding factors like grain orientation, density, moisture content, and elasticity, we can make informed choices about wood selection and treatment for different functions. This orthotropic behavior also contributes to wood's acoustic properties, where density and microfibril alignment within the cellular structure affect sound transmission and vibration. In musical instrument construction, this knowledge allows luthiers to select specific wood types for tonal qualities, ensuring each instrument's unique sound signature.

Furthermore, the aesthetics of wood, from color variations and grain patterns to intricate textures, adds another layer of value. These visual qualities arise from complex interactions within the wood's cellular structure and are influenced by external factors like light, heat, and oxygen exposure. The presence of extractives, such as tannins and phenolic compounds, gives certain woods their color and resistance to environmental wear, enhancing both their functionality and beauty. In addition, wood's workability is highly valued, as it allows artisans to shape, join, and finish it according to specific design needs.

Wood's durability, essential for both functional and aesthetic longevity, is impacted by factors like moisture content, decay resistance, and structural defects. Woods like cedar, oak, and redwood, with natural resistance to decay, are often chosen for applications exposed to environmental stress, while other species may require chemical treatments for enhanced durability. This aspect is critical in the creation of musical instruments, where longevity and consistent performance are essential.

In considering wood's role in instrument making, certain species, such as spruce and cedar for soundboards, maple and mahogany for backs and sides, and ebony and rosewood for fingerboards, have become traditional choices. Each type contributes a specific tonal quality and aesthetic, selected for its capacity to produce desired sounds, resonate efficiently, and sustain beauty over time. Understanding the structural, chemical, and acoustic characteristics of these woods aids in crafting instruments that are not only functional but also artful expressions of sound.

In conclusion, wood's multifaceted properties—its structural complexity, natural durability, workability, and acoustic potential—make it irreplaceable in many applications, particularly in the art of musical instrument making. By exploring wood science and anatomy, we appreciate its versatility and learn to harness its qualities to create lasting, high-performance, and aesthetically appealing products. In-depth knowledge of wood science allows us to bridge tradition and innovation, preserving the natural heritage of wood while continually finding new applications and treatments to enhance its inherent qualities.

References

1. G. Almeida, R.E. Hernández, Changes in physical properties of tropical and temperate hardwoods below and above the fiber saturation point. Wood Sci. Technol. **40**, 599–613 (2006)
2. T.D. Almeida, D.H. Almeida, L.A. Marcolin, D. Gonçalves, A.L. Christoforo, F.A.R. Lahr, Correlation between dry density and volumetric shrinkage coefficient of three Brazilian tropical wood species. Int. J. Mater. Eng. **5**(1), 1–4 (2015)
3. R.K. Bamber, Sapwood and heartwood. Forestry Commission of New South Wales. Technical Publication. New revised, No. 2 (1987). ISSN 0155-7548
4. M.C. Barbu, K. Buresova, E.M. Tudor, A. Petutschnigg, Physical and mechanical properties of *Paulownia tomentosa* × *elongata* sawn wood from Spanish, Bulgarian and Serbian plantations. Forests **13**(10), 1543 (2022)
5. J. Barnett, J. Gril, P. Saranpää, Introduction, in *The Biology of Reaction Wood*, ed. by B. Gardiner, J. Barnett, P. Saranpää, J. Gril. *Springer Series in Wood Science* (Springer, 2014), pp. 1–11. https://doi.org/10.1007/978-3-642-10814-3_1
6. V. Benítez, J. Franco, Á. Camargo, P. Raimonda, C. Mantero, C.M. Ibáñez, Influence of initial wood moisture on decay process by two brown-rot fungi. Maderas. Cienc. Tecnol. **23** (2021)
7. J. Bodig, B. Jayne, *Mechanics of Wood and Wood Composites* (Van Nostrand Reinhold, 1982)
8. A. Boura, D. De Franceschi, Is porous wood structure exclusive of deciduous trees? C.R. Palevol. **6**(6–7), 385–391 (2007)
9. A.J. Bowling, K.C. Vaughn, Gelatinous fibers are widespread in coiling tendrils and twining vines. Am. J. Bot. **96**(4), 719–727 (2009)
10. B.G. Butterfield, Wood anatomy in relation to wood quality, in *Wood Quality and Its Biological Basis*. ed. by J. Barnett, G. Jeronimidis (Blackwell Publishing, 2003), pp.30–52
11. C.G. Carll, T.L. Highley, Decay of wood and wood-based products above ground in buildings. J. Test. Eval. **27**(2), 150–158 (1999)
12. M.R. Carreira, P.G. de Alcântara Segundinho, F.G. Gonçalves, J.B. Paes, A.A. Dias, P.N. de Medeiros Neto et al., Influence of moisture content on the dynamic modulus of elasticity. Res. Soc. Dev. **11**(11), e232111133687 (2022)

13. F. Cartenì, A. Deslauriers, S. Rossi, H. Morin, V. De Micco, S. Mazzoleni, F. Giannino, The physiological mechanisms behind the earlywood-to-latewood transition: a process-based modeling approach. Front. Plant Sci. **9**, 1053 (2018)
14. S. Chauhan, A. Sethy, Differences in dynamic modulus of elasticity determined by three vibration methods and their relationship with static modulus of elasticity. Maderas. Cienc. Tecnol. **18**(2), 373–382 (2016)
15. B. Clair, B. Thibaut, Physical and mechanical properties of reaction wood, in *The Biology of Reaction Wood*, ed. by B. Gardiner, J. Barnett, P. Saranpää, J. Gril. *Springer Series in Wood Science* (Springer, Berlin, Heidelberg, 2014), pp. 171–200. https://doi.org/10.1007/978-3-642-10814-3_6
16. J.D. Cotton, J.D. Wolodko, Identification of sustainable tonewoods for acoustic guitars using materials selection software. AM&P Techn. Art. **181**(5), 13–18 (2023). https://doi.org/10.31399/asm.amp.2023-05.p013
17. D.J. Cown, Comparison of the Pilodyn and torsiometer methods for the rapid assessment of wood density in living trees. NZ J. Forest. Sci. **8**(3), 384–391 (1978)
18. J.V. Degraff, S.S. Agard, Defining geologic hazards for natural resources management using tree-ring analysis. Environ. Geol. Water Sci. **6**(3), 147–155 (1984)
19. F. Divos, F.S. Kiss, Strength grading of structural lumber by portable lumber grading-effect of knots, in *Proceedings of the Final Conference of COST Action E53. The Future of Quality Control for Wood & Wood Products*, Edinburgh, Great Britain, 4–7 May 2010
20. J. Fang, J.A. Lutz, H.H. Shugart, X. Yan, A physiological model for predicting dynamics of tree stem-wood non-structural carbohydrates. J. Ecol. **108**(2), 702–718 (2020)
21. U. Fischer, M. Kucukoglu, Y. Helariutta, R.P. Bhalerao, The dynamics of cambial stem cell activity. Annu. Rev. Plant Biol. **70**(1), 293–319 (2019). https://doi.org/10.1146/annurev-arplant-050718-100402
22. O.M. González, A. Velín, A. García, C.R. Arroyo, H.L. Barrigas, K. Vizuete, A. Debut, Representative hardwood and softwood green tissue-microstructure transitions per age group and their inherent relationships with physical-mechanical properties and potential applications. Forests **11**(5), 569 (2020). https://doi.org/10.3390/f11050569
23. L. Gurau, M.C. Timar, M. Porojan, F. Ioras, Image processing method as a supporting tool for wood species identification. Wood Fiber Sci. **45**(3), 303–313 (2013)
24. S.A. Harding, C.J. Tsai, Cell wall pectins in tree growth and woody biomass utilization, in *Advances in Botanical Research*, vol. 89 (Academic Press, 2019), pp. 235–256
25. S. Hirata, M. Ohta, Y. Honma, Hardness distribution on wood surface. J. Wood Sci. **47**(1), 1–7 (2001). https://doi.org/10.1007/BF00776637
26. ISO 13061-2, Physical and Mechanical Properties of Wood—Test Methods for Small Clear Wood Specimens—Part 2: Determination of Density for Physical and Mechanical Tests (2014)
27. ISO 13061-13, Physical and Mechanical Properties of Wood—Test Methods for Small Clear Wood Specimens—Part 13: Determination of Radial and Tangential Shrinkage (2024)
28. ISO 13061-14, Physical and Mechanical Properties of Wood—Test Methods for Small Clear Wood Specimens. Part 14: Determination of Volumetric Shrinkage (2024)
29. L. Karlinasari, Y. Fredisa, U. Adzkia, S. Fauziyyah, F. Dwiyanti, I.Z. Siregar, Use of a pin-penetration wood density meter to determine the density of 25 Indonesian species. BioResources **16**(2), 3032–3045 (2021). https://doi.org/10.15376/biores.16.2.3032-3045
30. E. Khosravi, M. Roohnia, A. Lashgari, A. Jahanlatibari, A. Tajdini, Evaluation of pin penetration probing technique for the assessment of basic density in air-dried wood. BioResources **16**(4), 6577–6586 (2021). https://doi.org/10.15376/biores.16.4.6577-6586
31. S. Leal, V.B. Sousa, H. Pereira, Within and between-tree variation in the biometry of wood rays and fibres in cork oak (*Quercus suber* L.). Wood Sci. Technol. **40**, 585–597 (2006). https://doi.org/10.1007/s00226-006-0073-x
32. M. Libralato, A. De Angelis, O. Saro, M. Qin, C. Rode, Effects of considering moisture hysteresis on wood decay risk simulations of building envelopes. J. Build. Eng. **42**, 102444 (2021)

33. W.J. Lucas, A. Groover, R. Lichtenberger, K. Furuta, S.R. Yadav, Y. Helariutta et al., The plant vascular system: evolution, development and functions. J. Integr. Plant Biol. **55**(4), 294–388 (2013)
34. E.W. Meier, *WOOD! Identifying and Using Hundreds of Woods Worldwide.* Wood Database. New edition 2023 (2015)
35. M. Merela, K. Čufar, Density and mechanical properties of oak sapwood versus heartwood in three different oak species. Drvna Ind. **64**(4), 323–334 (2013)
36. R.B. Miller, Characteristics and availability of commercially important woods, in *Wood Handbook: Wood as an Engineering Material* (USDA Forest Service, Forest Products Laboratory, Madison, WI, 1999), pp. 1.1–1.34, 113
37. P. Obara, Verification of orthotropic model of wood. Arch. Civ. Eng. **64**(3) (2018)
38. T. Ozyhar, S. Hering, P. Niemz, Moisture-dependent elastic and strength anisotropy of European beech wood in tension. J. Mater. Sci. **47**, 6141–6150 (2012)
39. G.D.M. Palermo, J.D.F. Latorraca, A.M. De Carvalho, F.W. Calonego, E.T.D. Severo, Anatomical properties of *Eucalyptus grandis* wood and transition age between the juvenile and mature woods. Eur. J. Wood Wood Prod. **73**, 775–780 (2015)
40. S. Rahimi, K. Singh, D. DeVallance, D. Chu, M. Bahmani, Drying behavior of hardwood components (sapwood, heartwood, and bark) of red oak and yellow-poplar. Forests **13**(5), 722 (2022)
41. P.D.O. Rodrigues, A.G. Corrêa, M.A. Baffi, D. Pasquini, Potential applications of hemicellulose, in *Handbook of Biomass*, ed. by S. Thomas, M. Hosur, D. Pasquini, C. Jose Chirayil (Springer, Singapore, 2024), pp. 697–727. https://doi.org/10.1007/978-981-99-6727-8_27
42. M. Roohnia, Wood: vibration and acoustic properties, in *Reference Module in Materials Science and Materials Engineering* (Elsevier, Oxford, 2019), pp. 1–13. https://doi.org/10.1016/B978-0-12-803581-8.01996-2
43. M. Roohnia, T. Yilmaz Aydin, Acoustic characterization vs. luthier's senses to select local substitutes for exotic famous tonewoods, in *Proceedings of ICONST LST 2023, International Conferences on Science and Technology. Life Science and Technology*, 30 Aug–1 Sept 2023, Budva, Montenegro
44. H. Rošić, A. Bjelić, A. Hodžić, M. Merhar, Optimization of solid wood workability parameters in the planing process. IOP Conf. Ser. Mater. Sci. Eng. **1298**(1), 012012 (2023). https://doi.org/10.1088/1757-899X/1298/1/012012
45. F. Ruffinatto, A. Crivellaro, *Atlas of Macroscopic Wood Identification: With a Special Focus on Timbers Used in Europe and CITES-Listed Species* (Springer Nature, 2019)
46. S. Sandoval-Torres, W. Jomaa, F. Marc, J.R. Puiggali, Causes of color changes in wood during drying. Forest. Stud. China **12**, 167–175 (2010)
47. A.G. Silveira, E.J. Santini, M. Talgatti, L.C. Cancian, R. Trevisan, T. Baldin, Spectroscopy and rockwell hardness to characterize the deterioration of wood treated with natural product. Floresta **50**(1), 1041–1046 (2020). https://doi.org/10.5380/rf.v50i1.60871
48. W. Simpson, A. TenWolde, *Physical Properties and Moisture Relations of Wood* (The Encyclopedia of Wood, 1999)
49. M.R. Suzuki, C.G. Hunt, C.J. Houtman, Z.D. Dalebroux, K.E. Hammel, Fungal hydroquinones contribute to brown rot of wood. Environ. Microbiol. **8**(12), 2214–2223 (2006)
50. N. Toghraie, Wood identification, vol. 1, in *Modern Woods*, ed. by H.R. Yazdani-Moghaddam (Jihad University Press, 2013) (in Farsi). ISBN: 9786001024658
51. M.C.A. Torbenson, D.W. Stahle, J. Villanueva Díaz, E.R. Cook, D. Griffin, The relationship between earlywood and latewood ring-growth across North America. Tree-Ring Res. **72**(2), 53–66 (2016)
52. H. Unterwieser, G. Schickhofer, Influence of moisture content of wood on sound velocity and dynamic MOE of natural frequency- and ultrasonic runtime measurement. Eur. J. Wood Prod. **69**, 171–181 (2011). https://doi.org/10.1007/s00107-010-0417-y
53. F. Vahabzadeh, The music of the mulberry: wood science, know-how and symbolism in instrument-making in Khorassan (Iran) and Central Asia, in *Wooden Musical Instruments: Different Forms of Knowledge*, ed. by M.A. Pérez, E. Marconi. *Book of End of WoodMusICK COST Action FP1302* (Philharmonie de Paris, 2018), pp. 399–415

54. E.A. Wheeler, P.E. Gasson, P. Baas, Using the InsideWood web site: potentials and pitfalls. IAWA J. **41**(4), 412–462 (2020). https://doi.org/10.1163/22941932-bja10032
55. A.C. Wiedenhoeft, R.B. Miller, Structure and function of wood, in *Handbook of Wood Chemistry and Wood Composites* (2005), pp. 9–33
56. M. Wiemann, Characteristics and availability of commercially important woods, chap. 2 in *FPL-GTR-282, 2-1* (2021)
57. R.S. Williams, Finishing of wood, in *Wood Handbook: Wood as an Engineering Material* (USDA Forest Service, Forest Products Laboratory, Madison, WI, 1999). General technical report FPL; GTR-113, pp 15.1–15.37, 113
58. T. Yilmaz Aydin, A. Ozveren, Effects of moisture content on elastic constants of fir wood. Eur. J. Wood Wood Prod. **77**(1), 63–70 (2019)
59. B.J. Zobel, J.P. Buitjeenen, *Wood Variation: Its Causes and Control* (Springer-Verlag, 1989)

Chapter 3
The Physics of Sound and String Vibration: A Luthier's Perspective

Abstract This chapter explores the fundamental principles of vibration and wave motion, particularly in relation to sound and stringed musical instruments. It begins with an introduction to vibratory motion, emphasizing the distinction between matter transfer and wave propagation as mechanisms of energy transmission. A focus is placed on simple harmonic motion (SHM), its defining characteristics, and its mathematical representation in terms of displacement, velocity, and acceleration. The discussion then extends to wave propagation, differentiating between transverse and longitudinal waves, and highlighting their relevance to string vibrations and sound waves. The physics of standing waves, harmonic frequencies, and resonance are analyzed, particularly in the context of musical instruments. Through practical exercises and examples, the chapter provides insight into how tension, length, and mass density of a string influence its vibrational characteristics. The latter sections of the chapter explore the properties of sound, including its generation, propagation, and interaction with different media. The speed of sound, wave reflection, refraction, and interference patterns are examined to explain how sound behaves in various environments. Special attention is given to the role of the bridge in stringed instruments, detailing the forces exerted by vibrating strings and their effects on sound transmission. Finally, the chapter delves into the perceptual characteristics of sound—loudness, pitch, and timbre—exploring their physical underpinnings and psychological interpretations. The impact of the missing fundamental effect on auditory perception is discussed, providing insights into how different listeners may perceive the same sound differently. By bridging theoretical physics with practical applications in instrument making, this chapter serves as a foundational resource for luthiers and acoustic engineers seeking to optimize sound quality through precise control of vibrational behavior.

M. Roohnia, *Properties and Treatment of Tonewood for String Instrument Construction*, Springer Series in Materials Science 355, https://doi.org/10.1007/978-3-032-00824-4_3

3.1 Concept of Vibration

Vibration is the back-and-forth motion of an object about a central point or equilibrium position. It's caused by a disturbance or force that displaces the object from its resting state.

3.1.1 Vibratory Motion and Waves

Vibration and waves are fundamental to countless natural phenomena. Waves, in particular, are efficient carriers of energy from one point to another. There are two considerable methods of energy transfer:

- **Matter transfer**: Energy is carried by moving matter, such as a bowling ball transferring the energy of a throw to bowling pins. This type of energy transfer involves the physical movement of matter from one place to another, carrying its kinetic and internal energy with it or just as blowing on a candle extinguishes its flame, wind propels a sailboat.
- **Wave propagation**: Energy is transmitted through a medium without the physical movement of the medium itself [15]. This phenomenon is exemplified by the sun's energy reaching us in light waves, or by the sonic boom of a jet breaking the sound barrier, which can shatter windows with its energy transmitted by wave.

While there are various types of waves, the underlying principle of energy transmission remains consistent. Waves are generated by the relative movement of matter and propagate through a material medium, carrying energy from the source of vibration. Although electromagnetic waves can travel through a vacuum, we won't pay attention into this extraordinary case here, since it is not related to acoustic science. It's important to note that both vibration and waves are fundamental concepts in physics, providing insights into phenomena like sound, light, heat, microwaves, and even the subatomic world.

Any motion that repeats itself identically at regular intervals is known as periodic motion. If this repetition occurs rapidly, we often refer to it as oscillatory motion [1]. Common examples of simple periodic motion in our daily lives include:

- The swinging of a pendulum in a grandfather clock.
- The rotation of a watch's balance wheel.
- The vibration of a tuning fork

3.1.2 Simple Harmonic Motion (SHM)

Simple harmonic motion is a type of periodic motion where a particle moves to and from along a straight line, with an acceleration that is always directed towards a fixed point on the line and whose magnitude is proportional to the displacement from the fixed point [1]. Key characteristics of SHM are:

- **Periodic motion**: The motion repeats itself after a fixed interval of time, called the period (T).
- **Restoring force**: The force acting on the particle is always directed towards the equilibrium position and is proportional to the displacement from this position. This force is often described by Hooke's Law [10] as also illustrated in Chap. 2 (Fig. 2.12):

$$F = k\delta \tag{3.1}$$

where F is the force, k is the spring constant, and δ is the displacement (Chap. 2—Fig. 2.12).

- **Sinusoidal motion**: The displacement, velocity, and acceleration of the particle vary sinusoidally with time.

  Examples of SHM are (Fig. 3.1):

- **Mass-spring system**: A mass attached to a spring oscillates back and forth when displaced from its equilibrium position.
- **Pendulum**: A simple pendulum swings back and forth under the influence of gravity. For small angular displacements, the pendulum's arc-like path can be approximated as a straight line. This linear approximation is a fundamental condition for simple harmonic motion.
- **Vibrating strings**: The vibrations of every individual mode a guitar string or a violin string are examples of SHM.
- **Seismic waves**: The propagation of seismic waves through the Earth's crust involves SHM.

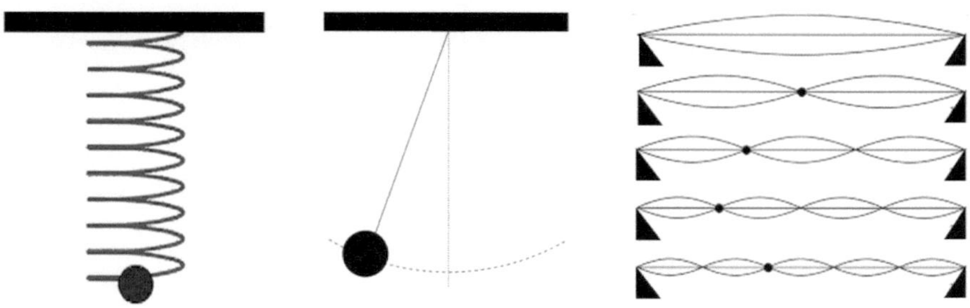

Fig. 3.1 Examples of simple harmonic motion. Three panels show (left) a mass attached to a spring oscillating vertically, (middle) a pendulum swinging in an small arc resembeling a linear movement, and (right) a series of images depicting the fundamental and higher harmonic modes of a vibrating string fixed at both ends

Simple harmonic motion is a fundamental concept in physics with numerous applications in various fields of science and engineering. Understanding SHM helps us analyze and predict the behavior of many natural and man-made systems.

3.1.3 Equations of Motion, Velocity, and Acceleration in Oscillations

As introduced, simple Harmonic Motion is a type of periodic motion where a particle moves to and from along a straight line, with an acceleration that is always directed towards a fixed point on the line and whose magnitude is proportional to the displacement from the fixed point [1]. The equation of motion for a particle undergoing SHM can be expressed as:

$$x(t) = A \cdot \sin(\omega t + \phi) \tag{3.2}$$

where:

$x(t)$: Displacement of the particle at time t.

A: Amplitude (maximum displacement).

ω: Angular frequency.

φ: Phase constant.

The velocity of the particle can be found by taking the derivative of the displacement equation with respect to time:

$$\mathrm{v}(t) = \frac{dx}{dt} = A\omega \cdot \cos(\omega t + \phi) \tag{3.3}$$

The acceleration of the particle can be found by taking the derivative of the velocity equation with respect to time:

$$a(t) = \frac{d\mathrm{v}}{dt} = -A\omega^2 \cdot \sin(\omega t + \phi) \tag{3.4}$$

Angular Frequency (ω) related to the period (T) and frequency (f) of oscillation:

$$\omega = \frac{2\pi}{t} = 2\pi f \tag{3.5}$$

Phase Constant (φ) determines the initial position and direction of motion of the particle.

Maximum Velocity (v_{max}) occurs at the equilibrium position:

$$v_{max} = A\omega \quad (3.6)$$

Maximum Acceleration (a_{max}) occurs at the extreme positions:

$$a_{max} = A\omega^2 \quad (3.7)$$

By understanding these equations and their physical interpretations, we can analyze and predict the behavior of various oscillating systems, from simple pendulums to complex mechanical vibrations in musical instruments.

To visualize simple harmonic motion, consider a reference circle with a radius equal to the amplitude of the oscillation (Fig. 3.2). A point on this circle moves with a constant angular velocity, completing one full revolution in a time period T. The projection of this point onto a diameter of the circle traces out simple harmonic motion. As the point moves around the circle, its projection oscillates back and forth along the diameter. At the endpoints of the diameter, the projection's velocity is zero, and its acceleration is maximum. At the midpoint, the velocity is maximum, and the acceleration is zero.

The frequency, f, of the oscillation is the number of complete cycles per second [1]. It is related to the period T by the equation:

$$f = \frac{1}{T} \quad (3.8)$$

The unit of frequency is the Hertz (Hz), named after the scientist Heinrich Hertz. One Hertz corresponds to one oscillation per second.

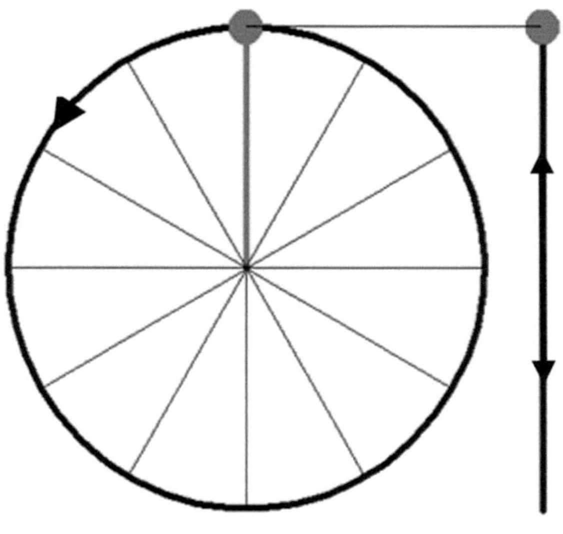

Fig. 3.2 Visualization of simple harmonic motion using a reference circle. A point moving with constant angular velocity on the circle's perimeter projects to a diameter, demonstrating the relationship between circular and simple harmonic motion. The projection's velocity and acceleration are indicated

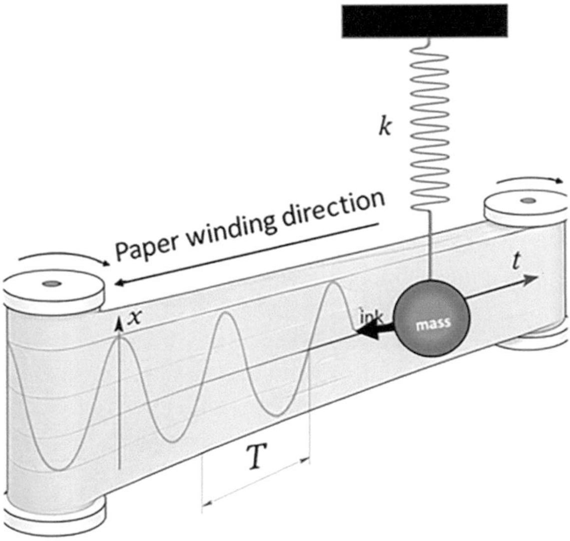

Fig. 3.3 Schematic of an oscillation system. A mass (labeled "mass") is attached to a spring (spring constant k). An ink pen ("ink") attached to the mass records oscillations on a paper drum rotating at a constant speed. The resulting trace on the paper measures displacement (X) and period (T). The "Paper winding direction" and the support structure are also shown

Imagine an oscillating particle coated in ink, tracing its path on a moving sheet of paper (Fig. 3.3). The resulting pattern is a wave-like graph, known as a sinusoidal curve. At the equilibrium point of oscillation (where the particle starts), the sine function in the trigonometric circle is also zero [i.e. $\sin(0°) = 0$]. At this same point, the particle's velocity in repeating oscillations reaches its maximum. This is because the cosine function, which represents velocity, is at its peak [i.e. $\cos(0°) = 1$] at this point (Fig. 3.4).

This connection between the particle's motion and trigonometric functions explains why the waveform is called sinusoidal or cosinusoidal. These two waveforms are essentially identical, differing only in their starting phase. Typically, we begin our analysis at the equilibrium point, where velocity is maximum, leading to a cosine-shaped velocity graph.

While a deep understanding of the sinusoidal nature of simple harmonic motion, including its velocity and acceleration, isn't crucial for most instrument makers, a basic qualitative understanding is beneficial. Recognizing the oscillatory patterns and the changes in velocity is sufficient for practical purposes.

The mathematical complexities of vibration, involving differential equations, are beyond the scope of this book and unnecessary for most instrument makers. However, it's worth noting that vibration analysis is a fundamental topic in mechanical engineering. Should you need to go deep into the intricate mathematical details of wave motion, velocity, and acceleration in musical instruments, consulting with a mechanical engineer would be advisable.

A sudden disturbance to an oscillator e.g., fingerpicking a guitar string by a fingernail can produce a complex response, exhibiting multiple vibrations and degrees of freedom, unlike the simple harmonic motion of a pendulum [12]. This complexity is characteristic of wave and vibration phenomena. Scientists use mathematical tools,

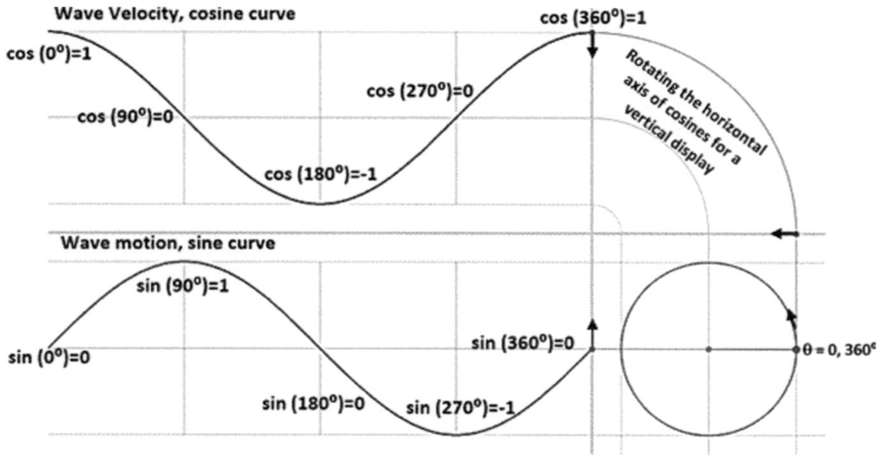

Fig. 3.4 Demonstrating the relationship between sine and cosine in wave motion. The top graph shows a cosine wave (velocity) with peak value at 0°, while the bottom graph shows a sine wave (position) starting at zero. Annotations highlight key values at different angles, explaining how velocity (cosine) is maximized when displacement (sine) is zero, corresponding to the equilibrium point of an oscillation

such as the Fast Fourier Transform, to decompose these complex vibrations into their individual components, enabling separate analysis and prediction of each oscillation (Fig. 3.1 right).

This book will guide you in understanding and controlling these vibrations in your instrument-making workshop. By avoiding excessive and uncontrolled vibrations, you can ensure the optimal performance of your instruments.

Let's continue our discussion of simple harmonic motion, focusing on the example of a pendulum.

A simple pendulum consists of a weight suspended from a fixed point by a string [1]. Its frequency of oscillation depends solely on its length and the local acceleration due to gravity. The mass and material of the weight do not influence the pendulum's frequency.

$$f = \frac{1}{2\pi} \cdot \sqrt{\frac{g}{L}} \tag{3.9}$$

where, g is the gravity acceleration, typically equal to 9.81 m/s^2 on Earth at sea level. L is the length of the pendulum (Fig. 3.5).

Now, I want you to produce a metal blade clamped at one end and free at the other, with a weight attached to the free end. When disturbed, the blade oscillates [43]. The oscillation frequency of the blade is influenced by several factors. Check these:

- **Blade length and thickness**: Longer and thinner blades tend to vibrate at lower frequencies.

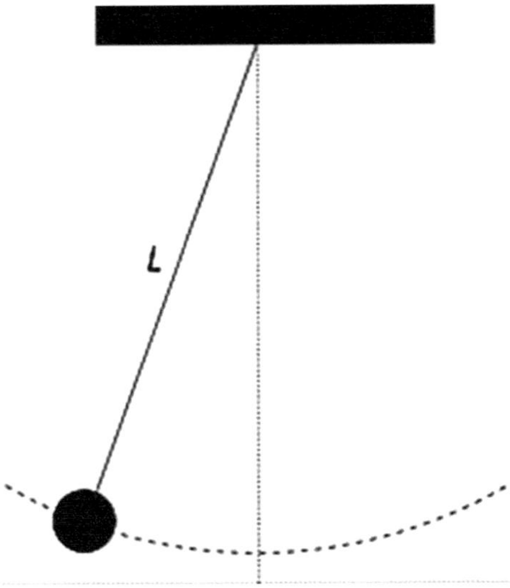

Fig. 3.5 A simple pendulum consists of a weight suspended from a fixed point by a string. Its frequency of oscillation depends solely on its length, *L*, and the local acceleration due to gravity, *g*. The mass and material of the weight do not influence the pendulum's frequency

- **Weight position**: Placing the weight closer to the free end reduces the frequency.
- **Weight mass**: Heavier weights decrease the frequency.
- **Material properties**: The density and stiffness of the blade material also impact its vibrational behavior.

A more advanced version of this vibrating blade is the tuning fork, which consists of a pair of U-shaped cantilever beams mounted on a common base (Fig. 3.6).

The tuning fork is a diapason, accepted as a standard tool for musicians to produce precise musical pitches. Similar to a metal blade, the longer and thinner the prongs of a tuning fork, the lower its fundamental frequency (pitch). Additionally, the material's stiffness and density also influence the tuning fork's frequency [7].

Pointing to "fundamental frequency" reminds that when you strike a tuning fork, it vibrates in multiple ways, producing a range of frequencies. While the fundamental frequency is the primary sound we hear, the tuning fork also vibrates in other modes, emitting higher frequencies known as overtones or harmonics. These additional frequencies contribute to the overall complexity and richness of the sound. This phenomenon will be explored in greater detail later in the book, which is dedicated to string musical instruments.

3.1.4 *Vibration and Waves*

Where a restoring force acts on an object, proportional to its displacement from an equilibrium position and in the opposite direction, the force continually pulls the object back towards the equilibrium position, causing it to oscillate back and forth.

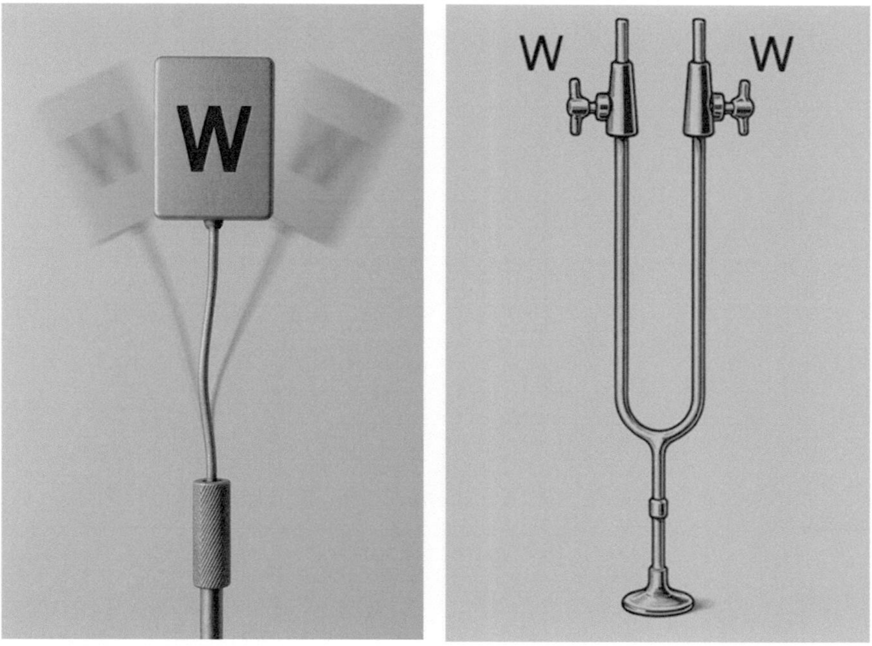

Fig. 3.6 To demonstrate oscillation: (left) a stiff metal bar clamped at one end with a weight, shown in three stages of vibration; (right) a tuning fork, consisting of two U-shaped prongs on a base, described as a more advanced version of the vibrating blade with movable weights, W, designed to produce a specific frequency of sound. Handart images enhanced by AI

- **Vibration** is a more general term that encompasses any repetitive motion, but it often refers to oscillatory motion, including simple harmonic motion.
- **Waves** are disturbances that propagate through a medium, transferring energy without the net transfer of matter.
- **Generating and propagating wave motion**: Harmonic motion is the foundation of many wave phenomena. When a particle oscillates harmonically, it creates a disturbance that propagates as a wave. Waves are produced through vibration. Vibration occurs under two primary conditions:
  - **Restoring force**: A force must exist that acts to return the object to its equilibrium position when it is displaced. This force arises from the object's tendency to maintain its original shape.
  - **Energy conversion**: The object must be able to convert potential energy (stored energy) into kinetic energy (energy of motion) and vice versa.

Examples of objects that can produce mechanical waves through vibration include springs, vibrating blades, tuning forks, and strings.

Imagine tying one end of a long rope or spring to a fixed point and holding the other end in your hand. If you quickly raise and lower your hand, you'll create a disturbance in the rope that propagates as a single wave pulse (Fig. 3.7).

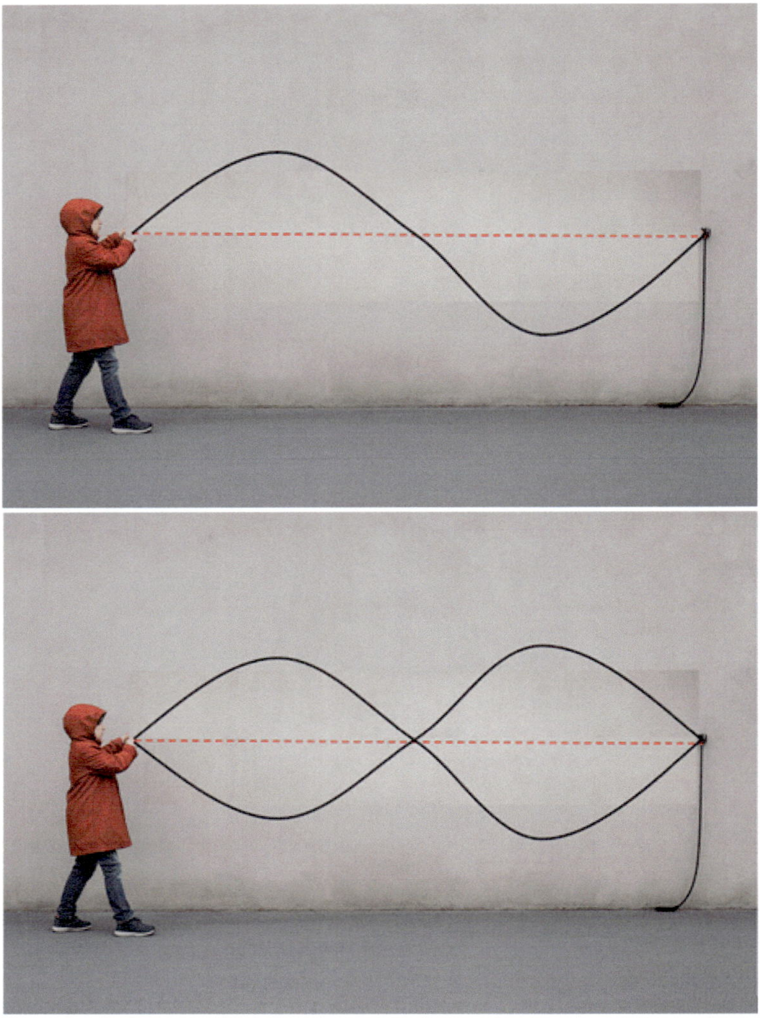

Fig. 3.7 A wave pulse traveling along a rope. A person on the left is shown creating the pulse by moving his hand, up and down, while the wave propagates toward a solid suport on the wall on the right. The image illustrates how a single disturbance travels through a medium and then reflects back, inverting its direction

As the disturbance travels along the rope, each point undergoes an up-and-down motion. Once the disturbance reaches the end of the rope and encounters an obstacle, it reflects back, inverting its direction.

If, instead of a sudden shock, we apply a continuous, simple oscillatory motion perpendicular to the rope with a period T, we generate a series of simple sinusoidal waves. These waves propagate along the rope, and each point on the rope begins

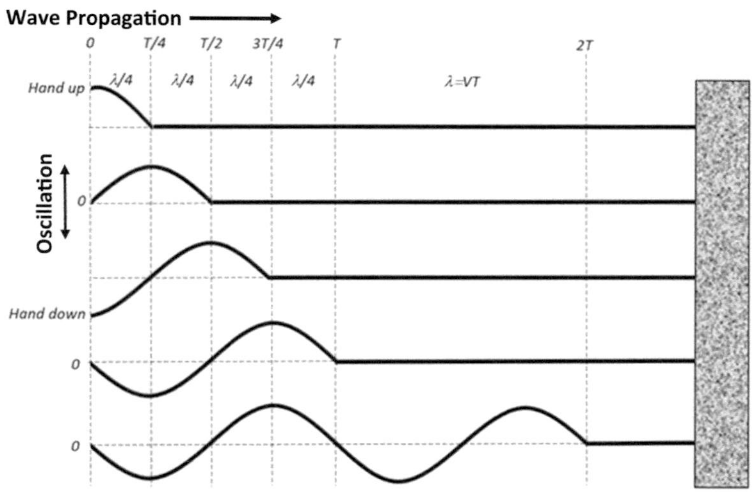

Fig. 3.8 Illustrating the generation and propagation of a wave on a rope, initiated by a hand motion (a hand motion (as depicted in Fig. 3.7). The sequence shows the wave at different time intervals (0, T/4, T/2, 3T/4, T, and 2T), demonstrating how the wave moves a quarter wavelength (λ/4) in a quarter of the wave period. The relationship $\lambda = VT$ is also shown

to vibrate sinusoidally after the wave reaches it [6]. Figure 3.8 illustrates this wave generation and propagation at different times.

- **Types of waves**: When we discuss sound and vibrating strings, we primarily encounter two types of waves [31]:
  - **Longitudinal waves**: In longitudinal waves, the particles of the medium vibrate parallel to the direction of wave propagation. Sound is a classic example of a longitudinal wave. When a sound source, like a vibrating string, produces sound, it creates compressions and rarefactions (regions of reduced air pressure) in the air. These compressions and rarefactions propagate through the air as a longitudinal wave.

    Imagine a spring being compressed and stretched repeatedly. The coils of the spring move back and forth parallel to the direction of the wave.
  - **Transverse waves**: In transverse waves, the particles of the medium vibrate perpendicular to the direction of wave propagation. When a string is plucked, it vibrates in a transverse manner. The particles of the string move up and down, perpendicular to the length of the string. This creates a transverse wave that travels along the string.

    As seen in Fig. 3.8, a wave on a rope is inserted by a vertical hand movement. The rope particles oscillate up and down, while the wave itself moves horizontally across the rope.

    There is a connection between the two. It means that while the vibrations of a string are transverse, the sound waves they produce in the air are longitudinal.

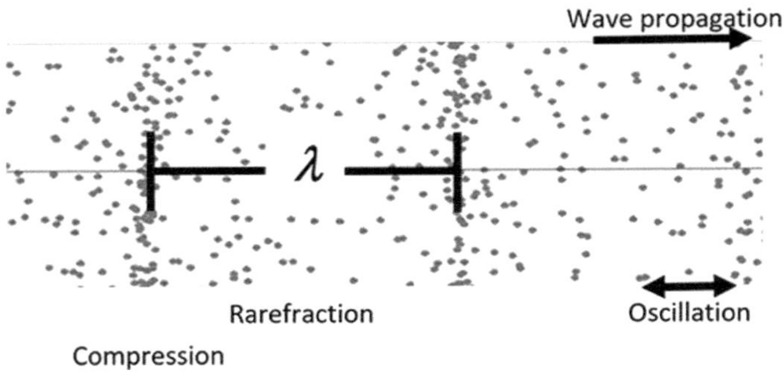

Fig. 3.9 Illustrating a longitudinal wave, such as a sound wave. The image shows particles of a medium (like air) being compressed and rarefied, with labels indicating "Compression" (high particle density) and "Rarefaction" (low particle density). Arrows show the direction of wave propagation and the parallel oscillation of individual particles. The wavelength (λ) is also indicated. Image from Wikimedia Commons, free resources (A Creative Commons Attribution License): https://commons.wikimedia.org/wiki/File:Molecule5.gif

The string's vibrations create pressure variations in the air, which propagate as longitudinal sound waves (Fig. 3.9).

- **Key points to remember**:
  - **Wavelength**: The distance between two consecutive corresponding points on a wave.
  - **Frequency**: The number of complete oscillations per unit time.
  - **Amplitude**: The maximum displacement of a particle from its equilibrium position.
  - **Oscillation range**: Total distance traveled in one complete oscillation (twice the amplitude). In essence, oscillation range is a broader measure that encompasses the entire extent of the motion, while amplitude focuses on the maximum displacement from a specific point.
  - **Speed**: The velocity or speed at which the wave propagates through the medium.
- **Other wave types**:

  Understanding longitudinal and transverse wave types is crucial in analyzing the behavior of sound and vibrating strings, and it is a fundamental concept in acoustics and music theory. However, it's worth exploring a few other wave types for a more comprehensive understanding.

  - **Torsional waves** involve the twisting motion of particles around an axis [28]. While not directly related to the primary vibration of a string, torsional waves can occur in the string itself, particularly in thicker strings or strings under high tension. These waves can influence the overall vibrational behavior and

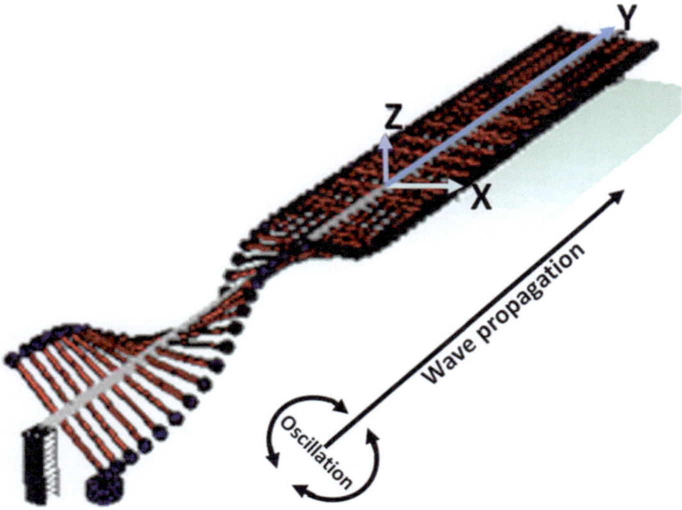

Fig. 3.10 Diagram of a torsional wave propagating along a rod. The markers on the rod show the circular motion of individual particles, perpendicular to the direction of wave propagation. The image visualizes the concept of twisting one end of a spring to create this type of wave. Image from Wikimedia Commons, free resources (A Creative Commons Attribution License): https://commons.wikimedia.org/wiki/File:Wavemachine.gif

sound quality of the string. It also occurs in plates of the sound boards but not as important as those above wave types.

To visualize torsional waves, imagine twisting one end of a spring back and forth. This twisting oscillation creates a wave that propagates along the spring, with each coil rotating around its axis. In this type of wave, the particles of the medium oscillate in a circular motion perpendicular to the direction of wave propagation (Fig 3.10).

– **Rayleigh waves** are surface waves that propagate along the surface of a solid material [31]. Rayleigh waves can probably influence the vibrational behavior of the soundboard of an acoustic instrument, particularly in the way it transfers energy from the strings to the air. But a limited number of studies have focused on applying Rayleigh solutions and methods to analyze soundboards. Notable examples include the work of Boutillon and Ege [5] and He et al. [14].

These waves involve a combination of longitudinal and transverse motion, causing particles within the solid to oscillate in elliptical paths. A key characteristic of Rayleigh waves is that the direction of particle rotation reverses with depth. Particles near the surface move in a counter-clockwise (retrograde) ellipse, while those deeper down move clockwise [35]. The importance of Rayleigh waves are in earthquakes, noting that although they travel slower than the other earthquake waves, they have larger amplitudes and cause the most damage due to their combined vertical and horizontal ground motion [8].

– **Water waves** are another example of the surface waves for a combination of longitudinal and transverse waves that propagate along the surface of water. The particles of water move in circular orbits as the wave passes [31]. While not directly applicable to string instruments, understanding water waves can provide insights into the behavior of waves in general, including sound waves and waves on strings.

By understanding these different wave types, we can gain a deeper appreciation for the complex interplay of vibrational phenomena in musical instruments.

Apart from surface waves, all other wave types i.e., longitudinal, transverse and torsional waves can propagate through solids to varying degrees. However, only longitudinal waves can penetrate deep into liquids and gases. This is because air and water molecules can only oscillate in the direction of wave propagation [31].

Transverse waves exhibit a unique characteristic that sets them apart from longitudinal and torsional waves. In longitudinal waves, vibration occurs parallel to the wave's direction of travel. In torsional waves, vibration is rotational around the wave's axis. Conversely, in transverse waves, vibration occurs perpendicular to the wave's direction of propagation. This allows for a multitude of vibration directions, as demonstrated by a rope oscillating in various directions perpendicular to its length (Fig 3.11).

When a transverse wave vibrates consistently in a single plane, i.e each of the blue, red or green planes of vibration (Fig 3.11), with both the vibration and propagation directions lying within that plane, we refer to it as a polarized wave [3].

- **Standing waves**: also known as a stationary wave, is a wave that oscillates in time but whose peak amplitude profile does not move. This phenomenon occurs when

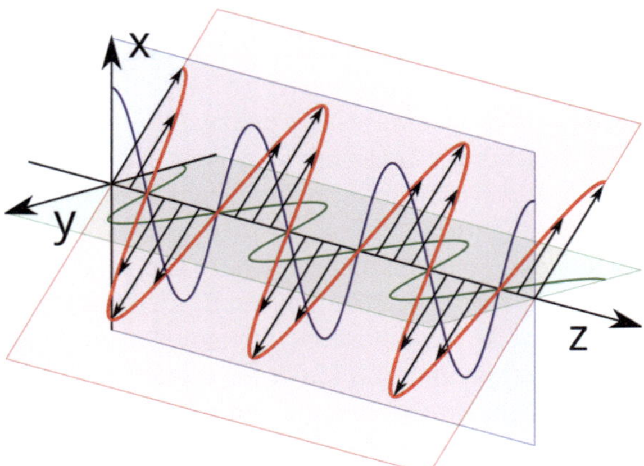

Fig. 3.11 This diagram illustrates a transverse wave vibrating in multiple planes. Arrows show various vibration directions, with sine waves in the planes, perpendicular to the wave's z-axis propagation. This multi-planar vibration is a key characteristic of transverse waves (Image from Wikimedia Commons, free resources; The copyright holder, released this work): https://commons.wikimedia.org/wiki/File:Linearly_Polarized_Wave.svg

two waves of identical frequency and wavelength travel in opposite directions through the same medium [42].

Imagine you have a long, taut string. If you pluck it, a wave travels down the string and then reflects back when it reaches the end. The incoming wave and the reflected wave will interfere with each other in a specific way. This means that when two identical waves traveling in opposite directions interfere, they create a pattern of alternating nodes and antinodes.

- **Nodes**: Points of zero displacement.
- **Antinodes**: Points of maximum displacement.
- **Boundary conditions**: The medium supporting a standing wave must be subject to boundary conditions, as exemplified by a string fixed at both ends. Standing waves arise from the superposition of two waves traveling in opposite directions. The boundary conditions dictate how these waves reflect and interfere with each other, leading to the formation of stationary nodes and antinodes characteristic of a standing wave.
- **Resonance**: This occurs when the frequency of the wave matches one of the medium's resonant frequencies. At these resonant frequencies, the wave reflects at the boundaries and interferes constructively with the incident wave, creating the stable pattern of a standing wave.
- **Resonant frequencies**: The resonant frequencies of a system determine the possible frequencies at which standing waves can form. For example, for a string fixed at both ends, the resonant frequencies are given by [13]:

$$f_n = n\frac{\mathrm{v}}{2L} \tag{3.10}$$

  where, f_n is the nth resonant frequency, v is wave speed in the medium and L is Length of the string between two supports, while fixed at both ends.

Standing waves are the basis for the sound production in many musical instruments, such as guitars, violins, and wind instruments.

3.1.5 Wave Propagation in a Homogeneous Medium

In a homogeneous medium, where the physical properties are uniform throughout, waves propagate at a constant speed in all directions. A common example is the circular ripple that spreads outward when a small spherical stone is dropped into a still pond.

The speed of wave propagation in a medium is largely unaffected by the wave's shape or amplitude, as long as the deformation caused by the wave remains relatively small [20]. The speed of wave propagation is primarily determined by the properties of the medium itself. For viscoelastic materials like wood, which exhibit a phase lag between stress and strain, the high speed testing methodologies e.g. very high ultrasonic frequencies can slightly affect the measured speed of sound [30]. This effect is often considered a minor measurement error and will be discussed for wood in the next chapter.

The speed of a transverse wave along a taut string can be calculated using the following equation [14]:

$$\mathrm{v} = \sqrt{\frac{F}{w}} \tag{3.11}$$

where:

v: Wave propagation speed (m/s).

F: Tensile force in the string (N).

w: Linear mass density of the string (kg/m).

This equation highlights the influence of string tension and linear mass density on the wave propagation speed.

Practical Exercise 3.1

A musical instrument string has a linear density of 0.5 g per meter. If the string is under 5 Newtons of tension, what is the predicted speed of wave propagation in meters per second? Assume the tension does not significantly elongate the string.

Solution

By substituting the values for tensile force and linear mass density into Eq. 3.11, we can readily calculate the speed of wave propagation in the string.

$$\mathrm{v} = \sqrt{\frac{F}{w}} = \sqrt{\frac{5}{0.0005}} = 100\ \mathrm{m/s}$$

As defined shortly earlier (Fig. 3.8), wavelength is the distance between two identical points in a wave, such as the distance between two consecutive crests or two consecutive troughs. It's often represented by the Greek letter lambda (λ).

In Eq. 3.10, the denominator, which is twice the length of the vibrating string between the supports, represents the wavelength of the fundamental frequency. This is the lowest frequency at which a standing wave can form on the string.

The wave speed is crucial in this context. By combining it with the wavelength, which is a multiple of the distance between the supports, we can predict the frequencies that the string can naturally produce.

The concepts of frequency, wavelength, and wave speed are important in understanding sound vibrations. There's a fundamental relationship between wavelength, wave velocity, and frequency [13, 20]. If we rewrite Eq. 3.10 for the fundamental frequency (n = 1), we get:

$$\lambda_n = \frac{2l}{n} \rightarrow\rightarrow \lambda_1 = \frac{2L}{1} = 2L$$
$$f_1 = 1 \times \frac{\mathrm{v}}{2L} = \frac{\mathrm{v}}{\lambda_1} \tag{3.12}$$

Since we haven't yet paid attention to the complexities of harmonic frequencies, we'll begin with a simplified model, focusing solely on the fundamental frequency to practice our calculations.

Practical Exercise 3.2

Calculate the fundamental harmonic frequency for Exercise 3.1, assuming a distance of 62.5 cm between the supports. Use the wavelength and wave propagation speed in your calculation.

Solution

In the previous exercise, we calculated the wave propagation speed in the string to be 100 m/s. Therefore:

$$\lambda_1 = \frac{2L}{1} = 2L = 2 \times 0.625 = 1.25 \text{ m}$$
$$f_1 = \frac{\mathrm{v}}{\lambda_1} = \frac{100}{1.25} = 80 \text{ Hz}$$

We'll also begin to explore the concept of harmonic frequencies, a complex topic we'll go deeper into when discussing sound color. When a vibrating string is plucked, it simultaneously produces multiple harmonics [9].

Table 3.1, illustrates the different modes of vibration, or harmonics, that can be produced in a string fixed at both ends. These standing waves are formed when the string vibrates at specific frequencies, resulting in points of zero displacement (nodes) and maximum displacement (antinodes) along the string. Each harmonic has its own unique wavelength and, consequently, a distinct frequency. So, if we rewrite Eq. 3.10 for the velocity, we get:

$$\mathrm{v} = \frac{\lambda_n f_n}{n} \tag{3.13}$$

As briefly mentioned earlier, harmonic vibrations exhibit an interesting relationship between wave speed and frequency. While wave speed influences frequency, changes in modal frequencies do not affect wave speed. This is because the wavelength adjusts proportionally in higher modes, maintaining a constant wave speed (Table 3.1). Wave propagation speed is primarily determined by the medium's properties. However, in viscoelastic materials like wood, where stress and strain are out of

Table 3.1 Illustration of harmonics in a string fixed at both ends

No.	Node/Antinode	λ	f	v
Fundmental	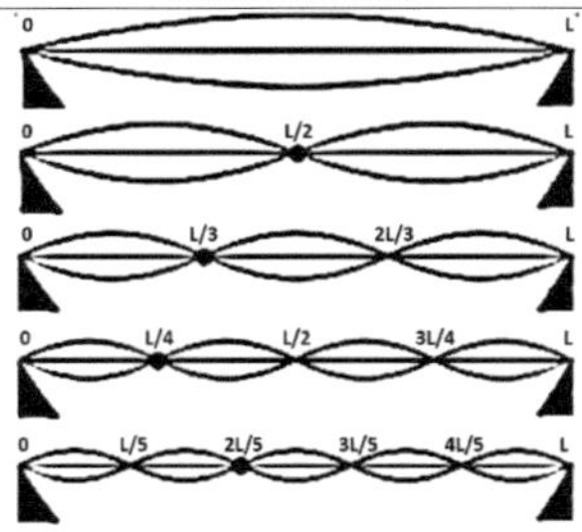	2L	f_0	v
2nd harmonic		L	$2f_0$	v
3rd harmonic		2L/3	$3f_0$	v
4th harmonic		L/2	$4f_0$	v
5th harmonic		2L/5	$5f_0$	v
nth harmonic	–	2L/n	nf_0	v

The table details the relationship between the mode of vibration (harmonic number), the pattern of nodes and antinodes, wavelength (λ), frequency (f), and wave speed (v). It demonstrates how wavelength adjusts proportionally with frequency in higher modes to maintain a constant wave speed

phase, high-speed testing (e.g., ultrasonic frequencies) can slightly alter the measured sound speed [30]. This can introduce errors in calculations of other mechanical and acoustic parameters dependent on sound speed. Since no ultrasonic examples have been considered thus far, we assume, according to Eq. 3.13, that wave speed remains constant across harmonic vibration modes. This topic will be explored further in the next chapter, where ultrasonic techniques will be also illustrated.

Look again at these key elements illustrated in Table 3.1:

- **String length**: The string is represented by the horizontal line, and its length is denoted by "L".
- **Nodes**: The points on the string where the displacement is always zero are indicated by small circles.
- **Antinodes**: The points on the string where the displacement is maximum are indicated by the peaks of the wave pattern.
- **Harmonics**: Each row in the image inserted in Table 3.1, represents a different harmonic, labeled from the fundamental (1st harmonic) to the 6th harmonic. The number of harmonics continues to infinity. For the fundamental harmonic (1st harmonic), the string vibrates as a whole with one antinode in the middle. The wavelength (λ) is twice the length of the string ($2L$). The frequency (f) is the fundamental frequency (f_1). For the second harmonic (2nd harmonic), the string vibrates with two antinodes and one node in the middle. The wavelength is equal to the length of the string (L). The frequency is twice the fundamental frequency ($2f_1$). It continues for the third harmonic (3rd harmonic). The string vibrates with three antinodes and two nodes. The wavelength is two-thirds the length of the string ($2/3L$). The frequency is three times the fundamental frequency ($3f_1$). This continues until the infinity.

As you move up the harmonic numbers, the number of antinodes and nodes increases, and the wavelength decreases. The frequency, on the other hand, increases. This pattern can be summarized as follows for a both ends clamped vibrating string:

The nth harmonic has n antinodes and (n + 1) nodes.
The wavelength of the nth harmonic is $\lambda_n = 2L/n$.
The frequency of the nth harmonic is $f_n = nfL$.

Understanding these standing wave patterns is essential in various fields, including music, acoustics, and physics. For example, in music, the different harmonics contribute to the unique sound of an instrument. In physics, these patterns are fundamental to the study of wave phenomena and resonance.

Practical Exercise 3.3

Calculate the higher harmonic frequencies (n = 2 to n = 4) for Exercise 3.2.

Solution

In the previous exercise, we applied the wave propagation speed in the string to be 100 m/s. Therefore:

$$\lambda_2 = \frac{2L}{2} = L = 0.625 \text{ m} \Rightarrow f_2 = \frac{\text{v}}{\lambda_2} = \frac{100}{0.625} = 160 \text{ Hz}$$
$$\lambda_3 = \frac{2L}{3} = \frac{2 \times 0.625}{3} \approx 0.42 \text{ m} \Rightarrow f_3 = \frac{\text{v}}{\lambda_3} = \frac{100}{0.42} = 240 \text{ Hz}$$
$$\lambda_4 = \frac{2L}{4} = \frac{2 \times 0.625}{4} = 0.3125 \text{ m} \Rightarrow f_4 = \frac{\text{v}}{\lambda_4} = \frac{100}{0.3125} = 320 \text{ Hz}$$

Practical Exercise 3.4

Tune a string of your instrument (e.g., violin string 1) to the fifth octave, E note (approximately 600 Hz). Given a linear density of 0.443 g per meter and a string length of 328 mm, calculate the string tension.

Solution

To determine the wave velocity on this string, we'll use the following formula:

$$f_1 = \frac{\text{v}}{2L} \Rightarrow\Rightarrow \text{v} = 2\text{Lf}_1$$
$$\text{v} = 2 \times 0.328 \times 600 = 393.6 \text{ m/s}$$

Then the string tension would be calculated:

$$F = v^2 w = 393.6^2 \times 0.000443 = 68.6 \text{ N}$$

While this remains a simplified calculation, it is important to acknowledge that real-world factors, such as changes in string length due to stretching, can influence the actual string tension.

As previously noted, changes in string length due to stretching can now affect the actual string tension. To compensate for this, we must first determine the effective vibrating length of the string, which is the length of the stretched string in the desired tuning. This effective length, along with the weight per unit length of the string, must be considered in our calculations.

To achieve precise string calculation, follow these steps:

- **Accurate tuning**: Ensure the string is tuned to the desired pitch.
- **Marking the nut and bridge**: Use a marker to mark the exact positions of the nut and bridge on the string itself.
- **Measuring string length**: Completely detune the string and measure the distance between the two marks. This is the relaxed string length.
- **Calculating elongation ratio**: Divide the string's length under tension (actual distance between nut and bridge) by its relaxed length (between the two marks) to determine the elongation ratio (*ER*).
- **Weighing the string**: Weigh any section of the string to obtain its mass.
- **Calculating tensioned length**: Multiply the measured string length by the elongation ratio to find the length it will have under tension (*EL*).
- **Estimating weight per meter**: Divide the string's weight by the tensioned length to calculate the weight per meter for that specific note.

Practical Exercise 3.5

Tune a violin string to the E note in the fifth octave (approximately 600 Hz). Given that a 450 mm piece of the string weighs 0.23 g and the original length between the bridge and nut is 328 mm, determine the actual tension when the string length between the bridge and nut markings after a complete detuning is shortened to 310 mm.

Solution

The elongation ratio (*ER*):

$$ER = \frac{328}{310} = 1.058$$

The length the relaxed string piece (*EL*) will have under tension:

$$EL = 450 \times 1.058 = 476.1 \text{ mm}$$

The actual linear density (w_a), kilograms per meter, for stretched string for this specific note:

$$w_a = \frac{0.00023}{0.4761} = 4.83 \times 10^{-4} \text{ Kg/m}$$

Then the wave velocity would be:

$$\text{v} = 2 \times 0.328 \times 600 = 393.6 \text{ m/s}$$

The actual string tension (F_a) would be calculated now:

$$F_a = \text{v}^2 w_a = 393.6^2 \times 0.000483 = 74.8 \text{ N}$$

Why not experiment with this practical exercise on other strings, notes, and even different string instruments?

3.1.6 Estimating Bridge Force in String Instruments: A Force Analysis Perspective

The constituent parts of a stringed instrument, exemplified by the Persian setar, are visualized in Fig. 3.13. The bridge, a crucial component of a string instruments, plays a pivotal role in transmitting string vibrations to the instrument's soundboard. Understanding the forces exerted on the bridge is essential for instrument makers, repair technicians, and researchers. This section presents a simplified approach to estimate the bridge force using basic force analysis principles.

The tension in each string, determined by tuning, is a primary factor influencing the force on the bridge. Higher tension, results in greater force on the bridge.

The angle at which the strings intersect the bridge affects the distribution of forces. A steeper angle generally leads to a higher vertical component of force on the bridge.

Friction between the strings and the bridge can influence the force transmission. While often neglected in simplified models, friction can affect the accuracy of force estimates.

The tension in each string acts along the string's direction. The bridge exerts reaction forces on the strings to balance the tension forces.

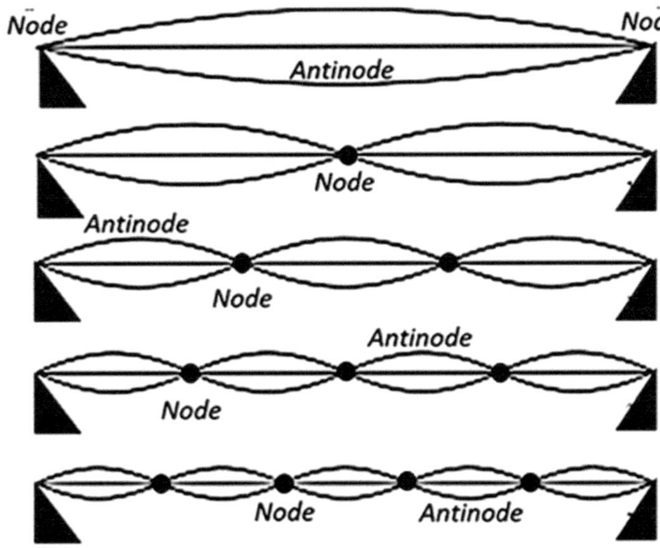

Fig. 3.12 Standing waves on a string, with clear labels marking the nodes (points of zero displacement) and antinodes (points of maximum displacement). The image emphasizes the stationary appearance of the string despite it vibrating in a standing wave pattern

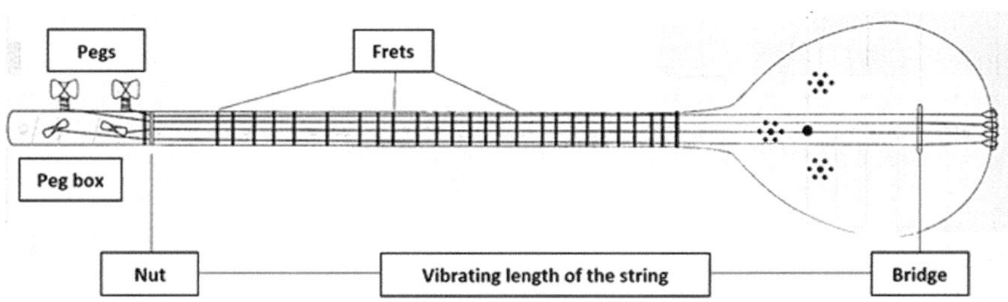

Fig. 3.13 Diagram of a long-necked plucked string instrument (a Persian Setar), with labeled parts: peg box (containing tuning pegs), nut, frets along the neck, the vibrating length of the string, and the bridge connecting the string to the instrument's body. The image provides a clear visual representation of the instrument's basic structure and string path

Ideally, in static equilibrium, the tensile force within a tuned string is constant throughout its length, irrespective of the angle at which it contacts the pole. This principle stems from fundamental physical laws. Newton's Third Law says that every action has an equal and opposite reaction. In the context of the string, this implies that forces exerted on the string are balanced by internal forces within the string itself.

In a uniform string, the tensile force is distributed evenly along its length. For the string to be in static equilibrium, the resultant force acting on any point within the string must be zero.

Why does this question arise?

While the ideal scenario assumes a constant tensile force, real-world conditions may introduce minor variations. Factors like friction between the string and the bridge, variations in string thickness, and the angle of contact can subtly influence the force distribution. However, these effects are generally negligible and do not significantly alter the overall tensile force within the string. In essence, the angle of contact with the bridge has a minimal impact on the string's internal tensile force in most practical scenarios.

- **Resolve forces**:

If we carefully assume equal tension for the continuous string on both sides of the bridge ($F' = F'' = F$), we can then decompose the tension forces into their vertical and horizontal components (Fig. 3.14):

- **Vertical component**: The vertical component of the string tension contributes to the downward force on the bridge (Fig. 3.14).

$$
\begin{aligned}
F'_{\mathrm{v}} &= F' \sin\left(\frac{3\pi}{2} - \beta\right) = -F' \cos\beta \\
F''_{\mathrm{v}} &= F'' \sin\left(\frac{3\pi}{2} + \alpha\right) = -F'' \cos\alpha
\end{aligned}
\tag{3.14}
$$

Reference to Fig. 3.14 provides a clear definition of F, F', F'', α, and β. The negative sign observed in the resultant vector calculation is a consequence of the downward orientation of the vertical components of both F' (located in the third quadrant) and F'' (located in the fourth quadrant).

- **Horizontal component**: The horizontal component of the string tension is balanced by frictional forces and the bridge's resistance to stay standing (Fig. 3.14).

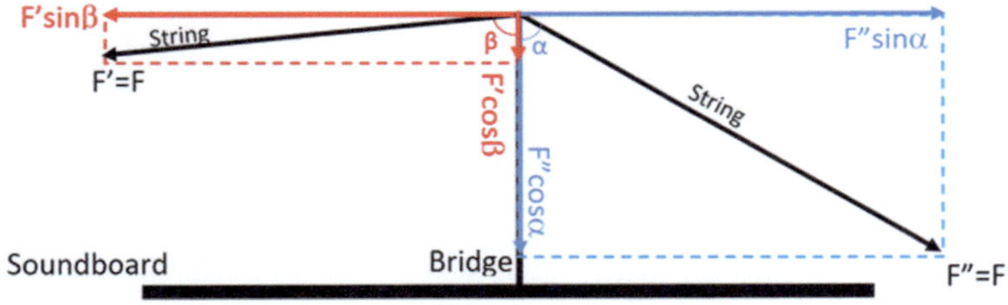

Fig. 3.14 Schematic of a string passing over a bridge, showing how the tension forces (F', F'') are resolved into vertical and horizontal components. The assumption of equal tension ($F' = F'' = F$) simplifies the analysis, allowing the calculation of forces acting on the bridge based on the angles α and β

$$
\begin{aligned}
F'_{\mathrm{H}} &= F' \cos\left(\frac{3\pi}{2} - \beta\right) = -F' \cos\beta \\
F''_{H} &= F'' \cos\left(\frac{3\pi}{2} + \alpha\right) = F'' \cos\alpha
\end{aligned}
\tag{3.15}
$$

The negative sign observed in the resultant vector calculation is a consequence of the leftward orientation of the horizontal component of F′ (located in the third quadrant). The horizontal component of F″ (located in the forth quadrant) is rightward and remained positive.

- **Estimate bridge force**: By summing the vertical components of the string tensions, we can obtain an estimate of the total downward force on the bridge (The negative sign was not considered).

$$
\begin{aligned}
&F_{\mathrm{v}} = F'_{\mathrm{v}} + F''_{\mathrm{v}} = F' \cos\beta + F'' \cos\alpha \\
&\text{Since} \Rightarrow F = F' = F'' \\
&\Downarrow \\
&F_{\mathrm{v}} = F(\cos\beta + \cos\alpha)
\end{aligned}
\tag{3.16}
$$

This estimate can be refined by considering the bridge angle and frictional effects, but these factors can be complex to model accurately, which will only be useful for academic research and is not very relevant in workshop instrumentation.

To maintain the balance of a bridge, the horizontal components of the string tension on both sides must be equal. An imbalance in these forces will pull the bridge towards the side with greater tension or potentially cause it to overturn. Fortunately, slight discrepancies are typically compensated by the friction between the string and the bridge, and between the bridge and the soundboard. Understanding the interplay of factors like friction, material flexibility, bridge height, and string tension seems to be necessary in predicting the behavior of the vibrating string.

Practical Exercise 3.6

A four-string instrument has two distinct string angles (Fig. 3.14) rested on the bridge: 84° (alpha) and 89° (beta). When tuned, the strings tensile forces were 74 N, 55 N, 40 N, and 38 N, respectively. Determine the total force exerted on the bridge by the string tension.

Solution

The vertical components of each srting:

$$F_{v1} = 74(\cos 89^\circ + \cos 84^\circ) = 9\ \text{N}$$
$$F_{v2} = 55(\cos 89^\circ + \cos 84^\circ) = 6.7\ \text{N}$$
$$F_{v3} = 40(\cos 89^\circ + \cos 84^\circ) = 4.9\ \text{N}$$
$$F_{v4} = 38(\cos 89^\circ + \cos 84^\circ) = 4.6\ \text{N}$$

Then a total vertical force exerted on the bridge legs by the strings tensions is:

$F_v = 25.2$ N.

The presented analysis is a simplified model that neglects factors like string stiffness, bridge flexibility, and the dynamic nature of string vibrations. Actual bridge forces can also slightly vary due to differences in instrument design, string gauge, tuning, and playing style. For example, putting a finger on a string on fingerboard, invreases the angle of β, raising the force a bit on bridge.

For more accurate analysis, advanced techniques like finite element analysis (FEA) can be employed to model the complex interactions between the strings, bridge and soundboard that as an academic approach could be out of the scope of this book for luthiery.

While the simplified force analysis presented here provides a basic understanding of the forces acting on the instrument's bridge, it's important to recognize its limitations. For more precise insights, a combination of theoretical analysis, experimental measurements, and advanced modeling techniques is necessary. By understanding the forces at play, instrument makers and repair technicians can make informed decisions to optimize instrument performance.

3.1.7 Wave Interference

Wave interference occurs when two or more waves meet while traveling through the same medium. The resulting wave pattern depends on the phase difference between the waves (Fig. 3.15) [13]. The standing wave that was introduced ealier this chapter is an example for the wave interfernce that the incoming wave and the reflected wave will interfere with each other in a specific way.

- **The principle of superposition**: states that when two or more waves overlap in space or in a medium, the resulting disturbance is equal to the algebraic sum of the individual disturbances. This means that the net effect of multiple waves at a given point is the sum of the effects of each individual wave at that point. For the waves

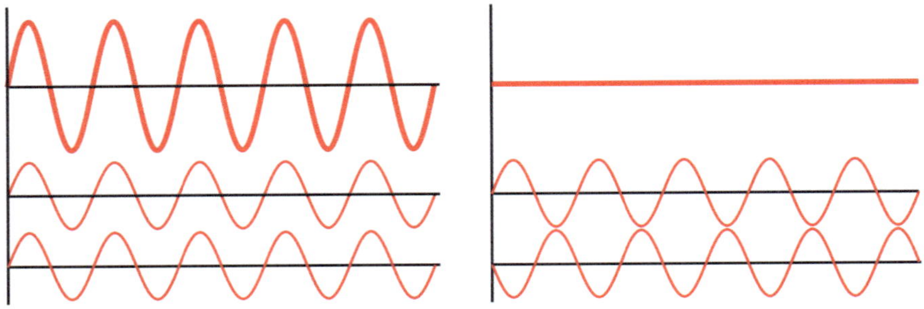

Fig. 3.15 The interference of two waves. When in phase, the two lower waves create constructive interference (left), resulting in a wave of greater amplitude. When 180° out of phase, they create destructive interference (right). Image from Wikimedia Commons, free resources (A Creative Commons Attribution License): https://commons.wikimedia.org/wiki/File:Interference_of_two_waves.svg

that originate from coherent sources, meaning they have the same frequency and a constant phase difference, these two kinds of interferences are expected [13]:

- **Constructive interference**: When the peaks of two waves coincide, their amplitudes add up, resulting in a wave with a larger amplitude. This is called constructive interference.
- **Destructive interference**: When the peak of one wave meets the trough of another, their amplitudes cancel each other out, resulting in a wave with a smaller amplitude or even no wave at all. This is called destructive interference. Noise-canceling headphones use destructive interference to reduce unwanted noise.

When waves of different frequencies interfere, the resulting pattern is more complex than the simple constructive and destructive interference seen with waves of the same frequency. This is because the phase relationship between the waves is constantly changing.

- **Beat frequency**: When two waves with slightly different frequencies interfere, they produce a phenomenon called "beats". The beat frequency is equal to the difference between the two original frequencies. This results in a periodic variation in amplitude, perceived as a pulsating sound (Fig. 3.16).
- **Complex interference patterns**: When two or more waves with significantly different frequencies interfere, the resulting pattern is more complex and can be difficult to predict. The interference pattern will change rapidly over time, making it difficult to observe a stable pattern.
- It's important to note that while the interference of waves with different frequencies can produce complex patterns, the underlying principle of superposition still applies. The resulting amplitude at any point is the sum of the amplitudes of the individual waves at that point.
- Now, imagine that the fundamental note and its higher harmonics, despite having different frequencies, combine harmoniously due to their regular frequency intervals, which are multiples of the fundamental frequency.

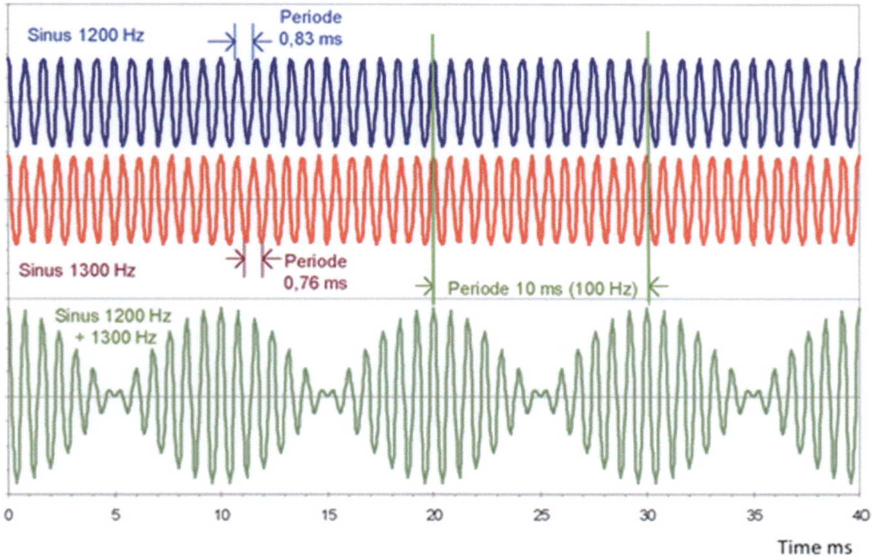

Fig. 3.16 Illustrating the phenomenon of beats. Two sine waves with frequencies 1200 Hz and 1300 Hz are shown separately and then combined. The resulting waveform demonstrates the periodic variation in amplitude, with the beat frequency (100 Hz, as indicated by the 10 ms period) clearly visible. The individual periods of the 1200 Hz (0.83 ms) and 1300 Hz (0.76 ms) waves are also marked. Image from Wikimedia Commons, free resources (The copyright holder of this work, released this work): https://commons.wikimedia.org/wiki/File:Kombinationston_Sinus.jpg

The resulting superimposed wave, while complex, has a distinct, repeating signature (Fig. 3.17). Conversely, the introduction of an unrelated frequency, not part of the harmonic series, can disrupt this harmony and create an unpleasant, noisy wave. We will move deeper into the superposition of harmonic and non-harmonic frequencies, as well as the concept of sound color, after further exploring the nature of sound and musical sounds.

3.2 Sound

Sound waves are mechanical waves generated by the vibration of elastic bodies and can propagate through gases, liquids, and solids. When these waves reach the sensitive eardrum, they are perceived as sound.

The science of acoustics encompasses not only the physics of sound but also the biology and psychology of hearing. While we will go deep into the physical aspects of sound, including its wave properties like frequency, wavelength, propagation speed, generation, propagation, reflection, and refraction, we will also briefly touch on sound detection and its impact on the human ear (Fig. 3.18).

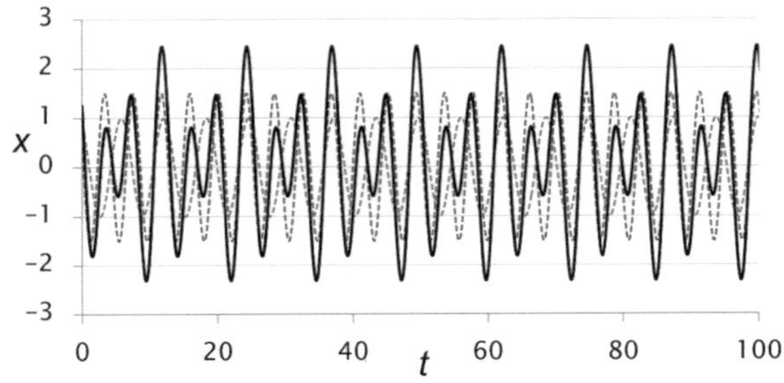

Fig. 3.17 Diagram of wave addition, explaining how waves combine harmoniously. Individual waves representing a frequency (blue wave) and a harmonic of it (red wave) are shown. Their sum (black wave) illustrates the principle of superposition, where the resulting amplitude is the sum of the individual wave amplitudes. Adding an unrelated frequency would create a noisy wave, destroying the existing harmony. Image from Wikimedia Commons, free resources (Under the Creative Commons Attribution License): https://commons.wikimedia.org/wiki/File:Superposition_of_two_simple_harmonic_motions,_rational_ratio_frequencies.png

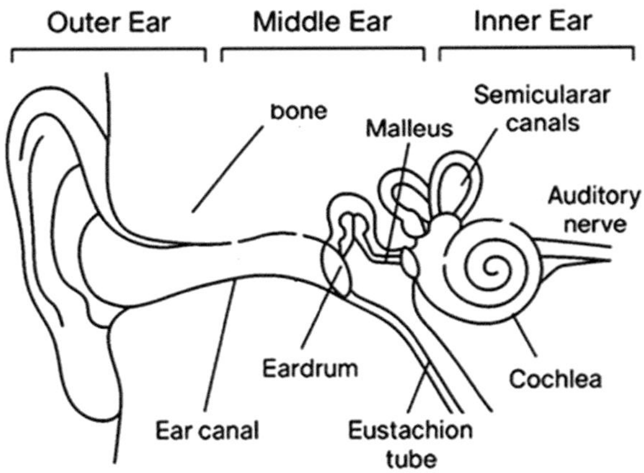

Fig. 3.18 Diagram of the human ear, illustrating the pathway of sound from the outer ear through the middle ear to the inner ear, highlighting the structures involved in hearing

3.2.1 Generation and Propagation of Sound

Sound is generated by the vibration of elastic objects. As we've discussed, many musical instruments produce sound through the formation of standing waves. The sound source can be a solid object, like a tuning fork or a stretched string, or a vibrating surface, such as a drumhead or a loudspeaker diaphragm. It can even be a mass of air, as in the case of sound produced by vocal cords.

The complexity of the sound produced depends on the nature of the vibration. A whistle with a circular lip, for instance, produces a simple sound due to its simple vibrations. In contrast, the complex vibrations of vocal cords result in complex sounds.

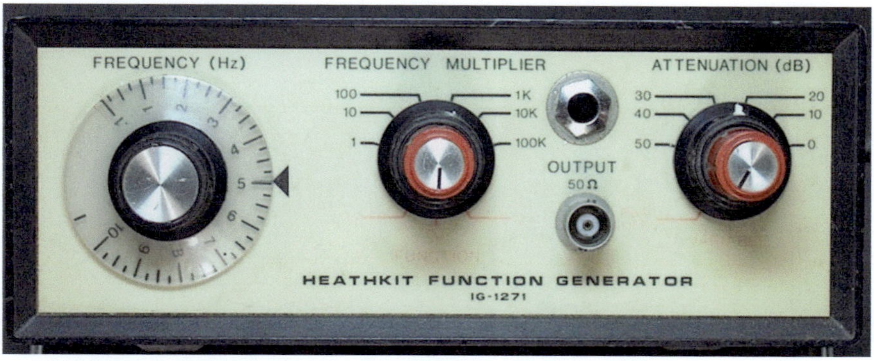

Fig. 3.19 Close-up of the control panel of an electrical function generator, showing dials for frequency, frequency multiplier, and attenuation. Free, powerful, web-based frequency generators are becoming widely available, reducing the need for these older, more expensive electrical equipments. Image from Wikimedia Commons, free resources (A Creative Commons Attribution License): https://commons.wikimedia.org/wiki/File:Signaalgenerator.jpg

Modern electronic oscillators allow us to generate electrical oscillations at specific frequencies, typically between 10 Hz and 100 kHz. These oscillations can be amplified and fed into a loudspeaker, where they are converted into mechanical vibrations of the loudspeaker diaphragm, producing sound waves.

Figure 3.19 depicts an older design of a device that generates electrical oscillations and converts them into various sound frequencies. Today, numerous user-friendly applications for cellphone devices or free online web applications can generate a wide range of frequencies.

The human ear can perceive sounds within a frequency range of approximately 20 Hz to 20,000 Hz [33]. Frequencies above 20,000 Hz are classified as ultrasonic or ultrasound, while those below 20 Hz are known as infrasonic or infrasound [21].

3.2.2 Speed of Sound Propagation

Sound doesn't propagate instantaneously. Instead, it travels through the air at a speed significantly slower than the speed of light. This propagation involves the transmission of vibrational motion from the sound source. The source emits waves that resemble those discussed in the previous section. These waves carry mechanical energy, specifically acoustic energy, derived from the mechanical energy of the vibration source.

Therefore, an elastic medium, such as air, is essential for the transmission of acoustic energy through longitudinal waves between the sound source and the ear.

The speed of sound in a material is directly related to its elastic stiffness. Generally, denser materials, which tend to be stiffer, are believed to have a higher speed of sound. However, this is a common misconception.

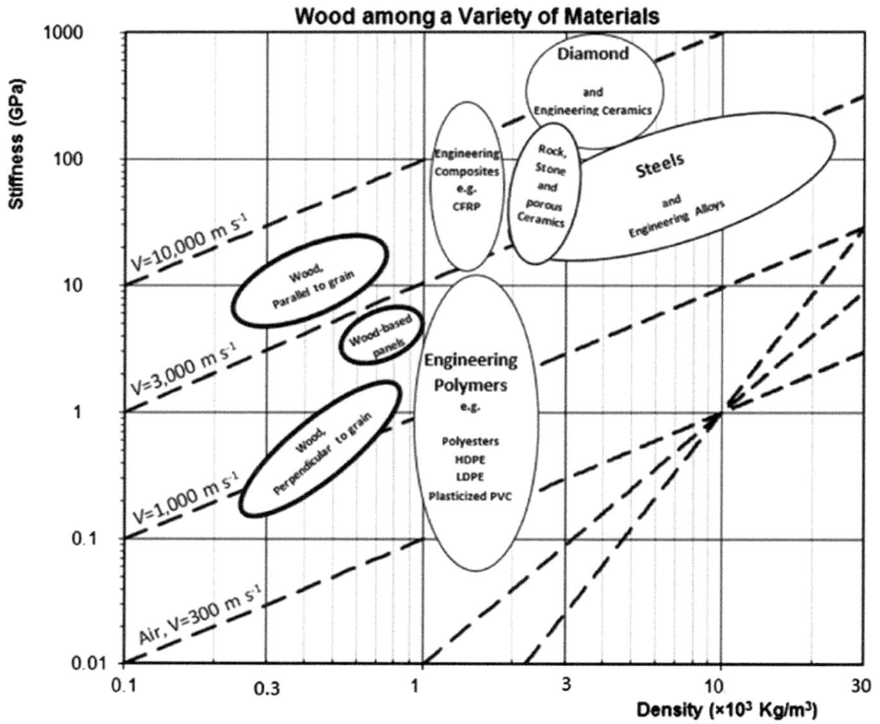

Fig. 3.20 Graph plotting Young's modulus (stiffness, GPa) against density (kg/m^3) for material selection. Wood, both parallel and perpendicular to the grain, is highlighted alongside other material groups (metals/alloys, ceramics, polymers, composites). Diagonal lines represent constant acoustic velocity (V = 10,000 m/s, 3000 m/s, 1000 m/s, and 300 m/s), illustrating performance trade-offs. Image generated by author, previously published [34]

As shown in Fig. 3.20, the speed of sound in wood, despite its lower density compared to steel, is surprisingly high [34]. This highlights that the ratio of stiffness to density is the crucial factor influencing the speed of sound in solids. Now you can imagine that any factor, such as temperature or humidity that alters a material's elasticity or density will also affect the speed of sound within that material. We will go deeper into this topic in the next chapter, focusing on wood.

3.2.3 Sound Reflection

When sound waves encounter an obstacle or a boundary between two media with different properties, a portion of the wave is reflected back. The remaining part may be transmitted into the new medium. If the new medium is significantly stiffer, the wave is primarily reflected [37].

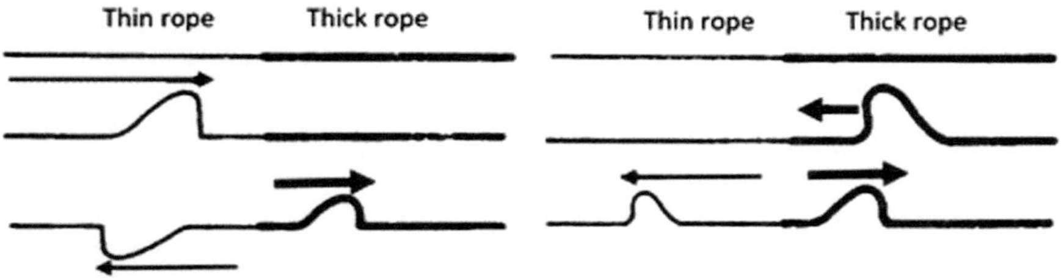

Fig. 3.21 Diagram explaining the concept of mechanical impedance mismatch using wave pulses on ropes of different (linear) densities. The illustration shows how a wave pulse is partially reflected and partially transmitted at the interface, with the key detail of wave inversion upon reflection from a higher impedance medium

To illustrate this, consider two ropes: one thin and light, the other thick and heavy (Fig. 3.21). If a disturbance is initiated on the thin rope, part of it is reflected at the junction with the heavier rope, while the rest is transmitted into the heavier rope. Notably, the reflected wave in the thin rope is inverted.

Conversely, if the disturbance originates on the heavy rope, some of it is transmitted to the thin rope, but the reflected wave in the heavy rope retains its original orientation.

If the end of a rope is fixed to a rigid wall, a wave traveling along the rope is completely reflected upon reaching the wall. The wall, being immovable, exerts a force on the rope, causing the reflected wave to invert.

This phenomenon, known as mechanical impedance mismatch (mechanical impedance is a measure of the total resistance to the vibration of a mechanical system), arises when two mediums with significantly different mechanical properties interact. This mismatch can lead to inefficient energy transfer between the mediums, resulting in a loss of sound energy.

In the next chapter, we will pay further attention to the concept of mechanical impedance, in terms of the elastic stiffness and the density, focusing on its relevance to wood. We will explore methods for calculating mechanical impedance and establish guidelines for acceptable levels of impedance mismatch between wood and conventional strings. This knowledge is important for optimizing the energy transfer between the string and the instrument's body, ultimately enhancing the sound quality.

In stringed instruments, the bridge acts as a boundary for the vibrating string. Imagine the string as a wave traveling through a relatively low-density medium. When this wave encounters the bridge, a denser structure, a portion of the wave is reflected back along the string.

This reflected wave interacts with the ongoing vibrations of the string, influencing its amplitude and sustaining the sound. Simultaneously, a significant portion of the vibrational energy is transmitted through the bridge to the instrument's body (the soundboard). Within the soundbox, the transmitted energy sets the top plate and air cavity into complex vibrations. These vibrations, after a series of reflections and interactions within the soundbox, are coupled back to the bridge and re-enter the string, further reinforcing its vibrations.

Fig. 3.22 Illustrating the reflection of waves and sound. The left panel shows a wave pulse reflecting from a rigid wall, with labels indicating "incident wave", "reflected wave". The right panel shows multiple sound waves reflecting off a surface, similar to light rays, with labels indicating "incident sound wave," "reflected sound wave," and the angles of incidence (i) and reflection (r)

This intricate interplay of reflections and energy transfer between the string, bridge, and soundbox creates a resonant system. The soundbox effectively amplifies and sustains the initial vibrations of the string, producing the rich and resonant tones that characterize stringed instruments.

Sound waves, which propagate spherically in a homogeneous medium, undergo similar reflection principles. To simplify the analysis, we often consider the direction of wave propagation using rays, which are lines perpendicular to the wavefronts. Similar to light rays, sound rays can be reflected (by obstacles with different impedance) (Fig. 3.22), focused (by concave reflectors) (Fig. 3.23), and refracted (when entering new environments) (Fig. 3.24).

Acoustic concave reflectors, often seen in science and amusement parks, demonstrate a fascinating principle of physics. These paired reflectors allow two people standing a distance apart to whisper to each other: one person whispers at the focal point of the first reflector, and the other, positioned at the focal point of the second, directly opposite reflector, clearly hears the whisper (Fig. 3.23).

However, these reflectors are more than just novelties. Historically, concave acoustic mirrors have served practical purposes, from amplifying weak sounds like birdsong and forest noises or detecting approaching enemy aircraft before the advent of radar—even for eavesdropping [39]. The back of many musical instrument resonance boxes is also parabolic or somewhat concave. When constructed from suitable

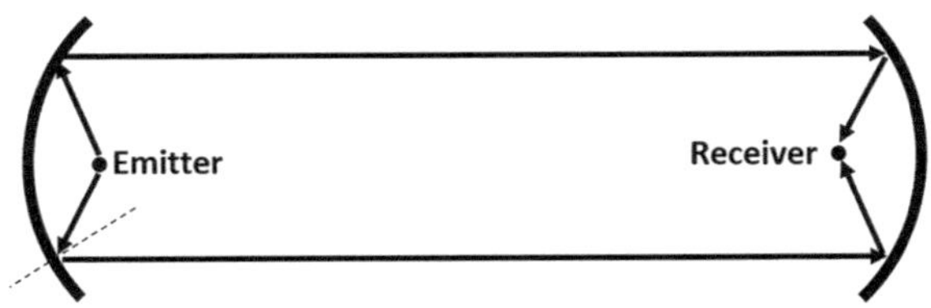

Fig. 3.23 Ray diagram illustrating the principle of sound wave focusing using concave reflectors. A sound source at emitter focal point, emits sound waves that are reflected by a concave surface and directed towards a second concave reflector. These rays are then focused at receiver point, demonstrating how sound intensity can be increased by concentrating the wave energy

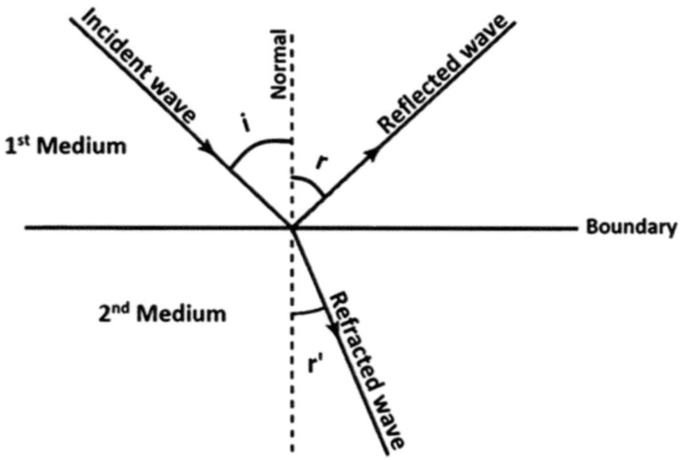

Fig. 3.24 Illustrating the refraction of a sound wave as it travels from one medium to another. The image shows the incident wave, the reflected wave, and the refracted wave, with labels indicating the angles of incidence (i), reflection (r) and refraction (r′). This refraction occurs due to a change in wave speed when moving between media of different densities (rarer to denser or vice versa)

wood, these curved surfaces, much like acoustic concave mirrors, reflect and focus sound waves, directing them towards the focal point and allowing for precise control of sound intensification value by the luthier.

3.2.4 Sound Refraction

When a periodic wave transitions from one medium to another, its speed and wavelength change, but its frequency remains constant. This relationship is expressed as:

$$\frac{v_1}{v_2} = \frac{\lambda_1}{\lambda_2} \tag{3.17}$$

This equation indicates that the ratio of the wavelengths, λ, in the two media equals the ratio of the wave speeds, v [11].

If a sound wave moves from a denser to a rarer medium, its speed decreases, causing the wave to condense at the interface. Consequently, little to no sound enters the rarer medium. A portion of the sound that does enter the rarer medium deviates from its original path, a phenomenon known as sound refraction. This refraction also occurs when a sound wave moves from a rarer to a denser medium.

3.2.5 Sound Characteristics

Researches reveal that sound categorization strategies vary depending on the specific sound type. For instance, people might categorize the sound of a car differently than the sound of rain, perhaps focusing on the distinct engine noise in the former and the continuous, textural quality of the latter. These findings offer valuable insights into how we perceive and process environmental sounds. This understanding is necessary for developing more accurate and robust sound classification models, which have applications in various fields, including automated sound recognition systems, environmental monitoring, and even improving accessibility for individuals with visual impairments. Further investigation into these diverse categorization strategies could lead to more nuanced and human-like sound recognition algorithms [4]. At this point, we can categorize sounds into two broad types:

- **Non-musical sounds**: These sounds are generally unpleasant or disharmonious, lacking a clear and defined pitch. Examples include the sounds of a gunshot, a file scraping metal, or a motorcycle exhaust. Non-musical Sounds often originates from irregular and unpredictable vibrations.

  The distinction between musical and non-musical sounds is not always clear-cut. While a regularly repeating sound like a clock ticking might lack the aesthetic qualities typically associated with music, certain non-musicals can be artistically integrated into musical compositions.

  For example, in the musics from Pink Floyd band, for instance, sounds like breaking glass or a cash register opening and closing are intentionally incorporated to create unique and evocative effects. Similarly, the rustling sound of metal rings on a daf (a frame drum) seems to be noisy but they can add a distinct character and beauty to the music (Fig. 3.25).

  Ultimately, our categorization of a sound as 'musical' or 'non-musical' is subjective and depends on its context, artistic intent, and cultural perception.
- **Musical sounds**: These sounds are characterized by a pleasing and organized quality, typically produced by musical instruments. They arise from the regular and consistent vibration of elastic bodies, with a smooth and predictable change in amplitude.

  Musical sounds are characterized by three primary perceptual qualities [25]:

  - **Loudness**: This subjective perception corresponds to the intensity or amplitude of the sound wave. Higher intensities are generally perceived as louder sounds.
  - **Pitch**: This subjective perception corresponds to the frequency of the sound wave. Higher frequencies are generally perceived as higher pitches.
  - **Timbre**: This subjective perception corresponds to the waveform or complexity of the sound wave. Timbre allows us to distinguish between different instruments playing the same note at the same loudness.

  Perceptual characteristics of sound are subjective and influenced by individual hearing sensitivity. This explains why different listeners may have varying opinions on the sound quality of the same instrument.

Fig. 3.25 Photograph of a daf (frame drum) used in Persian music. The image focuses on the rings attached to the drum's frame, which create a rustling sound described as both noisy and beautiful. The photographer and performer: Dr. Fariborz Hashemidizaji, Tonewood Scientist

Conversely, measurements of physical sound parameters provide more objective and reliable data for classifying sound quality. These measurements offer a consistent basis for analysis, minimizing the influence of individual perception.

Each of these characteristics of sound will be further clarified in the following sections.

3.2.6 Loudness Characteristic

Loudness is a perceptual characteristic of sound, describing how intense a sound appears to the human ear. While closely related to sound intensity, they are not synonymous.

Sound Intensity is a physical quantity representing the amount of sound energy passing through a unit area (typically 1 square meter) perpendicular to the direction of wave propagation per unit time.

Loudness is a physiological sensation influenced by sound intensity but also significantly impacted by the sensitivity of the human ear.

Factors that influence sound intensity include [19]:

- **Amplitude of vibration**: Larger vibrations generally result in greater sound intensity. Calculation demonstrates that the energy of vibrational motion is proportional to the square of its amplitude. This principle holds true for all vibrations, including sound waves. Consequently, the intensity of a sound wave is also proportional to the square of its amplitude. For instance, doubling the amplitude of vibration results in a fourfold increase in sound intensity, while tripling the amplitude leads to a ninefold increase.

- **Distance from the source**: Sound intensity decreases with increasing distance from the source. Calculation and experimental evidence demonstrate that sound intensity diminishes with the square of the distance from the sound source. This principle is known as the inverse square law of sound (Fig. 3.26).

  Consider a sound source, S, emitting sound waves. The amount of sound energy passing through a unit area at a distance of one meter from S will also pass through larger areas at distances of two meters and three meters, respectively. Since the surface areas of these spheres increase proportionally to the square of the distance (1:4:9), the sound energy passing through a unit area at each distance decreases accordingly (1:1/4:1/9). Consequently, sound intensity, which represents the power per unit area, decreases inversely with the square of the distance from the source.
- **Properties of the propagation medium**: The density and other properties of the medium affect how efficiently sound waves travel. The propagation of sound through a medium involves the absorption of sound energy, which is converted into heat. This energy absorption results in sound attenuation or muffling. The extent of sound absorption by a medium is frequency-dependent. Higher frequency sounds are generally more readily absorbed by air than lower frequency sounds.
- **Human ear sensitivity and hearing chart**: The loudness of a sound is determined not only by its energy but also by the sensitivity of the human ear. This sensitivity varies significantly with the frequency of the sound. Sounds with the

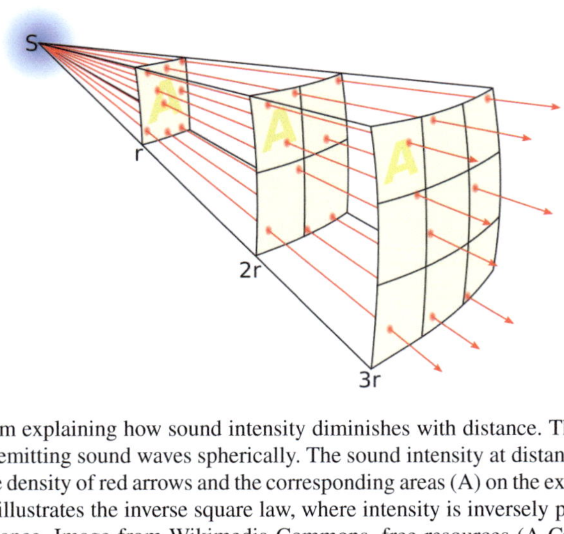

Fig. 3.26 Diagram explaining how sound intensity diminishes with distance. The image shows a sound source (S) emitting sound waves spherically. The sound intensity at distances r, 2r, and 3r is represented by the density of red arrows and the corresponding areas (A) on the expanding spherical wavefronts. This illustrates the inverse square law, where intensity is inversely proportional to the square of the distance. Image from Wikimedia Commons, free resources (A Creative Commons Attribution License): https://commons.wikimedia.org/wiki/File:Inverse_square_law.svg

same intensity but different frequencies can be perceived as having vastly different loudnesses.

Experiments have demonstrated that below a certain intensity, known as the hearing threshold, a sound becomes inaudible. Conversely, increasing the intensity eventually reaches the pain threshold, causing discomfort or pain in the ear [38, 44].

Extensive research has revealed that the human ear exhibits maximum sensitivity within the frequency range of 2000–4000 Hz [40]. This means that sounds within this range can be detected at lower intensities compared to sounds at much higher or lower frequencies. For instance, hearing sounds at the extreme ends of the audible range (around 20 Hz and 20,000 Hz) requires significantly higher intensities compared to those within the 2000–4000 Hz range [17, 26].

These findings are typically represented graphically in a hearing chart [17], which illustrates the variation of hearing threshold across different frequencies (Fig 3.27).

The most famous unit of loudness is the decibel (dB), which represents a relative measure of sound intensity. Different measurement systems may use different reference levels for zero decibels. Consequently, when comparing sound intensity levels, it's crucial to consider the specific reference point used. For instance,

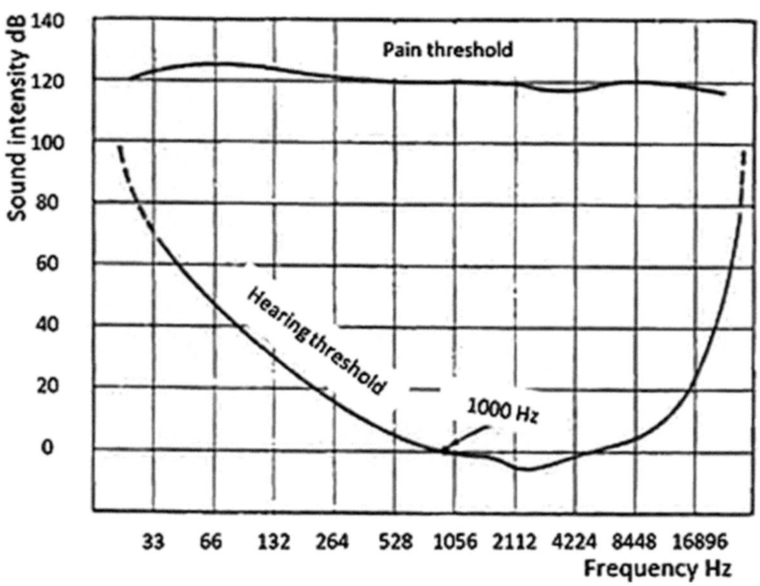

Fig. 3.27 Audiogram illustrating the frequency-dependent nature of human hearing. The graph plots sound intensity in decibels (dB) against frequency in Hertz (Hz), showing the thresholds of hearing and pain. The difference between these curves highlights how the ear's sensitivity to sound intensity changes across the frequency spectrum, explaining why sounds of equal intensity but different frequencies can be perceived as having different loudness

Table 3.2 Sound pressure levels in decibels, providing examples of common sounds and their corresponding dB values

Hearing threshold 0	0
Rustle of leaves 10	10
Conversation 60	60
Rock concert 110	110
Pain threshold 120	120
Jet engine 130	130

The table ranges from very quiet sounds like rustling leaves to extremely loud sounds like a jet engine, illustrating the broad dynamic range of human hearing

the zero point could be set at the threshold of human hearing at a particular frequency (Table 3.2) [23, 41].

The decibel (dB) is a logarithmic unit, whereas sound intensity is linearly measured as the sound power in watts. The average sound energy flowing through a specified point is also known as sound intensity or the sound power per square meter (W/m^2). However, for general use, and particularly among luthiers, the decibel scale is more commonly employed. We'll need to use both the linear and the logarithmic units when we get to Chap. 4, where we cover the bandwidth method for measuring damping capacity. We'll go into more detail there. But for now, let's stay on track.

3.2.7 Pitch Characteristic

Pitch is a perceptual attribute of musical sounds that allows us to distinguish between "high" and "low" notes. It is closely related to the physical property of frequency [18].

Higher frequency generally corresponds to a higher pitch. If two sounds have the same frequency, they will typically have the same pitch. These sounds are considered to be consonant.

It's important to understand that pitch is a perceptual quantity, influenced by both physical and psychological factors. While frequency is a purely physical property, our perception of pitch is usually (not always) directly proportional to frequency.

Simple experiments, such as playing a series of notes on a musical instrument or listening to a range of frequencies, can demonstrate the relationship between pitch and frequency. However, factors like timbre and loudness can influence our perception of pitch.

When you hear a pure tone, such as a whistle with a cicular lip, the perceived pitch directly corresponds to the frequency of that sound, as defined in physics. However, when multiple sounds are played together at harmonically related frequencies (forming a harmonic series), the perceived pitch often corresponds primarily to the fundamental frequency of that series, even if the fundamental itself is weak or even absent in the sound. This phenomenon is known as the missing fundamental effect [22]. The human ear can perceive the fundamental frequency based on the presence of its harmonic overtones, even if the actual fundamental tone is not physically present in the sound. This highlights the complex interplay between physical sound properties and human auditory perception. This suggests that the relationship between pitch and frequency is not always straightforward, and our perception of pitch can be influenced by other factors.

The findings of Ladd et al. [22] indicate that individuals exhibit stable differences in their perception of missing-fundamental tones. Some listeners perceive pitch based on the fundamental frequency, even when it is physically absent, while others rely more on the harmonic components present in the sound. These differences not only vary between individuals but can also shift within the same person depending on the frequency range.

Based on these findings, we can conclude that differences in opinions regarding the sound quality of a musical instrument are not solely a matter of personal taste. They also stem from individual variations in auditory perception and the way the brain reconstructs missing harmonics. One person may perceive the same instrument's sound as rich and resonant, while another might hear it as thin and lacking depth, even though both are listening to the same physical sound source.

This presents a significant challenge for instrument makers. If they rely solely on subjective human judgments, they will face contradictions and a wide range of opinions. These inconsistencies arise not just from differences in preference but also from inherent variations in auditory processing. Therefore, the use of precision measurement technology is essential for classifying and evaluating instrument sounds objectively. Such technology can provide measurable, standardized criteria that go beyond subjective human perception, helping to establish a more consistent approach to defining sound quality.

We will now extend this topic to a discussion of timbre.

3.2.8 Timbre—Sound Color

Just as visible light is composed of different frequencies that our eyes perceive as colors, sound is also composed of different frequencies that our ears interpret as pitches. However, the concept of "sound color" goes beyond simple pitch. It refers to the overall timbre or signature of a sound, influenced by the interplay of various frequencies and their intensities [18, 34].

Imagine a rainbow. Each color represents a specific frequency of light. Similarly, a musical note, say a middle C played on a piano (Fig. 3.28), is not just a single frequency but a complex mixture of frequencies, including the fundamental frequency and various overtones (harmonics). These overtones, or higher frequencies, determine the timbre or "color" of the sound. The superposition of multiple frequencies in a complex sound perturbs a sine wave. If these frequencies are harmonic, the perturbation creates a distinct, repeating pattern that serves as the sonic signature of the sound source (Fig. 3.29). So, a middle C played on a guitar will sound different from a middle C played on a violin because of the unique blend of overtones produced by each instrument. Several factors contribute to the perceived "color" of a sound:

- **Frequency spectrum**: The distribution of frequencies present in the sound. A sound rich in high-frequency overtones might be perceived as "bright" or "sharp," while a sound dominated by low-frequency overtones might be perceived as "dark" or "dull."

  Analyzing a simple, monochromatic sound, like a whistle with a circular lip, is straightforward. The sine wave representation allows us to easily determine key properties such as wavelength, period, frequency, wave speed, and attenuation. We can even glean information about related sound parameters by observing how the sine wave decays. However, complex sounds, often described as "colored," present a significant challenge. These sounds are composed of multiple frequencies, some harmonically related and others not, originating from the sound source itself and its overtones, or potentially from multiple distinct sources (Fig. 3.30).

  The principle of superposition dictates that these waves interact, leading to interference patterns that make it impossible to directly extract individual sound components. Therefore, to analyze a complex sound, we need a method to isolate the constituent frequencies and examine their individual loudness and duration within the frequency domain. Consider a sound composed of just three distinct frequencies like presented in Fig. 3.31. An oscilloscope or sound recording software would initially display a complex, seemingly chaotic waveform, as illustrated on the left view of the image (Fig. 3.31), similar to what was illustrated in Fig. 3.29.

  This waveform, a result of the combined attenuation and amplification of the three frequencies at their intersection points, lacks the regularity of a simple sine wave. While it exhibits a repeatable pattern that gives the sound its unique signature or timbre, already said that visually separating the individual frequencies is impossible. However, by transforming this time-domain representation into the frequency domain, as shown on the right view of the image, the three constituent frequencies become clearly distinguishable, each resembling a distinct simple sine wave (Fig. 3.31).

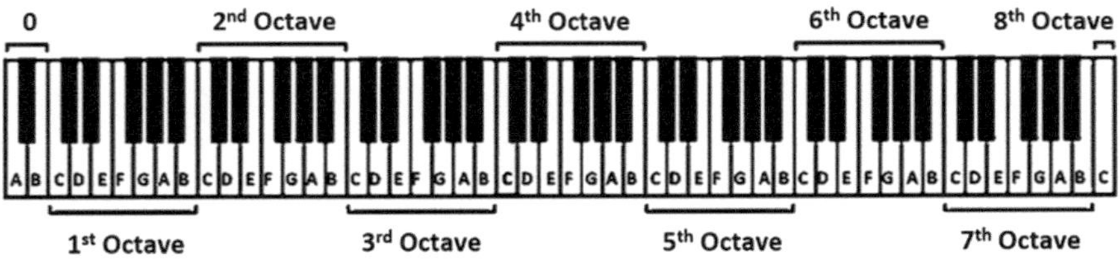

Fig. 3.28 Full piano keyboard diagram with each key labeled with its corresponding note (A–G) and octave number. Middle C, located in octave 4. The diagram spans from octave 0 to octave 8

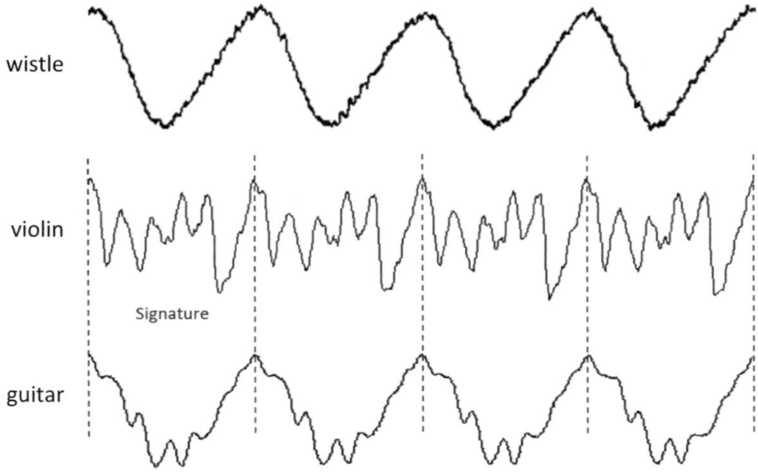

Fig. 3.29 Waveforms comparing the sound of middle C (262 Hz) played on a wistle with a circled lip, an amateur guitar and an amateur violin. The whistle's sound approximates a simple periodic sine wave (top). The violin waveform (middle) exhibits a more complex, less regular pattern with more high-frequency components, indicating a richer harmonic content and a brighter timbre. The guitar waveform (bottom) shows a simpler, more periodic pattern. Noise introduced during the sound generation or recording process is responsible for the irregular disturbances observed on these actual waves. The repeating pattern in each waveform is described as the instrument's "signature."

These tests can be performed with a wide variety of frequency spectrum analyzers; for now, Audacity is a free software option for sound analysis (Fig. 3.32).

This transformation, the separation of individual frequencies from a complex sound, is precisely what frequency analysis tools accomplish. Today, a variety of software programs utilize the mathematical technique of the Fast Fourier Transform (FFT) to decompose complex sounds into their constituent monochromatic frequencies. This allows us to analyze each frequency component and their ampltudes or their dampening behaviours, individually, providing a deeper understanding of the complex sound as a whole.

This technique allows for the identification of constructive and destructive interference patterns across different frequencies. By analyzing these patterns, a luthier can make informed decisions about instrument modifications. This includes addressing unwanted noises or destructive interference of specific frequencies, ultimately aiming to achieve a richer, more complex, and desirable sound.

While the intricate mathematical underpinnings of the Fast Fourier Transform (FFT) are not typically essential for a luthier's practical work, its application in acoustic analysis is invaluable. Therefore, we will omit the detailed mathematical derivations here. Fortunately, numerous software tools, many available at no cost e.g. Audacity (https://www.audacityteam.org/) software, employ FFT algorithms

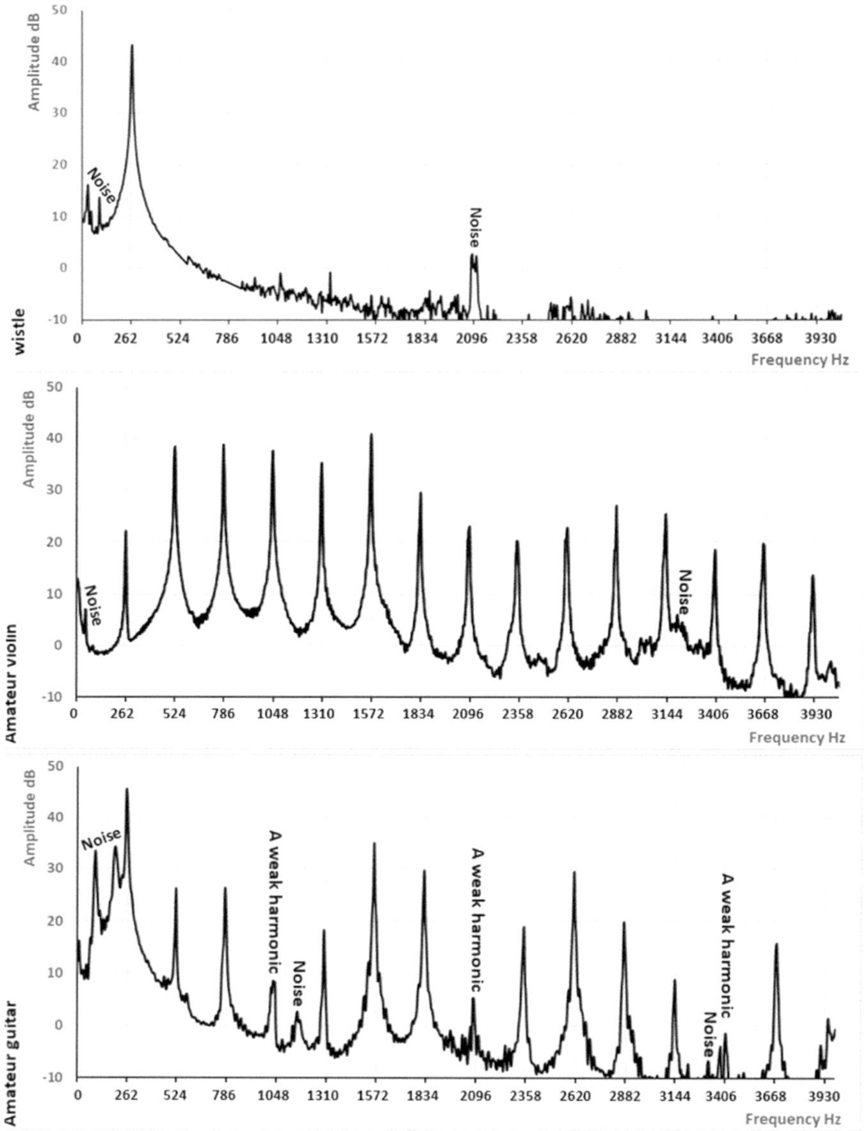

Fig. 3.30 Frequency domain analysis of amateur audio recordings, comparing the spectral characteristics of a whistle, violin, and guitar. The graphs highlight differences in harmonic content, noise (unwanted modes) levels, and overall spectral complexity between the three instruments

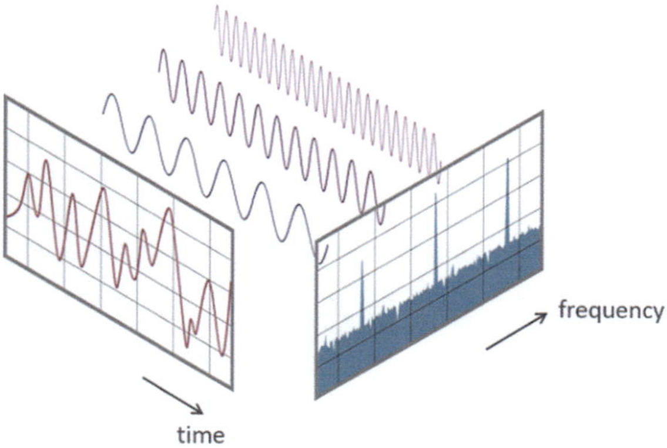

Fig. 3.31 Visual representation, illustrating how a complex waveform in the time domain can be decomposed into its constituent frequencies in the frequency domain. The image highlights the connection between the two representations and the concept of harmonic content [29]. Image from Wikimedia Commons, free resources (A Creative Commons Attribution License): https://commons.wikimedia.org/wiki/File:FFT-Time-Frequency-View.png

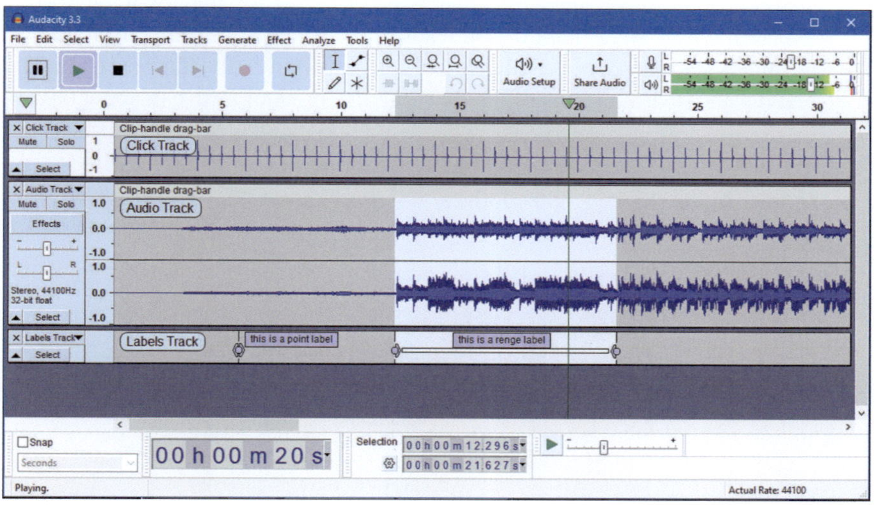

Fig. 3.32 A screenshot of Audacity 3.3 displays its interface with a timeline ruler across the top and multiple audio tracks showing waveforms. Basic playback, recording, and editing controls are visible in the toolbar. A labels track is also present. This illustrates Audacity, a free software used for sound analysis. Image from Wikimedia Commons, free resources (A Creative Commons Attribution License): https://commons.wikimedia.org/wiki/File:Audacity_version_3.3.png

to convert acoustic signals from the time domain to the frequency domain. This transformation provides luthiers with critical information, presented visually as spectra or numerically as frequency component magnitudes, facilitating detailed analysis of instrument sound. We will extensively utilize FFT spectra in Chaps. 4 and 5 to explore various acoustic characteristics of Tonewood.

- **ADSR envelope**: The overall shape of the sound wave over time, including the attack, sustain, and decay phases [24] is called "ADSR Envelope". It defines the way a sound starts and stops. A sudden, sharp attack can make a sound seem "bright," while a gradual decay can make it seem "soft" or "smooth."

  Let's break down the ADSR envelope in a way that's easy for a luthier to understand. Imagine you're watching a string being plucked or bowed. The ADSR envelope describes how the loudness of the sound changes over time, from the very first moment it's made until it fades away. It's like a map of the sound's volume. ADSR stands for four stages (Fig. 3.33):

  - *Attack*: This is how quickly the sound gets loud. Think of plucking a guitar string sharply—that's a fast attack. Now imagine slowly drawing a bow across a violin—that's a slower, more gradual attack. It's the time it takes for the sound to go from silence to its loudest point. For a piano or a drum, the attack is very quick, almost instant. For bowed instruments, the player has much more control over the attack, making it a gradual increase in volume. The next chapter will also discuss the mechanical impedance and viscoelastic behavior of tonewoods, explaining how stress application leads to a delayed strain response, to affect the attack section of the ADSR. Sound can attack instantly, like when you hit a piano key or a drum (think of it happening in a blink of an eye in about 0.01 s). Or, it can build up slowly, like when someone playing a violin or blowing a horn gradually gets louder over a few seconds.
  - *Decay*: Right after the attack, the sound might get a little quieter, even if the player is still bowing. This is the decay. It's a quick dip in volume before

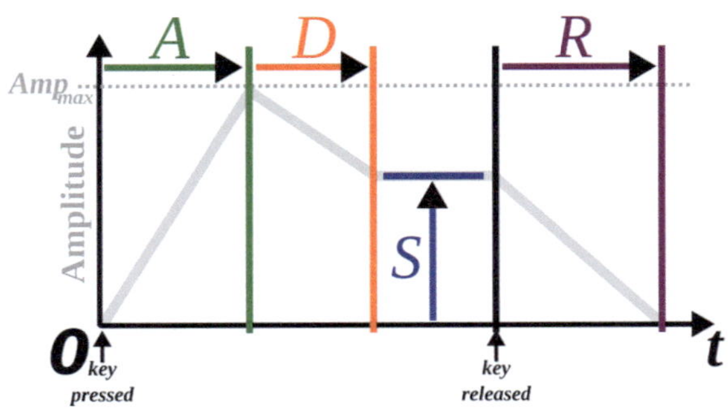

Fig. 3.33 Schematic graph of a sound's amplitude envelope, showing the four stages of Attack, Decay, Sustain, and Release. The attack phase (green) shows a rapid increase in amplitude. The decay phase (orange) shows a decrease to the sustain level. The sustain phase (blue) maintains a constant amplitude. The release phase (purple) shows the final decrease to zero amplitude. The graph is annotated with "key pressed" and "key released" to denote the triggering events for the envelope, which may represent the initiation and termination of the violin string's bowing. Image from Wikimedia Commons, free resources (A Creative Commons Attribution License): https://commons.wikimedia.org/wiki/File:ADSR_parameter.svg

the sound settles into the sustain phase. Sometimes, especially with bowed instruments, the decay is very short or even non-existent—the sound goes straight to the Sustain [24]. Think of a plucked string. It reaches its peak volume (the attack), then immediately starts to fade (the decay).

- *Sustain*: This is the part where the sound holds steady. When bowing a violin with a steady motion, after the decay, the sound maintains a consistent loudness—this is sustain. It represents the period of relatively constant volume. While a plucked string has a short sustain due to energy loss, bowed instruments offer player control over sustain through bow technique. Sustain is less about the precise duration in seconds and more about the consistency and steadiness of the sound's intensity. While sustained notes are characteristic of instruments such as electric guitars, violins, and wind instruments, the decay and sustain regions of plucked string instruments frequently exhibit similar slopes, rendering them indistinguishable. Therefore, these phases should be considered together, taking their combined duration into account.
- *Release*: This is how long it takes for the sound to fade away to nothing. When you stop bowing or plucking, the sound doesn't disappear instantly; it gradually gets quieter. The release is the time it takes for the sound to go from the sustain level back to silence. A plucked string has a longer release than, say, a struck drum.

  As a luthier, understanding ADSR helps you think about how your instrument sounds and how it responds to the player. The type of wood you use affects the ADSR. The way you join the wood, the arching of the top and back, all these affect how the sound surrounds the listener. The bridge, soundpost, and strings all influence the attack, decay, sustain, and release. A poorly fitted soundpost could make the attack dull.

  Essentially, by understanding ADSR, you can better craft instruments that give musicians the control they need over the sound, from the initial attack to the final fade. You can fine-tune your instruments to produce sounds with the desired character, whether it's a quick, percussive attack or a long, singing sustain.
- **Harmonics**: The richness or "color" of a sound also depends on how strong and noticeable the extra tones (called overtones or harmonics) are compared to the main tone. This parameter has been illustrated sufficiently.
- **Noise**: Sounds sometimes contain low-level, uncorrelated frequencies. These frequencies can change the sound, kind of like adding texture or color. Sometimes these additional sounds are desirable, like the twangy sound (ringing tone) of a flamenco guitar. Other times it's bad, like when you hear annoying buzzing or humming. We call these uncorrelated frequencies "noise," and they can change the overall sound color in both good (ringing tone) and bad (beat phenomenon and/or guitar fret buzz) ways.

The concept of sound color is particularly important in music, where it is used to create a wide range of timbres and textures. Composers and musicians carefully select instruments, playing techniques, and recording techniques to achieve specific sound colors that recall particular emotions or atmospheres.

Beyond music, sound color is also relevant in fields like acoustics, audio engineering, and even psychology. For example, researchers have studied how sound color can affect our mood, attention, and cognitive performance [2]. Having established the similarities between the creation of sound color and the color of light, it's worth noting that scientists are exploring sound color from another perspective, distinct from the concept discussed earlier. Given that the aesthetic dimension of music can be a fruitful area of study, let's briefly touch upon it.

Synesthesia is a fascinating phenomenon where stimulation of one sense triggers another. For example, someone with synesthesia might see a color when they hear a sound, or taste a flavor when reading a word. It's as if the senses are blended.

In a study focused on a specific type called "musical note-color synesthesia," where individuals see distinct colors when they hear musical notes (like C, D, E, etc.). Researchers aimed to understand how this connection between notes and colors is processed in their brains [16].

People in this type of synesthesia were asked to describe the colors they experienced when hearing each musical note. The results revealed a fascinating pattern. The notes corresponded to a spectrum of colors, with C being red, D yellow, E green, and so on, progressing to B as purple. Furthermore, higher-pitched notes corresponded to lighter, less saturated colors.

Interestingly, the color association was more strongly linked to the name of the note than its actual sound. For instance, C# (which sounds like Db) was associated with a color closer to red (the color of C) than Orange (near the yellow color of D). This suggests that the brains of these individuals respond more to the cognitive label of the note than its auditory properties.

The study concluded that two dimensions of music—note name (or "note class") and pitch—are connected to two dimensions of color: hue and brightness, respectively. In essence, the note's name determines its color, while its pitch influences its brightness.

To put it simply, imagine seeing the color red when you hear the sound "C." That's synesthesia. This study demonstrated that these color associations aren't random but follow a specific pattern. Crucially, it showed that the brain responds more strongly to the name of the note (like "C") than to the sound itself, as if the name is the key to unlocking the associated color.

It's true that humans make mistakes when relying solely on their five senses. That's why we've developed more accurate technologies and methodolodies to understand and categorize what we observe. This is especially evident in fields like medicine. Nowadays, even the most experienced doctors wouldn't prescribe medication without using tools like ultrasounds, X-rays, CT scans, MRIs, or lab tests to get a clear picture of what's going on.

It is observed that, currently, individuals exhibiting a high degree of self-absorption tend to exhibit a marked confidence in their own sensory perception coupled with a distrust of emerging diagnostic and classification technologies. So, It would be interesting to highlight that we often see what we want to see, a phenomenon explained by cognitive dissonance theory [27]. This theory posits that individuals tend to interpret reality through the lens of their desires and preferences, prioritizing beliefs that align with what they want to be true, even if those beliefs are at odds with reality. Cognitive dissonance theory explores the strategies people use to manage these internal conflicts and reduce the resulting psychological tension, making it a highly influential concept in social psychology.

The need for modern diagnostic tools, as previously discussed, stems from the inherent limitations of human perception. Individual taste, perceptual variations between and even within individuals, and the mind's tendency to reconstruct reality in different contexts all contribute to this. Cognitive dissonance and synesthesia further complicate subjective judgments. Therefore, luthiers relying solely on personal opinion and the perception of others, without accurate measurement tools, will inevitably encounter inconsistent and confusing results.

To return to the main topic of our discussion, we point out that sound color is a complex and multifaceted phenomenon that enriches our auditory experience. By understanding the factors that contribute to sound color, we can better appreciate the nuances of music and the world around us. Timbre and sound color are key topics explored in the following chapters, with a focus on tonewood and stringed instruments.

3.2.9 Sound Resonance

Resonance occurs when an object capable of oscillating at a specific natural frequency is subjected to periodic forces that match its natural frequency. This synchronization causes a significant increase in the amplitude of the object's oscillations.

While some energy is inevitably lost due to friction (converted into heat), the amplitude of the oscillations can remain constant if the applied forces continuously replenish the lost energy.

A simple example of resonance is a swing. The swing-person system forms a pendulum with a natural oscillation period determined by its length. When the frequency of the pushes applied to the swing matches the pendulum's natural frequency, the amplitude of the swing's motion gradually increases (Fig. 3.34).

Figure 3.35 illustrates another example of resonance. Two pendulums, R and E, are suspended from a horizontal steel wire. Pendulum E, with a heavier weight, acts as the driver, while pendulum R is the resonator. Initially, pendulum E has a longer length and thus a lower oscillation frequency than pendulum R.

When pendulum E is set in motion, pendulum R exhibits irregular oscillations. It undergoes a few small swings, then stops, and this pattern repeats. This occurs because the impulses transmitted from the driver pendulum to the resonator through the elastic base are not synchronized with the resonator's natural oscillation frequency. Each impulse may initiate a swing in the resonator, but the subsequent impulse tends to counteract this motion, preventing sustained oscillations.

Fig. 3.34 A small boy pushes an adult man on a swing set. Image from Wikimedia Commons, free resources (A Creative Commons Attribution License): https://commons.wikimedia.org/wiki/File:Boy_Pushing_Man_on_swing.jpg

However, as the length of the driver pendulum is gradually reduced, increasing its frequency, the amplitude of the resonator's oscillations begins to grow. When the frequencies of the two pendulums approach and eventually become equal, the resonator's oscillations reach maximum amplitude. This phenomenon is known as resonance.

Resonance is observed in various acoustic scenarios. For example, if two strings are tuned to the same pitch and one string is plucked, the other string will also vibrate sympathetically, producing the same note.

The illustration in Fig. 3.36 depicts a classic demonstration of resonance. Two identical tuning forks, mounted on identical resonance boxes, are used. When one tuning fork is struck, it begins to vibrate. This vibration causes the air within the first resonance box to vibrate as well. These air vibrations, in turn, act as a sound wave, traveling to the second resonance box.

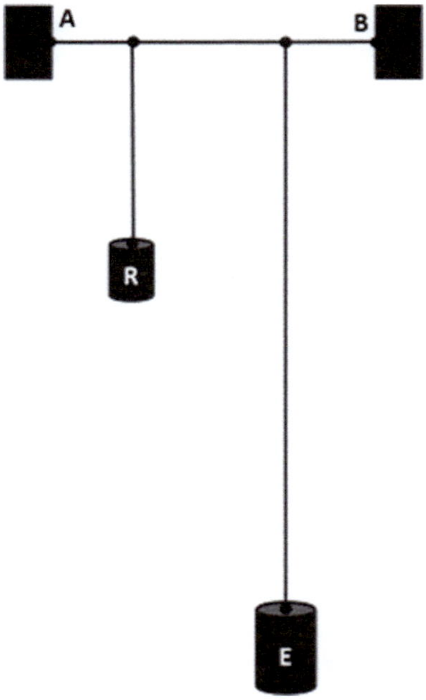

Fig. 3.35 Diagram of two pendulums, R (resonator) and E (driver), suspended from a horizontal steel wire. The image shows the initial setup where E has a longer length and lower frequency than R, resulting in irregular oscillations in R when E is set in motion. As the length of E is reduced and its frequency increases, R's oscillations grow, reaching maximum amplitude when the frequencies match, demonstrating resonance

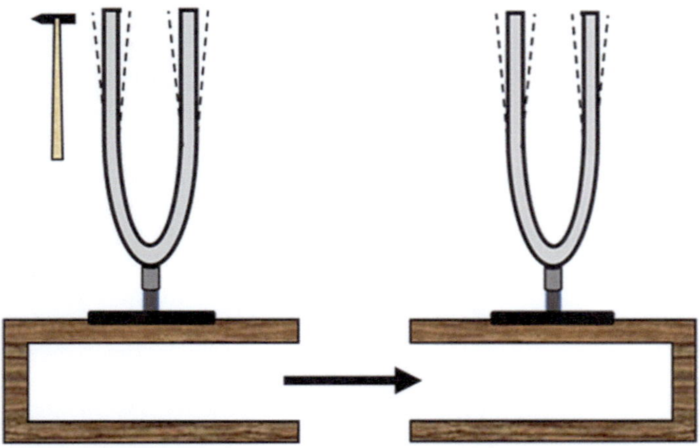

Fig. 3.36 Schematic of a system demonstrating acoustic resonance. A tuning fork is shown above an open box, implying that when the tuning fork is struck, the air column inside the box resonates with the tuning fork's frequency, amplifying the sound

The air within the second box, resonating with the incoming sound waves, sets the second tuning fork into sympathetic vibration. Even if we stop the first tuning fork, the second one continues to vibrate and emit sound.

For this resonance to occur effectively, both tuning forks and their corresponding boxes must have identical dimensions and, consequently, identical resonant frequencies.

Interestingly, the tuning forks boxes act as sound tubes in this experiment, a concept we will explore in more detail later.

Resonance is a fundamental phenomenon in musical instruments. In a stringed instrument, for example, exciting a string initiates vibrations that travel to the top plate through various pathways, including airborne sound and direct transmission through the bridge.

Resonance occurs when the frequencies of the vibrating string (including its fundamental frequency and higher harmonics) match or closely align with at least one of the natural resonant frequency modes of the mounted top plate. This causes the plate to vibrate sympathetically, amplifying the sound significantly.

The amplified sound waves then travel through the air within the instrument's body, reaching the back plate. This energy transfer can occur through air-borne sound waves, direct air transmission, or a combination of both.

The back plate, like a reflecting surface, reflects the sound waves back towards the top plate. This feedback loop further amplifies the vibrations of the top plate, which in turn reinforces the vibrations of the string. This rapid cycle of energy transfer and amplification occurs within fractions of a second, sustaining the sound until the next note is played.

Following resonance within the instrument's soundbox, the interaction between the vibrating plate and the plucking causes the string to continue oscillating. A continius regular plucking the string, excites vibrations primarily parallel to the plane of the plate, while the plate itself undergoes complex up-and-down motions. The instrument's top plate vibrates in multiple, complex modes, not just simple up-and-down motions. These modes are primarily bending, but also include twisting and other intricate movements (Fig. 3.37), all contributing to the richness and complexity of the sound produced. These two methods in this figure will be introduced and illustrated later in Chap. 5.

3.3 Sources of Musical Sounds

Musical sounds are characterized by their harmonious and pleasing effect on the human ear, whether played individually, sequentially (melody), or simultaneously (harmony). Two important aspects of musical sounds were demonstrated to be pitch and timbre. Pitch is primarily determined by the frequency of the sound source, while timbre is influenced by the presence and relative intensities of harmonics. The specific set of harmonics produced by an instrument depends on its type and the method of excitation.

Musical instruments must be capable of producing a wide range of musical notes with precisely defined intervals [11]. They are generally categorized into three groups [36]:

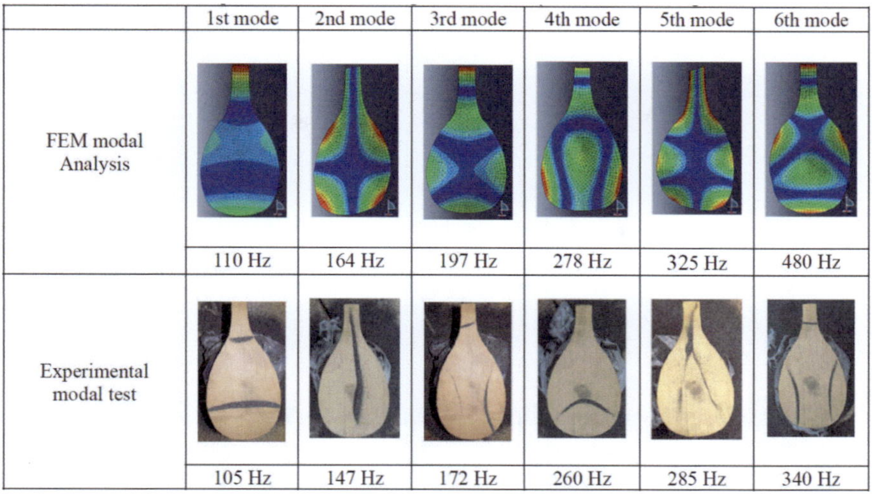

Fig. 3.37 Diagram of an instrument top plate vibrating in multiple modes (free boundary condition) using two methods of Finite Element Analysis (upper row) and Chladni experimental test (lower row). The image displays six distinct modes of vibration, each with a specific frequency (110–480 Hz). The modes are visualized using color-coded patterns, showing the complex up-and-down motions, including bending and twisting, that contribute to the instrument's rich sound. These vibrations are excited by plucking the string, which primarily vibrates parallel to the plate. Image from Pedrammehr et al. [32], Semantic Scholar, freely licensed (A Creative Commons Attribution License): https://doi.org/10.21595/jve.2018.19505

- **Stringed instruments**: or chordophones, are musical instruments that produce sound from vibrating strings. The sound is created when a performer strums, plucks, strikes, or sounds the strings in various ways. This category includes a wide variety of instruments such as the violin, viola, cello, guitar, and harpsichord, among others.
- **Wind instruments**: or aerophones are a class of musical instruments in which sound is produced by the vibration of air. This category includes woodwind instruments like flutes and clarinets, as well as brass instruments such as trumpets and trombones. Essentially, any instrument that creates sound by causing air to vibrate is considered an aerophone.
- **Percussion instruments**: Percussion instruments are musical instruments that produce sound when struck, scraped, or shaken. They can be categorized primarily into two groups:
  - Idiophones, which generate sound through the vibration of the instrument itself.
  - Membranophones, which produce sound from the vibration of a stretched membrane. Examples of percussion instruments include drums, cymbals, maracas, and tambourines.

Table 3.3 Table showing the speed of sound in air at different temperatures

Air temperature (°C)	Sound velocity (m/s)
0	331
15	340
20	343
28	348

The table has two columns: "Air Temperature (°C)" and "Sound Velocity (m/s)." The listed temperatures are 0 °C (331 m/s), 15 °C (340 m/s), 20 °C (343 m/s), and 28 °C (348 m/s). This data illustrates the positive correlation between air temperature and sound speed

While we've focused on stringed instruments, it's important to consider wind instruments, too, particularly in relation to the resonance boxes of stringed instruments. The air movement within a stringed instrument's soundbox exhibits similarities to the airflow in wind instruments like flutes, facilitating the transfer and amplification of sound energy.

3.4 Air Columns: Sound Tubes

While seemingly unrelated to vibrating strings, understanding sound tubes provides valuable insights into the resonance phenomena observed in stringed instruments, particularly those with soundboxes.

A sound tube is essentially a column of air that produces sound through vibration. In this context, we consider tubes with lengths 'L' significantly greater than their diameter to facilitate predominantly unidirectional air vibrations.

The formulas for calculating the frequencies of harmonics in open and closed tubes are illustrated here [45]. For open tubes, all harmonics are present, while closed tubes produce only odd harmonics. Here, 'K' represents the harmonic number. 'V' is the speed of sound in air and increases with the increase of the air temperature (Table 3.3). As an example, the speed of sound in air with a conventional desired temperature of about 28 °C is estimated to be approximately $V_{28\,°\mathrm{C}} = 348$ m per second [23].

- For an open tube:

$$f_k = K\frac{V}{2L} \tag{3.18}$$

- For a closed tube:

$$f_k = (2K - 1)\frac{V}{4L} \tag{3.19}$$

Practical Exercise 3.7

For a tube length of 65 cm (0.65 m), let's calculate the frequencies of the first four harmonics:

Solution

Open Tube:
1st Harmonic (K = 1): $f_k = 1 \times \frac{348}{2\times0.65} = 267.7$ Hz.
2nd Harmonic (K = 2): $f_k = 2 \times \frac{348}{2\times0.65} = 535.4$ Hz.
3rd Harmonic (K = 3): $f_k = 3 \times \frac{348}{2\times0.65} = 803.1$ Hz.
4th Harmonic (K = 4): $f_k = 4 \times \frac{348}{2\times0.65} = 1070.8$ Hz.
Closed Tube:
1st Harmonic (K = 1): $f_k = (2 \times 1 - 1)\frac{348}{4\times0.65} = 133.8$ Hz.
3rd Harmonic (K = 2): $f_k = (2 \times 2 - 1)\frac{348}{4\times0.65} = 401.5$ Hz.
5th Harmonic (K = 3): $f_k = (2 \times 3 - 1)\frac{348}{4\times0.65} = 669.2$ Hz.
7th Harmonic (K = 4): $f_k = (2 \times 4 - 1)\frac{348}{4\times0.65} = 936.9$ Hz.

We see that in closed tubes, only odd harmonics are produced. The first four harmonics in a closed tube are the fundamental frequency (1st harmonic), the third harmonic, the fifth harmonic, and the seventh harmonic, respectively.

Resonance boxes, like sound tubes, can be classified as open or closed depending on the placement of sound holes. However, unlike sound tubes, they have irregular shapes and a lower length-to-diameter ratio.

While relationships derived for sound tubes provide some guidance, they only account for a portion of the frequencies within a resonance box. Many other frequencies arise from bending and torsional vibrations within the resonator plate.

Nevertheless, these relationships offer valuable insights. Increasing the effective le ngth of the soundbox generally lowers the fundamental frequency, while shortening it raises it. This knowledge allows instrument makers to experiment with soundbox dimensions and soundhole placement to achieve desired tonal characteristics.

3.5 Final Words

Throughout this chapter, we have explored the intricate physics of vibration and wave motion, focusing on their role in sound production, wave propagation, and their practical applications in musical instruments. By understanding the fundamental principles governing these phenomena, we gain valuable insights into how vibrations shape the acoustic characteristics of stringed instruments, and more broadly, how sound interacts with the environment.

At its core, vibration is a universal phenomenon that extends far beyond the area of musical instruments. Whether in the oscillations of a pendulum, the resonance of a bridge, or the propagation of seismic waves, the principles of harmonic motion are deeply embedded in nature. Our discussion of simple harmonic motion (SHM) provided a foundation for understanding how mechanical systems, including strings and soundboards, behave when disturbed. By examining the mathematical relationships governing displacement, velocity, acceleration, and wave frequency, we have equipped ourselves with the tools necessary to predict and manipulate vibratory behavior in real-world applications.

One of the key takeaways from this chapter is the importance of resonance and harmonic frequencies in shaping the tonal qualities of musical instruments. When a string vibrates, it does not simply produce a single frequency but rather a complex mixture of fundamental and harmonic frequencies. These harmonics, in combination with the instrument's structural properties, define its unique sound signature. Luthiers and acoustic engineers can connect this knowledge to refine instrument design, adjusting parameters such as string tension, bridge positioning, and soundboard material to achieve optimal tonal quality.

The discussion of wave propagation highlighted the distinction between transverse and longitudinal waves, as well as the role of standing waves in musical acoustics. Understanding how wave motion is influenced by factors such as density, stiffness, and boundary conditions of the medium allows instrument makers to fine-tune the vibrational response of their designs. The concept of impedance matching, for example, is particularly relevant in ensuring efficient energy transfer between the string and the soundboard, thereby enhancing sound projection and sustain.

Moreover, the exploration of sound perception—including pitch, loudness, and timbre—emphasized the subjective nature of auditory experience. While frequency and amplitude are measurable physical properties, the way humans perceive sound is influenced by a variety of psychological and physiological factors. The phenomenon of the missing fundamental effect demonstrates how our brains interpret harmonic relationships to reconstruct pitch, even in the absence of a fundamental frequency. This insight underscores the complexity of sound perception and explains why different listeners may have varying preferences for the tonal qualities of an instrument.

Beyond the realm of luthiery, the principles covered in this chapter have implications across multiple scientific and engineering disciplines. In mechanical engineering, vibration analysis is important in designing stable structures and machinery.

In medical fields, acoustic waves are used in imaging technologies such as ultrasound. Even in seismology, understanding how waves propagate through the Earth's crust helps scientists predict and analyze earthquake behavior. This highlights the interconnected nature of wave physics, demonstrating that the study of vibration extends far beyond the scope of musical acoustics.

For instrument makers, an appreciation of these physical principles is invaluable in crafting instruments that not only meet aesthetic and structural criteria but also deliver exceptional acoustic performance. The subtle interplay between material selection, string tension, and body resonance determines the final sound produced by an instrument. By combining scientific understanding with artistic intuition, luthiers can continually refine their craftsmanship, achieving a balance between tradition and innovation.

In conclusion, the study of vibration and wave motion is not merely an academic pursuit—it is a gateway to deeper appreciation and control of the sounds that shape our world. Whether you are a physicist, an engineer, a musician, or a luthier, understanding the behavior of vibrating systems allows for the creation of more refined, expressive, and sonically rich instruments. The journey into the physics of sound does not end here; as we move forward, we will further explore the role of different materials, construction techniques, and environmental factors in shaping musical acoustics. The pursuit of perfect sound is an ongoing endeavor—one that bridges science, craftsmanship, and the art of music itself.

References

1. S. Alrasheed, Oscillatory motion, in *Principles of Mechanics. Advances in Science, Technology & Innovation.* (Springer, Cham, 2019). https://doi.org/10.1007/978-3-030-15195-9_10
2. J.M. Barbiere, A. Vidal, D.A. Zellner, The color of music: correspondence through emotion. Empir. Stud. Arts **25**(2), 193–208 (2007). https://doi.org/10.2190/A704-5647-5245-R47P
3. A. Brauchler, P. Ziegler, P. Eberhard, Examination of polarization coupling in a plucked musical instrument string via experiments and simulations. Acta Acust. **4**(3), 9 (2020). https://doi.org/10.1051/aacus/2020009
4. O. Bones, T.J. Cox, W.J. Davies, Sound categories: category formation and evidence-based taxonomies. Front. Psychol. **9**, 1277 (2018). https://doi.org/10.3389/fpsyg.2018.01277
5. X. Boutillon, K. Ege, Vibroacoustics of the piano soundboard: reduced models, mobility synthesis, and acoustical radiation regime. J. Sound Vib. **332**(18), 4261–4279 (2013). https://doi.org/10.1016/j.jsv.2013.03.015
6. B. Crowell, Vibrations and waves (Vol. 3). Light and Matter (2008). https://ufdcimages.uflib.ufl.edu/AA/00/01/17/30/00001/Vibrations.pdf
7. J.-M. Friedt, E. Carry, Introduction to the quartz tuning fork. Am. J. Phys. **75**(5), 415–422 (2007). https://doi.org/10.1119/1.2711826
8. A. Gantner, R.H.W. Hoppe, D. Köster, K.G. Siebert, A. Wixforth, Numerical simulation of piezoelectrically agitated surface acoustic waves on microfluidic biochips. Comput. Visual Sci. **10**(3), 145–161 (2007). https://doi.org/10.1007/s00791-006-0040-y
9. N. Giordano, On hearing the "shape" of a vibrating string. Comput. Sci. Eng. **4**(3), 100–105 (2002). https://doi.org/10.1109/5992.998647

10. M.J. Giuliodori, H.L. Lujan, W.S. Briggs, G. Palani, S.E. DiCarlo, Hooke's law: applications of a recurring principle. Adv. Physiol. Educ. **33**(4), 293–296 (2009). https://doi.org/10.1152/advan.00067.2009
11. D. Halliday, J. Walker, R. Resnick, *Fundamentals of Physics Extended*, 10th edn. (Wiley, Hoboken, 2014), p. 1450. https://ia601307.us.archive.org/15/items/fundamentals-of-physics/Fundamentals-of-Physics-Textbook-booksfree.org_.pdf
12. C.M. Harris, A.G. Piersol, *Harris' Shock and Vibration Handbook*, 5th edn. (McGraw-Hill, New York, 2002), 1568p
13. U. Hassan, M.S. Anwar, Vibrations on a String and Resonance, in *Lab Monograph Introductory Experimental Physics*, p. 190 (2010). https://physlab.lums.edu.pk/images/b/b6/Vibrationsv6.pdf
14. L. He, Y. Liang, L. Zhang, J. Zhou, R. Wang, Z. Liu, Measurement and analysis of the vibration responses of piano soundboards with different structures. Materials **17**(5), 1004 (2024). https://doi.org/10.3390/ma17051004
15. A. Ishimaru, Wave propagation and scattering in random media and rough surfaces. Proc. IEEE **79**(10), 1359–1366 (1991). https://doi.org/10.1109/5.104210
16. K. Itoh, H. Sakata, I.L. Kwee, T. Nakada, Musical pitch classes have rainbow hues in pitch class-color synesthesia. Sci. Rep. **7**(1), 17781 (2017). https://doi.org/10.1038/s41598-017-18150-y
17. R. Jeche, M.R. Maslin, M. Park, A mechanistic study of tinnitus and hyperacusis in relation to temporary conductive hearing loss. Master's thesis, University of Canterbury (2023). https://ir.canterbury.ac.nz/server/api/core/bitstreams/4e68e8ac-619c-4614-adb1-6da3b5c1e9b5/content
18. K. Jensen, Perceptual and physical aspects of musical sounds. J. Sangeet Res. Acad. 1–22 (2002). https://citeseerx.ist.psu.edu/document?repid=rep1&type=pdf&doi=cd5c35047bc63bc4a5718abf32cf4478c7b29496
19. N. Kaplanis, J. Van Velzen, Hearing through darkness: a study of perceptual auditory information in real rooms and its effect on space perception. Proc. Inst. Acoust. **34**(4), 114–125 (2012)
20. L.E. Kinsler, A.R. Frey, A.B. Coppens, J.V. Sanders, *Fundamentals of Acoustics* (John Wiley & Sons, 2000)
21. C. Koch, Hearing beyond the limit: Measurement, perception and impact of infrasound and ultrasonic noise, in *12th ICBEN Congress on Noise as a Public Health Problem*, Zurich (2017). https://docs.wind-watch.org/ICBEN-2017_Koch_4163.pdf
22. D.R. Ladd, R. Turnbull, C. Browne, C. Caldwell-Harris, L. Ganushchak, K. Swoboda, V. Woodfield, D. Dediu, Patterns of individual differences in the perception of missing-fundamental tones. J. Exp. Psychol. Hum. Percept. Perform. **39**(5), 1386–1397 (2013). https://doi.org/10.1037/a0031261
23. L. Learning, Sound interference and resonance: standing waves in air columns, in *Fundamentals of Heat, Light & Sound* (2021). https://pressbooks.nscc.ca/heatlightsound/chapter/17-5-sound-interference-and-resonance-standing-waves-in-air-columns/
24. F. Leccese, G. Salvadori, G. Bernardini, P. Bernardini, The bowed string instruments: acoustic characterization of unique pieces from the Italian lutherie. IOP Conf. Ser. Mater. Sci. Eng. **364**(1)(2018). https://doi.org/10.1088/1757-899X/364/1/012022
25. S. McAdams, P. Depalle, E. Clarke, Analyzing musical sound, in *Empirical Musicology: Aims, Methods, Prospects*, ed. by Smith, Jones (Oxford University Press, Oxford, 2004), pp. 157–196
26. H. McDermott, D. Baldwin, M. Nyffeler, The importance of perceptual bandwidth and how frequency compression extends it. Hear. J. **63**(5), 34, 36, 38–40 (2010). https://doi.org/10.1097/01.HJ.0000373448.52956.c1
27. M.K. Miller, J.D. Clark, A. Jehle, Cognitive dissonance theory (Festinger), in *The Blackwell Encyclopedia of Sociology*, ed. by G. Ritzer (Wiley Online Library, Hoboken, NJ, 2015). https://doi.org/10.1002/9781405165518.wbeosc058.pub2
28. A. Morales, J. Flores, L. Gutiérrez, R.A. Méndez-Sánchez, Compressional and torsional wave amplitudes in rods with periodic structures. J. Acoust. Soc. Am. **112**(5), 1961–1967 (2002). https://doi.org/10.1121/1.1509431

29. G. Nicoletti, Mastering the sound of the acoustic guitar. Julius Guitars. Printed in Poland by Amazon Fulfillment Poland Sp. z. o. o. Wrocław (2024). ISBN 979-10-415-1253-9
30. D. Ouis, On the frequency dependence of the modulus of elasticity of wood. Wood Sci. Technol. **36**(4), 335–346 (2002). https://doi.org/10.1007/s00226-002-0145-5
31. H.J. Pain. *The Physics of Vibrations and Waves*, 6th edn. (John Wiley & Sons, Hoboken, NJ, 2021)
32. S. Pedrammehr, N. Jafarzadeh Aghdam, S. Pakzad, M.M. Ettefagh, M. Homayoun Sadeghi, A study on vibration of Setar: stringed Persian musical instrument. J. Vibroeng. **20**(7), 2680–2689 (2018). https://doi.org/10.21595/jve.2018.19505
33. M.A. Persinger, Infrasound, human health, and adaptation: an integrative overview of recondite hazards in a complex environment. Nat. Hazards **70**(1), 501–525 (2014). https://doi.org/10.1007/s11069-013-0827-3
34. M. Roohnia, Wood: vibration and acoustic properties, in *Reference Module in Materials Science and Materials Engineering* (Elsevier, Oxford, 2019), pp. 1–13. https://doi.org/10.1016/B978-0-12-803581-8.01996-2
35. D.A. Russell, *Acoustics and vibration animations*. Graduate Program in Acoustics, The Pennsylvania State University (2016). https://www.acs.psu.edu/drussell/Demos/waves/Rayleigh-2016.gif
36. C. Schmidt-Jones, in *Classifying Musical Instruments*. cnx.org (2017). https://www.melissagonzalez.org/uploads/7/0/4/1/7041280/classifying_instruments.pdf
37. S.K. Şengören, R. Tanel, N. Kavcar, Students' difficulties about the wave pulses propagating on a rope. J. Turkish Sci. Educ. **6**(1), 50–59 (2009). https://www.tused.org/index.php/tused/article/view/106
38. H. Senses. *The Human Ear-Hearing, Sound Intensity and Loudness Levels* (2013). https://courses.physics.illinois.edu/phys406/sp2017/Lecture_Notes/P406POM_Lecture_Notes/P406POM_Lect5.pdf
39. V. Stojanović, Z. Milivojevic, *The Polar Characteristic of an Acoustic Parabolic Reflector*. Facta Universitatis, Series: Working and Living Environmental Protection, pp. 293–303 (2025). https://doi.org/10.22190/FUWLEP241003028S
40. C.W. Turner, D.A. Nelson, Frequency discrimination in regions of normal and impaired sensitivity. J. Speech Lang. Hear. Res. **25**(1), 34–41 (1982). https://doi.org/10.1044/jshr.2501.34
41. P.P. Urone, R. Hinrichs, S. Beceiro-Novo, *Sound intensity and sound level. In ISP209: The Mystery of the Physical World* (2012). https://openbooks.lib.msu.edu/collegephysics/chapter/sound-intensity-and-sound-level/
42. D.R. Ventura, P.S. de Carvalho, M.A. Dias, Standing waves in an elastic spring: a systematic study by video analysis. Phys. Teacher **55**(4), 232–234 (2017). https://doi.org/10.1119/1.4978723
43. W.S. Yoo, J.H. Lee, S.J. Park et al., Large oscillations of a thin cantilever beam: physical experiments and simulation using the absolute nodal coordinate formulation. Nonlinear Dyn. **34**(1), 3–29 (2003). https://doi.org/10.1023/B:NODY.0000014550.30874.cc
44. W.A. Yost, M.C. Killion, Hearing thresholds, in *Encyclopedia of Acoustics*, vol. 3, ed. by M.J. Crocker (1997), pp. 1545–1554. https://etymotic.com/wp-content/uploads/2021/05/erl-0096-1997.pdf
45. L. Zeng, C. Smith, G.H. Poelzer, J. Rodriguez, E. Corpuz, G. Yanev, Illustrations and supporting texts for sound standing waves of air columns in pipes in introductory physics textbooks. Phys. Rev. Special Top. Phys. Educ. Res. **10**(2), 020110 (2014). www.utrgv.edu https://doi.org/10.1103/PhysRevSTPER.10.020110

Chapter 4
Tonewood: Acoustic Identification

Abstract This chapter explores the identification and classification of tonewoods, focusing on their acoustic properties, anatomical characteristics, anwd historical significance in instrument making. Tonewood selection plays a critical role in the sound quality and performance of stringed instruments, with species such as spruce, maple, and mahogany being commonly used due to their superior mechanical and acoustic attributes. The discussion highlights the physical properties of these woods, including density, stiffness, and damping capacity, which collectively influence their suitability for specific instrument components. A significant section of the chapter delves into various methods of tonewood identification, encompassing both traditional and modern scientific techniques. Traditional methods rely on visual inspection, empirical knowledge, and tap-tone assessments, while advanced analytical techniques include microscopic examination, X-ray densitometry, and spectroscopy as well as the acoustic measurements. These methods provide valuable insights into the microstructure and mechanical behavior of wood, aiding luthiers and researchers in making informed material choices. Furthermore, the chapter discusses the role of geographical origin and environmental factors in determining wood quality. Growth conditions such as altitude, climate, and soil composition significantly impact the grain structure, density, and elasticity of tonewoods. These variations contribute to regional differences in sound characteristics, making provenance an essential factor in the selection of premium-quality woods. The historical and cultural aspects of tonewood use are also examined, tracing the evolution of wood preferences across different time periods and regions. The chapter highlights how historical luthiers such as Stradivari and Guarneri meticulously selected and treated their woods to enhance the tonal richness of their instruments. Additionally, the chapter addresses sustainability concerns related to the depletion of traditional tonewood sources, emphasizing the importance of responsible harvesting, certification programs, and alternative materials such as thermally modified woods and engineered composites. Lastly, the chapter presents emerging research in the field of tonewood analysis, including the application of machine learning and computational modeling in predicting wood performance. These advancements offer new possibilities for optimizing tonewood selection, improving consistency, and enhancing the overall quality of instrument craftsmanship. In conclusion, the identification and characterization of tonewoods are

M. Roohnia, *Properties and Treatment of Tonewood for String Instrument Construction*, Springer Series in Materials Science 355, https://doi.org/10.1007/978-3-032-00824-4_4

crucial aspects of instrument making, combining art, science, and tradition. By integrating historical knowledge with modern analytical techniques, the field continues to evolve, ensuring that high-quality, sustainable materials remain available for future generations of musicians and luthiers.

4.1 Understanding Tonewood: A Key to Instrument Sound

This chapter explores the important link between wood properties and musical instrument tone. Building upon the fundamentals of wood science and acoustics in Chaps. 2 and 3, we now go deep into the specific mechanical and acoustic parameters that define "tonewood". Acoustics is the science and engineering of sound. It encompasses the study of how sound is produced, transmitted, and received. By carefully selecting and working with wood, instrument makers can fine-tune the acoustic properties of their instruments to achieve specific tonal qualities. Key concepts like density, stiffness, sound velocity, and damping are explained, revealing how these properties shape an instrument's sound. We'll examine the unique acoustic demands of various instrument components, from the resonating soundboard to the structural bracing, and how wood selection addresses these needs. We'll touch upon the methods used to measure these vital parameters, from analyzing beams and plates to understanding how grain analysis and tree selection influence instrument quality. This exploration provides a framework for understanding why certain woods are prized for specific instruments and how luthiers optimize their material choices.

"Tonewood" refers to wood specifically selected for its acoustic properties and suitability for use in musical instruments. It's chosen as the name because it directly emphasizes the wood's primary function: to contribute to the tone or sound of the instrument. The term highlights that not all wood is created equal when it comes to musical applications. Tonewood possesses specific characteristics that make it ideal for resonating, projecting, and shaping sound, distinguishing it from wood used for other purposes like construction or furniture.

The journey from a standing tree in the forest to a finely crafted musical instrument is a complex one, fraught with variables that can drastically alter the wood's inherent properties. While the species of a tree provides a starting point for understanding its potential as tonewood, it's necessary to recognize that this is just the beginning of the story. Significant variations exist not only between different species but also within a single species, even between individual trees growing in the same location. Furthermore, the specific part of the tree from which the wood is harvested plays a critical role. Wood from the heartwood, sapwood, or even different sections along the trunk will exhibit distinct characteristics.

This inherent variability is compounded by a range of external factors that influence the wood's properties from the moment it's felled. The drying process, whether natural air-drying or kiln-drying, dramatically affects the wood's moisture content, which in turn influences its density, stiffness, and resonance. The conditions under which the wood is dried and stacked, including temperature, humidity, and airflow, all contribute to the final outcome. Even the duration of the drying process plays a role, as prolonged or improper drying can lead to warping, cracking, or other defects.

Beyond drying, the wood continues to interact with its environment. Weathering, exposure to sunlight, rain, and temperature fluctuations, can degrade the wood's surface and alter its internal structure. Biodegradation, caused by fungi, insects, or bacteria, poses a constant threat, potentially compromising the wood's structural integrity and acoustic properties. Even seemingly benign factors like storage conditions and the passage of time can subtly shift the wood's microscopic characteristics.

Therefore, relying solely on visual inspection or the assumed species of a piece of wood is a risky proposition, especially when crafting a musical instrument where tonal nuances are paramount. The wood's appearance, while informative, can be deceptive. Two pieces of wood from the same species might look similar but possess vastly different acoustic properties due to the aforementioned variables. A beautiful grain pattern doesn't necessarily guarantee exceptional tone. Similarly, the age of the wood, while sometimes considered a positive attribute, doesn't automatically equate to superior quality. In fact, improperly stored or degraded older wood might be far less suitable than a carefully selected and properly seasoned piece of younger wood.

While many factors influencing the musical properties of wood will be explored in the next chapter (Chap. 5), it's necessary to understand the potential for misjudgment when relying solely on external signs. We often form opinions about a piece of wood—optimistic or pessimistic—based on its appearance, but this can be misleading. Just how wrong can we be? That's a question we'll address indirectly. This chapter focuses specifically on identifying whether a piece of wood is suitable for musical instrument construction. We will concentrate on assessment and avoid, for the moment, any discussion of methods for treating, altering, or "tuning" wood. These techniques, crucial as they are, will be covered in detail in Chap. 5. So, this chapter emphasizes the critical importance of moving beyond superficial assessments and a deeper understanding of the wood's actual properties. Before undertaking any luthiery work, thorough measurements and evaluations are essential. These measurements, as detailed in this chapter, will reveal the wood's current density, stiffness, damping capacity, and other crucial parameters. Only through these rigorous assessments can a luthier confidently determine if a piece of wood is truly suitable for its intended purpose and predict its contribution to the instrument's final tone. Ignoring this critical step can lead to disappointing results, wasted materials, and ultimately, an instrument that fails to live up to its sonic potential. The following sections will equip you with the knowledge and tools necessary to make informed decisions, ensuring that the wood you select is not just visually appealing, but also acoustically exceptional.

It bears repeating: modern medicine wouldn't dream of prescribing treatment without the insights gained from tools like ultrasounds, X-rays, CT scans, MRIs, and lab tests. These diagnostic tools provide a clear understanding of a patient's condition, enabling informed decisions. It's puzzling, then, how some luthiers continue to rely solely on their five senses when selecting tonewood. Why forego the precision of measurement and analysis when the acoustic properties of wood are so complex and variable? This chapter argues for a more scientific approach, emphasizing the necessity of objective data to inform wood selection and ensure the creation of instruments that truly sing.

4.2 Acoustic Properties of Wood for Musical Instruments

In the scope of musical instruments, wood is primarily classified based on its solid acoustic properties within the audible frequency range. This classification emphasizes the sound produced by the wood, rather than solely its material composition or visual appearance. This chapter aims to clarify these acoustic properties, including, density, specific stiffness, acoustic velocity, acoustic radiation coefficient, internal dissipation or damping capacity, acoustic conversion efficiency, mechanical Impedance, timbre and/or sound color.

While the acoustic properties of wood are necessary for an instrument's sound production and clarity, their significance extends beyond musical performance. These properties also play a role in minimizing sound transmission between rooms and in the development of noise barriers for mitigating highway noise.

Furthermore, acoustic wave and vibration analyses are employed to assess wood strength, monitor processing variables during instrument construction, and provide a detailed understanding of wood characteristics. Each of these areas is complex, but they are all based on the fundamental principles discussed herein. For further information on the broader applications of wood acoustics, please refer to the resources provided at the end of this chapter. Within the context of this book, our attention will be directed exclusively to the acoustic properties of wood in musical instrument applications.

4.2.1 Density

Wood density, a physical quantity, was defined and its measurement limitations (due to environmental factors) and solutions were addressed in Chap. 2. We assume the reader is familiar with that material. The present discussion explores the influence of density on the acoustic and sound properties of wood.

In any vibrating medium, including wood, oscillatory motion exhibits a cyclical pattern of velocity and acceleration. At the equilibrium point (the center of the vibration), velocity is maximal, while acceleration is zero. As the medium moves toward the extremes of its displacement, its velocity decreases, and its acceleration increases (though it's decelerating). At the point of maximum displacement, velocity momentarily reaches zero, and acceleration is maximal. The medium then reverses direction, accelerating back toward the equilibrium point, reaching maximum velocity and zero acceleration once again. This process repeats, constituting a vibration cycle.

This oscillatory behavior, governed by Newton's laws of motion, is fundamentally linked to the mass of the vibrating medium. The continuous changes in acceleration, particularly the need to overcome inertia at the points of maximum displacement, highlight the importance of density. For sound-producing components in musical instruments, such as the soundboard of stringed instruments, density plays a crucial role. While reducing thickness can also contribute to lightness, there are inherent limitations for thin vibration. Therefore, selecting naturally light wood species is a primary consideration, with adjustments to thickness being a secondary factor. This latter point will be discussed further in Chap. 5.

However, in specific applications within musical instruments, notably the back plate of bowl-shaped instruments, the wood serves as a reflector of acoustic energy. To fulfill this function, the material should exhibit significant density (and rigidity), thereby minimizing its susceptibility to sympathetic vibration. This characteristic enables efficient reflection of incident sound waves, effectively redirecting the energy towards the soundboard and the external environment.

It's important to note that excessively high or low densities can be detrimental. Extreme densities may lead to undesirable exaggerated resonances or reflections, negatively impacting the instrument's intended sound. However, in certain instruments, or for specific tonal preferences, the ringing sound produced by the vibrating back plate can be a desirable characteristic. This preference can influence the acceptable density range for that instrument.

Following the work of Wegst [61], we will refer to her proposed ranges for acoustic parameters in musical instruments, beginning with density and continuing with other parameters in subsequent sections. These ranges provide a valuable starting point (Table 4.1). Luthiers can then make informed adjustments within these ranges to tailor the instrument's sound to meet diverse aesthetic preferences, while carefully monitoring the resulting sonic changes.

Following the overview of density parameters in Chap. 2, this chapter will address the measurement of apparent density. The impact of secondary factors on density, notably moisture content and the related processes of wood treatment and tuning for musical applications, will be considered in the subsequent chapter.

For a piece of raw wood intended for musical instrument construction that has a regular, easily definable geometric shape (or can be machined into one), determining apparent density is straightforward. The volume is calculated by multiplying the cross-sectional area by the length. The mass is then measured directly using a scale. Dividing the mass by the volume yields the apparent density of the wood at its current moisture content.

Practical Exercise 4.1

A luthier is selecting maple for a violin back from commercially available, aged and air-dried, rectangular beams. A specific beam measures 10 cm x 25 cm x 50 cm and weighs 9 kg. Is the density of this piece suitable for a violin back?

Table 4.1 Table of material property ranges for stringed instrument construction, suggested by Wegst [61], and highlighted later by Bucur [10] and Roohnia [50]

Property	Min		Max	
	Top-plate	Back and side	Top-plate	Back and side
Density (kg/m^3)	300	600	550	900
Elastic stiffness (GPa)	6	8	20	18
Sp. stiffness ($\times$ 10^6 m^2/s^2)	16	9	50	30
Longitudinal velocity (m/s)	4000	3000	7000	5500
Anisotropy (v_L/v_R)	6	2	12	5
Damping capacity (tan δ)	0.003	0.008	0.009	0.018
Acoustic characteristic impedance (kPa s/m)	1300	2200	3300	4000
Sound radiation coefficient (m^4/kg s)	9	5	16	8
Acoustic conversion efficiency	1000	250	5000	1000

The table provides a general guideline for acceptable minimum and maximum values of key sound quality parameters. These parameters include density, elastic stiffness (modulus of elasticity) and specific stiffness, sound velocity and anisotropy ratio, damping capacity, sound radiation coefficient, acoustic characteristic impedance and overall acoustic conversion efficiency. The data is categorized for sound boards and backs/sides, highlighting the different material requirements for each

Solution

To determine suitability, we need to calculate the density:

$$\begin{aligned}\text{Volume} &= \text{Length} \times \text{Width} \times \text{Height} = 10\,\text{cm} \times 25\,\text{cm} \times 50\,\text{cm}\\ &= 12{,}500\,\text{cm}^3 = 0.0125\,\text{m}^3\end{aligned}$$

$$\text{Density} = \text{Mass/Volume} = 9\,\text{kg}/0.0125\,\text{m}^3 = \mathbf{720\,kg/m^3}$$

The density of maple used for violin backs based on Table 4.1 ranges from about 600 to 900 kg/m^3. A luthier might consider this wood, but would likely assess other factors in conjunction with the density, before making a final decision.

However, determining the density of irregularly shaped wood pieces presents a significant challenge. Traditional volume measurement techniques are inapplicable to objects lacking regular geometric forms. For instance, accurately determining the density of a violin back after it has been carved and graduated is problematic. Destructive sampling, involving cutting and removing a geometrically regular portion, would be required, thereby ruining the instrument component. While immersion in a liquid could, in principle, determine volume displacement, the use of water or other penetrating liquids is unsuitable due to their interaction with the wood, leading to inaccurate volume measurements. Furthermore, the toxicity and expense of mercury preclude its use. For high-value samples, advanced techniques like radiodensitometry, available in specialized wood science laboratories, can provide density measurements. For antique and valuable instruments, the primary constraint is their preservation. Although non-destructive measurement of individual wood component density is technically achievable, the associated cost is frequently prohibitive and of diminished importance in the case of highly valuable specimens like antique musical instruments. Therefore, this density measurement challenge is most relevant to contemporary instrument makers, who require cost-effective methods for characterizing the wood used in their instruments. Researchers are also exploring non-destructive methods, such as analyzing the shift in vibrational mode frequencies upon the addition or removal of small weights, to infer density in cases where direct volume or mass measurement is difficult [31]. Efforts are also underway to evaluate density using semi-nondestructive methods, such as thin pin penetration depth measurements [28]. While promising, these approaches have yet to be widely adopted for musical instrument resonator components, though their successful implementations would offer a cost-effective solution.

Consequently, until these advanced techniques are refined and made accessible to the musical instrument industry, the most reliable and currently standard practice for density determination remains measuring the density of the raw wood prior to shaping. This direct approach offers the greatest accuracy.

To measure the density of a cylinder (Fig. 4.1), you'll need to determine its mass and volume. Here's a breakdown of the process:

- **Measure the mass**: Use a balance or scale to measure the mass of the cylinder in grams (g) or kilograms (kg). Record this value.

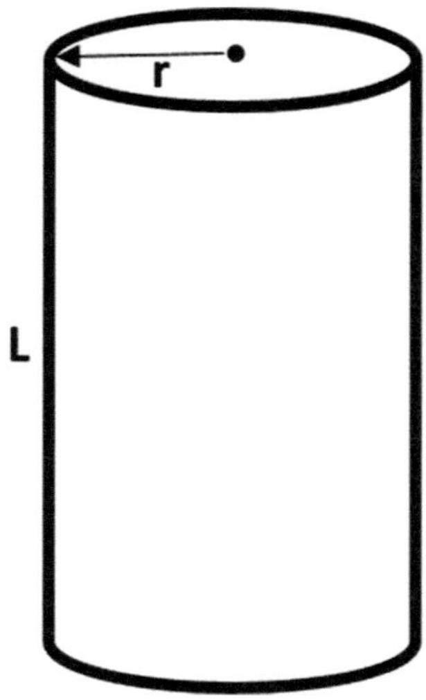

Fig. 4.1 A cylinder, illustrating the shape of a radial commercial log of a debarked tree. The diagram labels the relevant dimensions: length (L) and radius (r)

- **Measure the dimensions**: Measure the diameter of the circular base of the cylinder and divide it by two to get the radius. Measure the length of the cylinder. Record these measurements in centimeters (cm) or meters (m), ensuring consistency with your mass units.
- **Calculate the volume**: The formula for the volume (V) of a cylinder is: $V = \pi r^2 L$, where π (pi) is approximately 3.14159.

  Plug in the values for the radius and length you measured and calculate the volume. The volume will be in cubic centimeters (cm^3) or cubic meters (m^3), depending on your length units.
- **Calculate the density**: Divide the mass of the cylinder by its volume to get the density. The density will be in grams per cubic centimeter (g/cm^3) or kilograms per cubic meter (kg/m^3).

Practical Exercise 4.2
Let's say you have a cylinder with:

Mass (m) = 125 kg
Radius (r) = 25 cm
Length (h) = 100 cm

Solution

Volume: $V = \pi(25\ cm)^2(100\ cm) \approx 196{,}350\ cm^3$

Density: $\rho = (125{,}000\ g)/(196{,}350\ cm^3) \approx 0.637\ g/cm^3 \times 1000 = 637\ kg/m^3$

To measure the density of a Prism with trapezoidal cross-section:

As seen, the density of an object is calculated by dividing its mass by its volume. To find the density of a prism with a trapezoidal cross-section, you need to determine both its mass and its volume.

The volume of a prism is calculated by multiplying the area of its cross-section by its length (or height). In this case, the cross-section is a trapezoid. The area of a trapezoid is given by the formula:

$$A = \frac{1}{2}(a + b) \times h \tag{4.1}$$

where 'a' and 'b' are the lengths of the two parallel bases of the trapezoid, and 'h' is the perpendicular distance between these two bases (Fig. 4.2).

Once you have the area of the trapezoidal cross-section, multiply it by the length of the prism to get the volume. To find the mass of the prism, you can use a balance or a scale.

Now that you have both the mass and the volume of the prism, simply divide the mass by the volume to calculate the density.

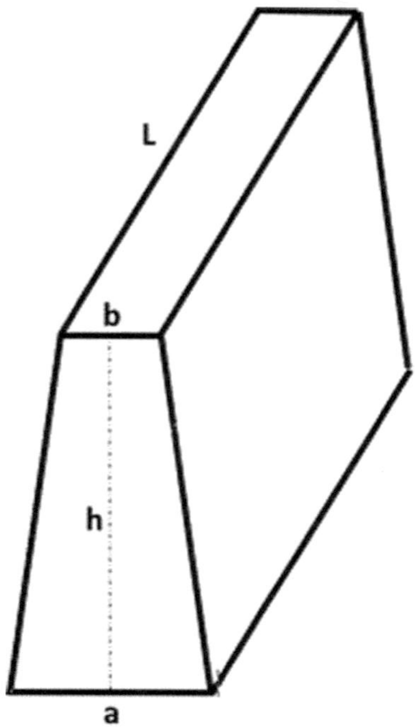

Fig. 4.2 A prism with a trapezoidal cross-section, illustrating the shape of a radial tonewood block used in violin making. The diagram labels the relevant dimensions: length (L), top width (b), height (h), and bottom width (a). This shape is typical for the initial block before the final concave profile of the violin top or back is carved

Practical Exercise 4.3

Let's say your selected spruce tonewood is a prism with a trapezoidal cross-section (wedge cut). The trapezoid has bases of 3 and 5 cm, and a height of 18 cm. The prism is 45 cm long. Its Mass is 1.4 kg. Calculate its density and determine its suitability for your instrument's top-plate.

Solution

$$\text{Area} = (0.5) \times (3\ \text{cm} + 5\ \text{cm}) \times 18\ \text{cm} = 72\ \text{cm}^2$$

$$\text{Volume} = 72\ \text{cm}^2 \times 45\ \text{cm} = 3240\ \text{cm}^3$$

$$\text{Density} = (1400\ \text{g})/(3240\ \text{cm}^3) \approx 0.432\ \text{g/cm}^3 \times 1000 = 432\ \text{kg/m}^3$$

A luthier might consider this wood for top-plate, but would likely assess other factors in conjunction with the density, before making a final decision.

A luthier's first and simplest concern must be apparent density, which should be measured using the simplest methods before proceeding.

Because wood is a layered material, its properties vary significantly between early and late growth seasons (Fig. 4.3). Early-season wood, with its thinner cell walls, is less dense, more porous, and mechanically weaker. Late-season wood, on the other hand, develops thicker cell walls and smaller cavities, resulting in greater density and

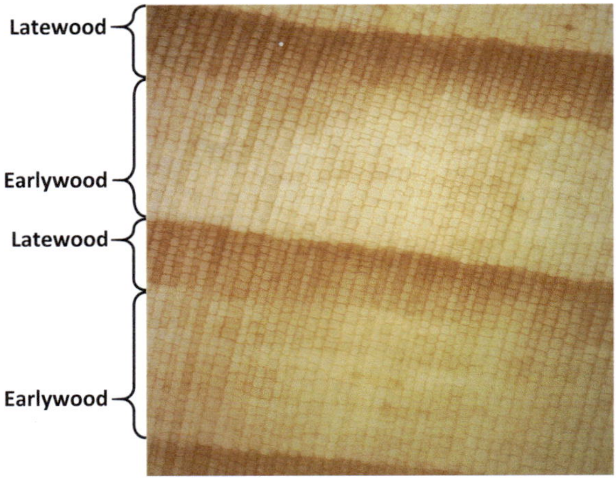

Fig. 4.3 Illustration of a conifer stem in cross-section (*Picea abies*), highlighting the differences between earlywood and latewood. The earlywood is described as consisting of large, thin-walled cells with bigger cellular cavities, while the latewood is composed of small, thick-walled cells and smaller cavities. Image adapted and improved from Wikimedia commons free resources. The copyright holder of this work, released this work: https://commons.wikimedia.org/wiki/File:Earlywood-latewood_PCAB.jpg

strength. Therefore, the ratio of early wood to total ring thickness is an important parameter for density (Chap. 2).

As previously mentioned, cellular constituents, including water and extractives, can contribute to wood weight and dramatically influence density.

4.2.2 *Elastic Stiffness*

As discussed in Chap. 2, which the reader is assumed to have reviewed, elastic stiffness quantifies a material's resistance to reversible deformation. A higher stiffness implies that a greater force is required to achieve a given elastic deformation. It is important to distinguish between stiffness, hardness, and toughness. While these terms are sometimes used interchangeably in common parlance, they have distinct definitions and applications in solid mechanics. This property (elastic stiffness) is characterized by two related parameters. The first, Young's modulus (E), also known as the modulus of elasticity, is an intrinsic material property independent of sample dimensions. It is measured in Pascals (Pa). The second, the springy constant (k), is an extrinsic property dependent on the sample's geometry. It is measured in Newtons per meter (N/m). For a constant applied force, a longer or thinner sample will exhibit a greater elastic deformation.

While the intuitive approach might be to compensate for low wood stiffness by increasing thickness, and vice versa, this simplification overlooks critical factors.

Merely increasing the thickness of a vibrating wood component increases its mass, thereby raising its inertia. This added inertia can hinder vibration and lead to premature damping. Therefore, acoustic properties of wood for musical instruments are governed by optimal ranges of modulus of elasticity and density, not simply springy constant and the mass alone. Reducing dimensions and thickness to lighten a component primarily decreases its stiffness constant (springy constant, k), which is generally undesirable. The interplay between stiffness and density, and how this balance influences the acoustic behavior of musical instrument components, will be explored further.

In our analysis of musical instrument acoustics, the initial calculations will be based on the wood's modulus of elasticity and density. Subsequent, minor adjustments to the dimensions will be considered only to effect subtle changes in the apparent springy constant of stiffness and the mass, thereby allowing for fine-tuning of the acoustic characteristics.

To understand how stiffness influences the musical sound produced by instruments, we must first acknowledge the orthotropic nature of wood. Its elastic stiffness varies significantly along the longitudinal, radial, tangential, and intermediate axes. The values presented in Table 4.1 are related to stiffness measured along the grain (longitudinal direction). Transverse stiffness, whether radial or tangential, falls outside this range. Consequently, instrument makers prioritize straight-grained wood with highly parallel grain alignment, ensuring that the longitudinal axis of the wood closely matches the direction of tree growth. This selection process optimizes the acoustic properties of the instrument's soundboard.

Furthermore, the dynamic nature of vibrations in musical instruments necessitates the use of the dynamic elastic modulus. However, it's necessary to understand the distinction between dynamic and static moduli. Both parameters are relevant in different contexts, and the numerical differences between them can be a source of confusion if not properly addressed. Therefore, we will first examine methods for measuring elastic stiffness under both dynamic and static conditions. Following this, we will explore the origin of these differences and their implications for instrument design.

- **Static bending method**:

  The static bending test is a well-established method for determining the modulus of elasticity of wood. This test, performed according to international standards, measures elastic stiffness, bending strength (or modulus of rupture), and work at proportional limit (toughness) in the longitudinal direction of the wood sample. Accurate evaluation of these properties requires careful sample preparation. For longitudinal modulus, the sample's longitudinal axis must align with the wood's grain direction. Similarly, radial and tangential moduli require samples whose longitudinal axes are aligned with the radial and tangential directions, respectively. In the context of acoustic and vibrational properties of wood relevant to musical instruments, we are primarily concerned with the elastic range and small, reversible deformations. Therefore, for determining the modulus of elasticity, established methods from static bending tests, as specified in standards like ISO 13061-4:2014 [26], are sufficient. While precise adherence to this standard may require specialized equipment, a reasonably accurate simulation can be achieved in a well-equipped luthier's workshop. The process of adapting the standard for workshop use can offer valuable insights and potentially lead to innovative approaches.

This standard specifies determining the modulus of elasticity by measuring the mid-span deflection of a simply supported beam under a gradually increasing transverse load, within the material's linear elastic region. A suitable testing machine must provide: (1) a rigid frame to support the specimen while allowing unrestrained deflection; (2) a loading head that applies the force without inducing stress concentrations; (3) a controlled, low loading rate (or loading head movement) to minimize viscoelastic effects and ensure accurate deflection measurements; and (4) a calibrated force-measuring device accurate to within 1%.

Experimental setups for material testing often utilize rigid base frames, supports, and specialized loading heads to prevent stress concentrations at the loading points. These components are readily available or can be manufactured in instrument workshops. As shown in Fig. 2.17, the stress–strain curve exhibits a linear relationship within the elastic limit. This constant slope enables the calculation of the modulus of elasticity by measuring the deformation resulting from a constant applied load (Fig. 4.4).

$$E = \frac{Fl^3}{4bh^3\delta} \tag{4.2}$$

where E is the modulus of elasticity in geometrically longitudinal axis (Pa), F represents the applied load (N), l is the distance between supports (m), b is the specimen width (m), h is the specimen height (m), and δ is the deflection at the mid-span (m). The effective bending length is defined by the distance between the supports, not the total specimen length (L). The overhanging portions are not considered in the bending analysis.

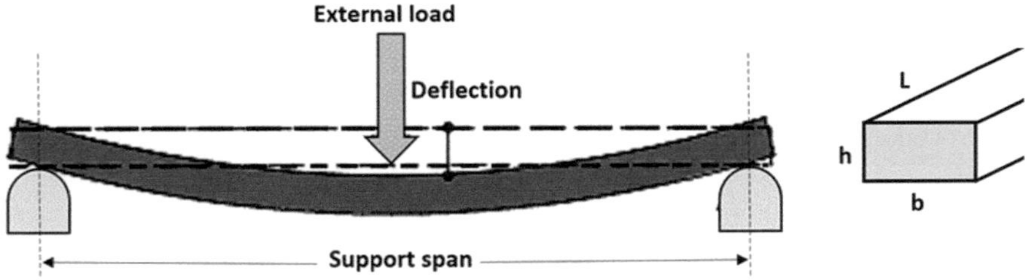

Fig. 4.4 Schematic of a simply supported beam under load, demonstrating deflection. The beam is supported at its two ends, and a load is applied at the center. The diagram includes labels for "external load," "deflection," and "support span," clarifying the key parameters measured in a bending test. The dashed line indicates the unloaded position of the beam

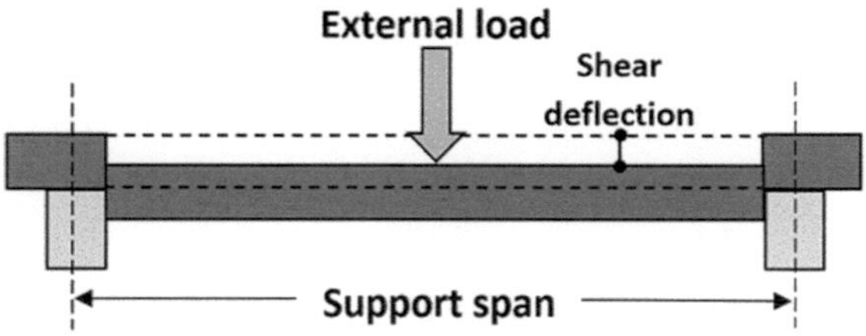

Fig. 4.5 Schematic of a beam illustrating the concept of shear deflection. While the image exaggerates the deflection near the supports for clarity, the shear stress occurs along the entire span, decreasing from the supports to the mid-span. The diagram includes labels for "external load," "shear deflection," and "support span"

While standardized static bending test equipment minimizes measurement errors, constructing a bending apparatus with basic workshop resources is feasible provided the potential for error is understood.

- **Shear deflection error**: A common source of error arises from shear deflection. When a wooden beam is loaded at mid-span, the total observed deflection is a composite of bending and shear deflections (Fig. 4.5). Treating the entire deflection as solely due to bending, as is often done, introduces error. Ideally, the shear deflection component, however small, should be subtracted from the total deflection before calculating the modulus of elasticity. Failing to account for shear deflection leads to an underestimation of the true modulus of elasticity (Eq. 4.2).

  Shear stress is not uniform along the beam's length. It is typically greatest near the supports and diminishes toward the mid-span. Thus, the measured deflection at any point along the beam includes a contribution from shear stress, varying in magnitude.

  The geometry of the support contact surfaces (Fig. 4.4) influences the magnitude of the shear deflection error. While a carefully designed support can mitigate this error, it cannot be entirely eliminated. The span length-to-beam height ratio (l/h) is another critical factor. Increasing this ratio reduces the relative contribution of shear deflection to the total deflection. A larger l/h ratio greater than 20, means that the bending deflection becomes dominant, and the shear deflection error becomes less significant. However, even with optimized supports and a high l/h ratio, shear deflection cannot be entirely eliminated. Careful consideration of these factors allows the shear deflection error to be minimized to a negligible level.
- **V-shaped deformation**: This standard applies to perfectly rectangular, regularly shaped small clear specimens. While this three-point bending test (with a single mid-span force and two support reactions) is a common method, it can lead to non-uniform bending in larger specimens, potentially causing a V-shaped deformation. Therefore, a four-point bending test, where two forces are applied at points one-third of the span length apart, is often employed to

create a region of constant bending moment and ensure more uniform stress distribution.

With careful consideration of factors such as uniformly distributed loading across the mid-span, this method can also be adapted for use with rectangular wooden plates of uniform thickness and lower widths. While the inherent imprecision of a self-designed and assembled device will result in greater measurement uncertainty compared to a standardized static bending apparatus, the results can provide an acceptable approximation of wood stiffness for comparative purposes within the musical instruments industry. Therefore, in situations where access to a standardized testing device is unavailable, constructing and utilizing this simplified tool is recommended (Fig. 4.6). Because it's better than nothing.

Characterizing wood stiffness, even with some small degree of uncertainty, offers significant advantages over neglecting this important material property. However, access to and utilization of calibrated, standardized testing equipment remains the most reliable approach for obtaining accurate and comparable results.

Practical Exercise 4.4

A static bending test was conducted on a rectangular aged and air-dried spruce wood plate. The grain direction of the wood was aligned with the longitudinal axis of the plate. A 5 kg mass applied at the midline of the

Fig. 4.6 Photograph of a static bending test apparatus. A wooden sample is supported at both ends, with a dumbbell applying a load at the center. A dial indicator is used to measure the resulting deflection (the effect of the dial indicator on the stiffness of wood is concerned and compensated). This setup, devised by Persian setar luthier, Mr. Kyomars Soozani, allows for stiffness measurements, acknowledging the need for calibration and precise control for acceptable results

42 cm span (l) between supports resulted in a 4 mm maximum deflection (δ). The plate's overall length (L) was 50 cm, with a thickness (cross-sectional height) of 5 mm (h) and a width of 15.5 cm (b) and its mass was 165 g. Determine the apparent density (D) and longitudinal modulus of elasticity (E) of this spruce specimen.

Solution

$$D = \frac{m}{L \times b \times h} = \frac{165}{50 \times 15.5 \times 0.5} \approx 0.426\,\text{g/cm}^3 \times 1000 = 426\,\text{kg/m}^3$$

To calculate the external load in Newton, we multiply the mass of the external weight in kilograms by the acceleration due to gravity, which is approximately 9.81 m/s^2. This calculation is based on Newton's second law of motion, which states that force is equal to mass times acceleration ($F = ma$). In this case, the force is the external load, the mass is the mass of the external weight, and the acceleration is the acceleration due to gravity.

$$E = \frac{Fl^3}{4bh^3\delta} = \frac{5 \times 9.81 \times 0.42^3}{4 \times 0.155 \times 0.005^3 \times 0.004}$$
$$= 11{,}722{,}633{,}548\,\text{Pa} = 11.72\,\text{GPa}$$

A luthier might consider this wood for top-plate, but would likely assess other factors in conjunction with the density and elastic stiffness, before making a final decision.

- **Specific stiffness and acoustic velocity**:
  As we have seen, both density and stiffness are effective for the vibration properties of wood. We said that for the continuity of vibration, it is better to use wood with a lighter density and a higher stiffness. Therefore, the combination of these two parameters can be considered as an indicator of the vibration capability of wood.

$$S = \frac{E}{D} \tag{4.3}$$

Specific stiffness (S) in wood is a measure of how stiff a material is relative to its density. It essentially tells you how much a material resists bending or deformation for a given weight.

The soundboard of a stringed instrument, like a guitar, oud, Turkish baglama or violin, is responsible for amplifying the vibrations of the strings. A high specific stiffness allows the soundboard to vibrate efficiently and project sound well. A high specific stiffness means that the soundboard can be lightweight yet stiff, making it more responsive to vibrations. This results in a clearer and more resonant sound.

Spruce is an example of a popular choice for soundboards due to its high specific stiffness and excellent acoustic properties. Cedar is another common wood for soundboards, known for its warm and mellow tone.

Specific stiffness is a factor in the design and construction of musical instruments, as it directly affects the instrument's sound quality and responsiveness.

Let's break down the unit of the specific stiffness (S) within the SI system. Elastic Modulus (E) Measured in Pascals (Pa), which is equivalent to kg/m s^2. Density (D) Measured in kilograms per cubic meter (kg/m^3). Therefore, the unit of specific stiffness (S) is calculated as $S = (\text{kg/m s}^2)/(\text{kg/m}^3) = \text{m}^2/\text{s}^2$. It means that in the SI system, specific stiffness (the ratio of elastic modulus to density) is measured in m^2/s^2, equivalent to the dimensions of velocity squared.

In solids, vibrational wave motion manifests as dynamic longitudinal stress waves. Analyzing the speed of sound propagation allows for a dynamic estimation of specific stiffness. Incorporating the material's density then yields the elastic stiffness of the wood sample.

- **Direct measurement of acoustic velocity**:

  A seemingly straightforward approach to measuring sound speed is the direct distance-over-time method, analogous to speed cameras and similar devices. However, while conceptually simple, this method presents practical challenges. Accurately measuring the transit time of sound through wood requires sophisticated equipment capable of precisely triggering a timer upon sound entry and termination, especially given the potential for error within such short time intervals. While accurate, such devices represent a significant investment.

  Despite the cost, properly calibrated direct measurement systems offer the most versatile and efficient means of determining acoustic parameters in wood for musical instruments. Critically, they can measure sound velocity in arbitrary directions (Fig. 4.7), unconstrained by the sample's geometry. This adaptability allows for testing samples of any shape, size, and (within the instrument's accuracy limits) diameter. In contrast, other established methods are often limited by the sample's geometry, required alignment with specific growth directions, and other constraints.

  Figure 4.7 shows how longitudinal wave velocity (v) is experimentally determined by measuring the distance (L) sound travels and the transit time (t) between the emitter and receiver. This measurement is based on the principles of stress wave propagation.

$$\text{v} = \frac{L}{t} \tag{4.4}$$

  The longitudinal wave speed in wood provides an estimate of specific stiffness. Assuming negligible influence from other factors, the relationship is approximated as:

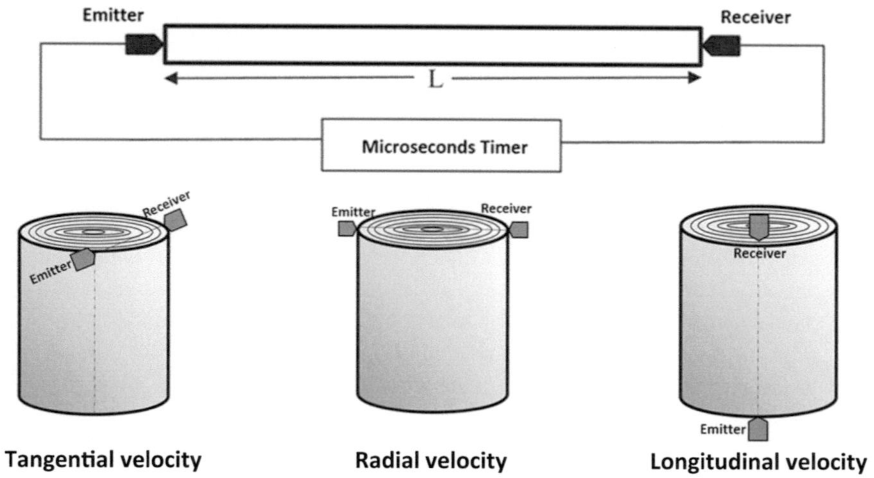

Fig. 4.7 Schematic of an apparatus for measuring sound velocity in wood. The diagram shows two transducers, one acting as an emitter and the other as a receiver. The emitter is excited by either a hammer impact or an ultrasonic pulse, generating a sound wave that travels through the wood to the receiver. The time taken for the sound wave to travel a known distance is measured using a microsecond timer, allowing for the calculation of sound velocity

$$S = \mathrm{v}^2 = \frac{E}{D} = \frac{L^2}{t^2}$$
$$E = D\mathrm{v}^2 \qquad (4.5)$$

Two distinct methodologies exist for the direct measurement of sound speed based on distance and time: one employing ultrasonic transducers to generate and detect acoustic pulses, and the other relying on the acoustic impulse generated by a mechanical impact, such as a hammer striking a sensor.

Both these two methods for determining the speed of sound in wood involve placing two opposing transducers on the wood sample at a known distance. One transducer acts as the emitter, and the other as the receiver. In one approach, a hammer impact generates the sound wave; in the other, an ultrasonic source is used. Upon excitation (hammer impact or ultrasonic pulse), the piezoelectric element within the emitter transducer generates a voltage pulse, triggering a microsecond timer. Simultaneously, the sound wave propagates through the wood to the receiver transducer. When the sound wave arrives, the receiver's piezoelectric element generates a second voltage pulse, which stops the timer. The elapsed time is recorded, and knowing the distance between the transducers, the speed of sound within the wood in selected growth direction can be calculated.

The speed or the velocity of sound along the grain of wood can exceed 5000 m/s. Accurately measuring the transit time of such a fast pulse requires extremely high precision to minimize measurement error. Achieving this level of accuracy requires state-of-the-art instrumentation, which comes at a significant cost. However, the directness and ease of obtaining results with these acoustic methods make them attractive to researchers and luthiers compared to alternative techniques.

Let us now consider the possible errors associated with this technology and the solutions, either realized or necessary, for their reduction or elimination.

- **Sensor-wood coupling**: The efficacy of piezoelectric sensors in acoustic measurements of wood is often compromised by imperfect contact between the sensor and the wood's surface. This imperfect coupling can lead to incomplete or delayed transmission of the acoustic pulse, introducing inaccuracies into the measurements [19]. While ultrasound coupling gels are frequently employed to improve contact, their water-based composition poses a significant drawback. Rapid penetration of the gel into the wood can alter the wood's acoustic properties due to moisture content changes, thus confounding the measurement. An alternative approach involves nail-shaped sensors, which achieve more complete contact by penetrating the wood (Fig. 4.8). However, this method with the highest precision compare to the other, particularly when applied to high-value woods used in musical instruments, is often considered invasive and undesirable. Therefore, a practical solution frequently adopted is a hybrid approach. This involves judicious selection of sensor type: flat sensors with coupling gel where appropriate, and minimally invasive nail-shaped sensors (with needle-like small diameters to limit damage) when permitted. This strategy allows luthiers to balance the need for accurate acoustic coupling with the desire to preserve the integrity of the wood sample.
- **Sensor delay compensation**: Sensor transit time and signal propagation delays introduce systematic errors in acoustic measurements. These delays, arising from the time required for sound to traverse the sensors and for electrical signals to travel through the connecting wires, respectively, contribute a constant offset

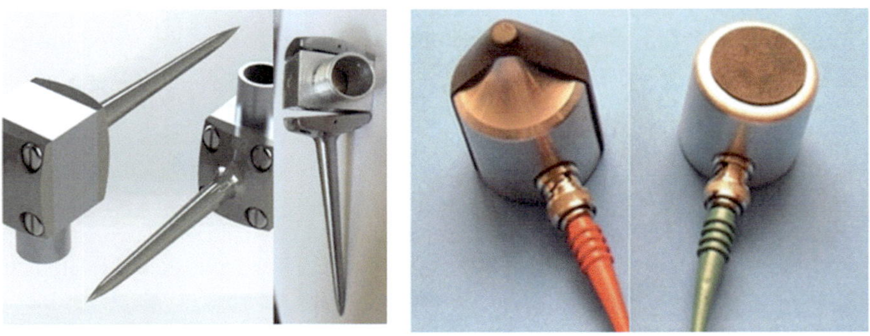

Fig. 4.8 Image of piezoelectric sensors for wood testing, including a close-up view of a nail-shaped transducer with visible prongs, a conic transducer with a pointed tip, and a flat transducer with a circular contact surface. Images of these sensors, the Lucchimeter (right) and my handmade Tonewood Analyzer (left, I inspired it from Hungarian FAKOPP tree tomography sensors and improved it for tonewood evaluation), are contributed to this book by the manufacturers

to the measured values. To mitigate these errors, measurements are repeated across multiple samples with varying dimensions and sensor distances. This multi-faceted approach allows for the identification and characterization of the constant error, which is subsequently compensated for through software or hardware calibration of the device.

- **Viscoelastic error**: The apparent elastic modulus of viscoelastic materials like wood is falsely frequency-dependent [40]. Because stress and strain are not perfectly in phase, deformation lags behind stress. At higher frequencies, the material doesn't have sufficient time to fully deform, resulting in a smaller observed deformation. This underestimation of deformation leads to an overestimation of the elastic modulus. Similarly, dynamic tests at higher frequencies yield smaller apparent deformations, thus inflating both the estimated oscillation frequency and the calculated elastic modulus. Ultrasonic testing, with its inherently high frequencies, produces the largest (and most exaggerated) dynamic modulus, deviating significantly from results obtained through standard static bending tests. To mitigate this issue, lower test frequencies are preferable. Consequently, hammer impact testing offers closer agreement with static bending test results compared to ultrasonic methods.
- **Local inhomogeneity and damages**: Damage and inhomogeneity within a sound propagation path lead to underestimations of acoustic velocity. This is because the sound waves are forced to deviate from a direct path, effectively circumventing the damaged or inhomogeneous regions. Consequently, velocity measurements become unreliable in global measurements. For instance, consider the radial velocity profile across the diameter of a log, as illustrated in Fig. 4.7.

  Sound traveling along a radial path must traverse both the juvenile wood and the mature wood, regions known to possess a weaker structure than the mature wood layers. To mitigate this issue and enable accurate, localized evaluation of the wood, nail-shaped transducers are employed. These transducers can be positioned at arbitrary locations within the medium, facilitating area-by-area or local analysis, thereby minimizing the influence of known damage and inhomogeneity's (Fig. 4.9).

  Sound stress waves in solids propagate as expanding wavefronts originating from the source. In homogeneous materials, these wavefronts are typically circular (or spherical in 3D). However, material inhomogeneity causes wavefront distortion; propagation is faster, and thus wavefronts elongate, in regions of higher sound speed [11]. Defects, voids, or damage impede propagation, locally reducing sound speed. Critically, the fastest path from emitter to receiver, regardless of wavefront shape, governs the arrival time and triggers the receiver sensor.

Although dynamic wood testing reveals a phase lag between stress and strain (due to viscoelasticity) at high frequencies, dynamic stiffness still remains the dominant effective elastic stiffness of vibrating wood in musical instruments. To minimize this

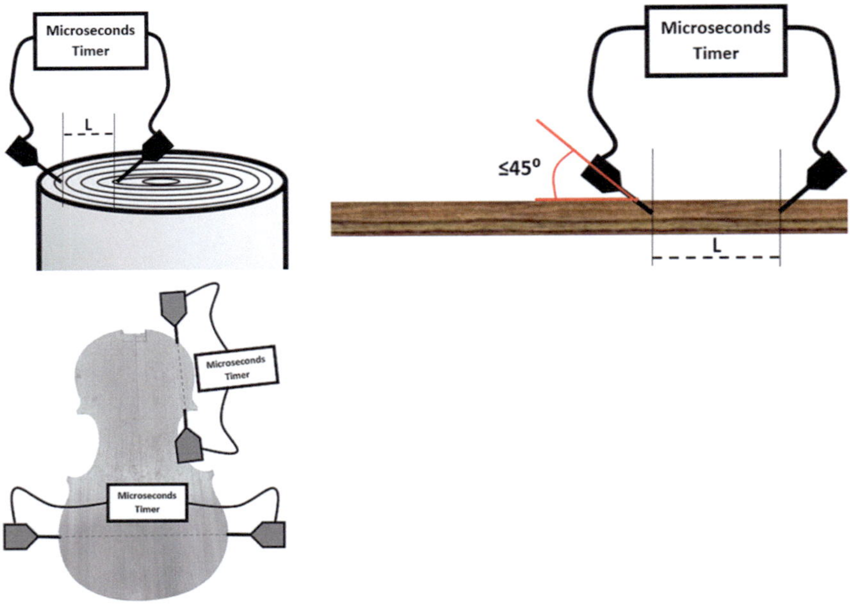

Fig. 4.9 Diagram highlighting the technical advantage of nail-shaped sensors in wood assessment. The illustration shows their ability to perform localized measurements, emphasizing the enhanced detail provided by nail-shaped sensors. Device manufacturers recommend keeping the straight needle angles or below 45° for reliable readings

viscoelastic error, it is advisable for luthiers to utilize dynamic testing equipment operating at lower frequencies, when feasible (Fig. 4.10).

Beyond impact-based methods, which employ a hammer strike rather than an ultrasonic source, more accurate techniques leverage the natural frequencies of beam and plate vibrations. These frequencies are significantly lower than those used in ultrasonic methods, allowing for more precise measurements of sound velocity and specific dynamic stiffness. While highly effective and a standard reference often used for calibrating other stiffness measurement tools, these modal vibration methods are currently limited by the regular geometric requirements of the test samples [50]. The following paragraphs describe methods that utilize the wavelength and natural modal frequencies of the wood specimen, rather than directly measuring the time of flight of sound.

Practical Exercise 4.5

A longitudinal stress wave ultrasonic velocity test was performed on a spruce specimen (50 cm length, 5 mm thickness, 15.5 cm width, 165 g mass), consistent with the rectangular dimensions outlined in Practical Exercise 4.4. **Sensors were firmly coupled to the specimen along the grain**

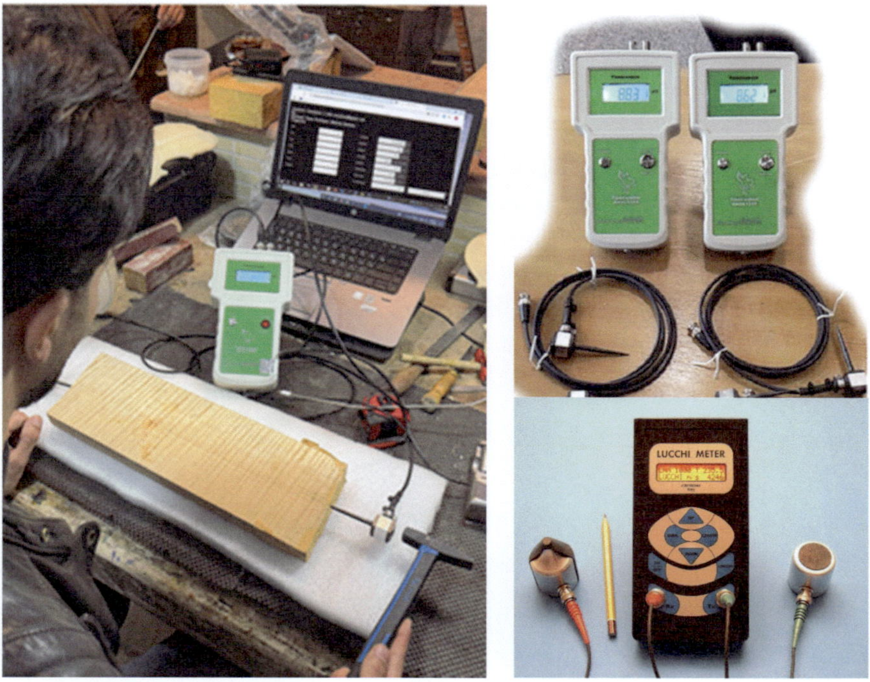

Fig. 4.10 Photograph showcasing two devices used for measuring sound velocity and acoustic parameters in wood. On the lowerright is a Lucchimeter, and on the upright is a handmade Tonewood Analyzer of mine. Both devices utilize the direct method of measuring distance and transit time of ultrasonic waves to calculate sound speed. On the left, luthier Mr. Mikaeil Ghazal is shown analyzing a wedge-cut maple tonewood for violin. Images of these tools, the Lucchimeter and Tonewood Analyzer, were gifted to this book by their respective manufacturers

direction. The dynamic modulus of elasticity (E) was then determined. The sound traveling time was 87 μs measure by a microseconds timenr device.

Solution

$$E = D\text{v}^2 = \frac{m}{bhL} \times \frac{l^2}{t^2} = \frac{0.165}{0.155 \times 0.005 \times 0.5} \times \frac{0.5^2}{0.000087^2}$$
$$= 14{,}064{,}158{,}132\,\text{Pa} \Rightarrow 14.06\,\text{GPa}$$

In this context, l represents the sensor tip-to-tip distance, while L denotes the total length of the test specimen. When employing surface-mounted sensors at both ends of the specimen, l is equivalent to L. However, when utilizing needle-tip sensors, which may partially penetrate the specimen, the effective tip-to-tip distance, l, must be accurately determined. This distance will typically be less than L due to sensor penetration.

Further Analysis

A hammer-impact excitation of the emitter sensor, replacing the ultrasonic source, yielded a sound transmission time of 90 μs (again, the sensor is a contact type without the sensor penetrating the wood, i.e. $l = L$). These dynamic stiffness results should be compared with the static bending test data from the preceding analysis.

$$E = Dv^2 = \frac{m}{bhL} \times \frac{l^2}{t^2} = \frac{0.165}{0.155 \times 0.005 \times 0.5} \times \frac{0.5^2}{0.000090^2}$$
$$= 13{,}142{,}174{,}432\,\text{Pa} \Rightarrow 13.14\,\text{GPa}$$

Analysis of Elastic Moduli in Wood: Static Versus Dynamic Methods

Three distinct elastic moduli were determined from a single wood specimen, revealing variations attributable to the measurement methodology. The static bending test yielded the lowest modulus, a result consistent with the anticipated influence of shear deflection, which inherently reduces the apparent stiffness. Conversely, the piezoelectric supersonic velocity test, employing direct measurement, produced the highest modulus. This overestimation is attributed to the viscoelastic behavior of wood at high frequencies, leading to an increased apparent stiffness under rapid sonic vibrations. The stress wave test, which primarily induces longitudinal waves, has a minimal shear deflection. Consequently, the modulus obtained from this method was approximately 20% greater than that derived from the static bending test. However, a modified piezoelectric stress wave test, utilizing a hammer impact on the emitter sensor without ultrasonic pulse transmission, effectively evaluates the speed of sound at lower frequencies. This approach yielded a modulus relatively closer to the static bending test result, although still approximately 12% greater.

The observed discrepancies between static and dynamic stiffness measurements in this exercise align with established research. Bell et al. [4] and Bodig and Jayne [5] reported a 10% underestimation of static stiffness attributed to shear deflection, a recognized systematic error. Dynamic stiffness, conversely, typically exceeds static stiffness by 10–25%, a variation influenced by wood species, vibration direction, and the viscoelastic behavior of wood. Given that musical instruments engage dynamic vibrations, the dynamic stiffness is exceptionally the most relevant parameter for this particular application.

Understanding the sources and magnitudes of these measurement errors is therefore essential for accurate characterization of wood's elastic properties in dynamic contexts. The observed discrepancies highlight the importance of selecting appropriate measurement techniques based on the frequency range and loading conditions relevant to the intended application.

In piezoelectric sensor-driven stress wave testing, where an impact hammer generates the input pulse, it's reasonable to expect that higher-order vibrational modes, characterized by ultrasonic frequencies, would propagate to the receiver more rapidly. This raises concerns regarding viscoelastic effects and high oscillation speed errors. While this hypothesis is logically sound, it's important to understand the spectral content of the impact-generated stress wave.

The higher-order modes induced by impact excitation, while present, typically exhibit significantly lower frequencies compared to those employed in conventional ultrasonic testing. Consequently, they do not produce the exaggerated dynamic stiffness values that would arise from high-frequency ultrasonic excitation. Furthermore, established calibration procedures mitigate systematic errors associated with impact-based stress wave testing.

These procedures involve extensive testing and comparisons with sensor-free modal longitudinal vibration analysis, which operates at significantly lower frequencies. This comparative approach allows for the identification, correction, and calibration of inherent errors. For instance, in the development of my tonewood analyzer devices for luthiers, I calibrate and equate the stress wave test results with those obtained from modal longitudinal vibration analysis, where the fundamental longitudinal mode frequency serves as the primary criterion for velocity and dynamic elastic modulus estimation.

- **Longitudinal free vibration of a free-free beam**:

  Longitudinal stress wave testing can be performed on wood without embedded sensors, though careful consideration of several factors is required [7, 29, 35, 50, 51, 62]. Longitudinal wave velocity (v) is related here to wavelength and longitudinal modal frequency. For the first mode of free vibration in a slender bar, its relationship simplifies to:

  $$\mathrm{v} = 2Lf \tag{4.6}$$

  where L is the bar length (m) and f is the fundamental longitudinal resonant frequency (Hz).

  A free-free wooden beam subjected to a sudden compressive impact along its longitudinal axis will undergo transient vibration (Fig. 4.11). Due to the wood's elastic properties, the beam will initially respond by vibrating predominantly at its fundamental longitudinal resonant frequency. Because the beam is unrestrained, higher-order longitudinal modes will also be excited.

  In this non-contact method illustrated in Fig. 4.11, an acoustic wave is generated by a light hammer impact at one end of the wood sample, and the resulting excitation is detected at the opposite end. The fundamental resonant frequency is then determined from the frequency spectrum obtained via Fast Fourier Transform (FFT) analysis of the recorded signal. A luthier must not be concerned about the

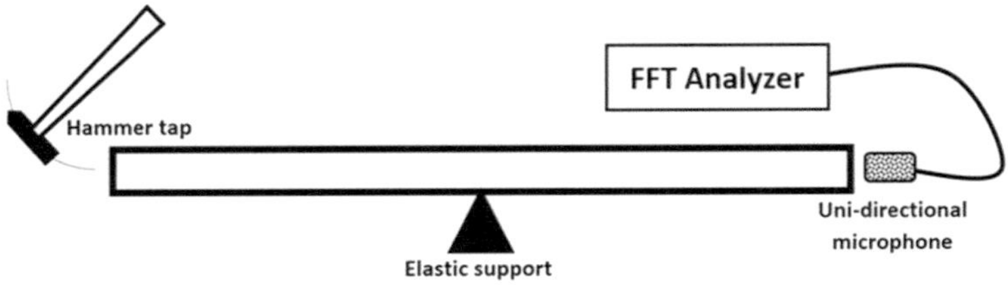

Fig. 4.11 Schematic of a wood sample undergoing longitudinal free vibration testing. The sample is supported at its nodal point with a soft, elastic support to simulate a free-free bar. A hammer impact generates vibrations, and a microphone connected to an FFT analyzer records the resulting sound. The diagram labels the key components of the setup

higher modes of vibration in this particular test method, but must know the nodal points and the wave length carefully.

In the fundamental mode of longitudinal vibration, the node, a point of zero displacement, is located precisely at the beam's midpoint. Positioning a thin, compliant material, such as a thin rubber cushion, at this node effectively isolates the beam's vibration from external constraints. This minimizes damping from the support, allowing the beam to vibrate as if it were unconstrained. For this fundamental mode, the wavelength (λ) is twice the beam's length (L), as expressed in Eq. 4.6.

Knowing both the acoustic velocity and the material density, the modulus of elasticity can be calculated (as illustrated in Eq. 4.5). This modulus is then used in the determination of other acoustical properties.

Let us now consider the possible errors associated with this technology and the solutions, either realized or necessary, for their reduction or elimination.

- **Viscoelastic error**: Dynamic stiffness measurements, whether obtained through piezoelectric sensor-based stress wave tests or sensorless vibration analyses, are susceptible to errors arising from the phase lag between applied stress and resulting deformation. This phase lag, exacerbated by high vibration frequencies, leads to an overestimation of dynamic stiffness. Consequently, lower vibration frequencies minimize this error. Given that the natural frequency of the first vibrational mode, utilized in this sensorless test, is significantly lower than ultrasonic frequencies and within the audible range, we anticipate reliable dynamic stiffness values. While inherent errors, acknowledged and understood, will still be present, their magnitude will be reduced. Therefore, provided specimen dimensions adhere to established standards, this sensorless method, based on first-mode natural frequency, is recommended for comparison and calibration of other dynamic vibration-based techniques.
- **Wave dispersion error**: Although incorporated into standards like ASTM D6874 [2], this simplified approach to stiffness estimation, which relies solely on longitudinal wave velocity and density, neglects the influence of changing cross-sectional dimensions. These dimensional variations can introduce inaccuracies in velocity-based stiffness evaluations due to the dispersive nature of

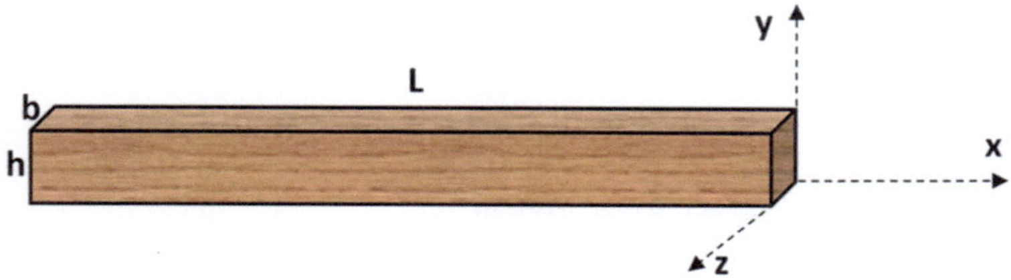

Fig. 4.12 Illustration of a rectangular prism representing a wooden beam, with dimensions labeled as L (length), b (width), and h (height). The image includes a 3D coordinate system (x, y, z) to define the orientation of the beam, suggesting it's used for scientific or engineering calculations

stress waves. Corrections for these cross-sectional dimensional effects, as well as the influence of Poisson's ratio, were initially addressed by Rayleigh [42] and subsequently refined for wood by Brancheriau [7], with further contributions simplified by Mirbolouk and Roohnia [35] and Roohnia [50]. So, the correct form of Eq. 4.2 for longitudinal vibration of a free-free beam presented as (Fig. 4.12):

$$\mathrm{v_x} = \sqrt{\frac{E_x}{D \cdot C_x}} \tag{4.7}$$

C represents Brancheriau's correction coefficient, simplified by Roohnia [50] for orthotropic materials, such as wood, specifically for impact excitation in the longitudinal (x) direction and defined as:

$$C_x = 1 + \frac{\pi^2}{AL^2}\left(I_z v_{xy}^2 + I_y v_{xz}^2\right) \tag{4.8}$$

where, A is cross section area in (m^2), v is Poisson's ratio (with regards to orthotropic coordinates, x, y and z), I_z and I_y are the moment of inertia in (m^4), for a rectangular beam would become as:

$$\begin{aligned} I_z &= \frac{1}{12} bh^3 \\ I_y &= \frac{1}{12} hb^3 \end{aligned} \tag{4.9}$$

Accurate determination of a wood's orthotropic Poisson's ratios is essential for reliable dynamic analysis. In practical luthiery, these ratios are often difficult to obtain due to time and resource limitations.

Consequently, free longitudinal vibration testing of free-end beams should be restricted to thin, long beams. For geometries such as planes or beams with large cross-sections, this method becomes unreliable without incorporating the specific orthotropic Poisson's ratios of the wood sample. Consequently, this methodology remains primarily confined to research laboratories and, in my assessment, is not yet suitable for general luthiery applications, except perhaps for slender beams. For this application, a beam should be considered slender when the length-to-cross-sectional-dimension ratio exceeds 20.

- **Flexural free vibration of a free-free beam**:

  The acoustic linear velocity in wood can be indirectly determined through flexural resonance analysis by leveraging the relationship between resonant frequencies, beam geometry, and material stiffness. To accurately simulate free-free boundary conditions, elastic supports, if necessary, must be positioned at the theoretical nodal points of the flexural excited beam. These nodal points, characteristic of the specific mode shape of a free-free beam, are tabulated in Table 4.2.

  By precisely identifying and supporting these nodal points, the beam can vibrate in its natural modes without significant external constraints. Excitation at an antinode, coupled with response measurement at another appropriate antinode, allows for the determination of the resonant frequencies associated with each mode [44–46, 49, 50]. These frequencies, in conjunction with the beam's dimensions and the mode-specific constants, enable the calculation of the flexural wave velocity.

  Any deviation from accurate nodal support placement introduces external friction and boundary condition errors. This leads to inaccuracies in the measured resonant frequencies and, consequently, in the calculated wave velocity. Furthermore, external frictional forces introduced by improper support placement directly affect the determination of internal friction.

Table 4.2 Position of nodal points for a free-free beam in flexural free vibration

L = 1000 mm	Mode number	Nodal points (mm from one end)							
	1	224	776						
	2	132	500	868					
	3	94	356	644	906				
	4	73	277	500	723	927			
	5	60	226	409	591	774	940		
	6	51	192	346	500	654	808	949	
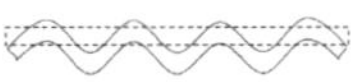	7	44	166	300	433	567	700	834	956

For a perfectly isotropic and homogeneous beam, theoretical modal frequency ratios are precisely defined. Specifically, if the fundamental (first mode) frequency is normalized to 1 Hz, the subsequent modal frequencies for flexural vibration will exhibit specific ratios: approximately 2.756 Hz for the second mode, 5.404 Hz for the third mode, 8.933 Hz for the fourth mode, and so forth. These theoretically predicted frequencies represent the ideal, or "mathematical," frequencies for a perfectly uniform material.

However, in real-world materials, particularly wood, deviations from perfect isotropy and homogeneity are common. These heterogeneities cause shifts in the observed modal frequencies compared to the theoretical values. The magnitude and direction of these frequency shifts serve as a sensitive indicator of the material's homogeneity. Larger deviations suggest a greater degree of inhomogeneity or defects, while closer agreement with the theoretical ratios implies a more uniform material structure [49].

Based on Euler–Bernoulli's Elementary theory, the flexural vibration of a beam formulates as (Fig. 4.13):

$$\alpha = \frac{I}{Al^2} \Rightarrow \left(\frac{E}{D}\right)_n = \left(\frac{4\pi^2 l^2 f_n^2}{\alpha \cdot m_n^4}\right) \tag{4.10}$$

where, n is the mode number, f is the modal frequency, I is the moment of inertia, A is the cross section area and m is the support's boundary condition constant. For the 1st and the 2nd mode of a free-free beam: m = 4.730 and 7.853, respectively, and when n > 2, is the *nth* result of the following equation [23]:

$$m_n = \pi \frac{2n+1}{2} \tag{4.11}$$

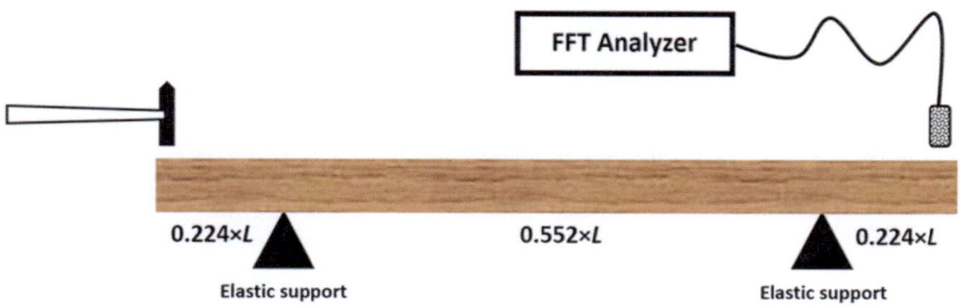

Fig. 4.13 Schematic of a free-free wooden beam undergoing flexural vibration testing. The beam is supported at 0.224 × L from each end with elastic supports, and excited by a hammer impact at one end. A microphone, typically placed at the other end, captures the sound, which is processed by an FFT analyzer to obtain the frequency spectrum. For the first mode, the microphone can also be placed at the mid-span nodal point of the 2nd mode to minimize interference

Due to the inherent influence of shear force and rotary motion on flexural deformation, the stiffness evaluated using Euler–Bernoulli beam theory consistently underestimates the stiffness determined through longitudinal vibration analysis. This discrepancy increases proportionally with the mode number.

Consequently, to minimize the impact of shear force and rotary motion errors, the flexural vibration analysis is typically limited to the first mode, with greater l/h ratios ($20 \leq l/h$, as already described for static bending), where the shear force and rotary motion effect is minimal.

Let us now consider the possible errors associated with this technology including the shear deflection error and the solutions, either realized or necessary, for their reduction or elimination.

- **Shear force and rotary motion error**: In the free flexural vibration analysis of a free-free beam, shear deformation, analogous to that observed in static bending tests, must be considered. When a beam is excited by an impact, as depicted in Fig. 4.13. The impulsive force can induce both flexural and shear wave propagation. This is because, as discussed in Chap. 3, the particle motion associated with transverse vibration is perpendicular to the wave propagation direction, inherently leading to periodical shear oscillation. While the contribution of shear deformation to the fundamental flexural mode for longer beams is typically negligible compared to bending deformation, its significance increases with higher mode numbers. As the bending deformation associated with higher modes diminishes, the relative influence of shear deformation becomes more pronounced and measurable. This phenomenon was recognized and studied by scientists such as Euler, Bernoulli, and later Timoshenko, who developed theories for the vibration of short beams where shear deformation effects are more substantial.

  With the advent of computational techniques, researchers like Bordonné [6] successfully applied Timoshenko's theory to estimate both the elastic modulus and shear modulus of wood. This was achieved by analyzing the progressive changes in apparent stiffness observed in higher flexural modes, which are directly influenced by shear deformation.
- **Viscoelastic error**: Reliable dynamic stiffness measurements necessitate minimizing the phase lag between stress and deformation, a challenge common to both longitudinal and flexural vibration techniques. Given that this phase lag is frequency-dependent, lower vibration frequencies are preferred. The first flexural mode, operating within the audible range and at frequencies considerably lower than those used in longitudinal testing, provides a practical means of achieving accurate dynamic stiffness values.

Today, standardized procedures, such as ASTM C1548-02 [1] or D6874 [2], provide guidelines for dynamic stiffness measurements based on flexural vibration analysis.

For luthiers, the accurate determination of shear modulus from higher vibration modes, while informative, is often considered less critical in the practical selection and utilization of wood for musical instruments. Therefore, this discussion will primarily focus on the Euler–Bernoulli beam theory for dynamic stiffness estimation. Readers interested in exploring Timoshenko beam theory and methods for shear modulus determination through comparative modal analysis are encouraged to consult the cited references, including Roohnia [50], Roohnia and Brancheriau [49], Roohnia and Tajdini [48], Roohnia et al. [45], Bordonné [6], and the relevant ASTM standards, C1548-02 [1] or D6874 [2].

Free flexural vibration techniques are also effectively applied to rectangular wood plates of uniform thickness, to evaluate the most important elastic constants of wood including the dynamic stiffness along the grain [12, 16, 22, 33].

While identifying vibration modes from Fast Fourier Transform (FFT) spectra generally necessitates training and experience, the task becomes significantly more complex when analyzing plate vibrations. This increased complexity arises from the intricate coupling of longitudinal, transverse bending, and torsional modes inherent in plate structures. Consequently, accurate modal identification requires meticulous control of experimental parameters. Specifically, mastery of plate excitation techniques for inducing bending vibrations, precise selection of support and sound recording locations, and refined interpretation of FFT spectra are essential. With sufficient training and practice in these areas, the determination of elastic parameters for plates can achieve a level of simplicity comparable to beam measurements.

Practical Exercise 4.6

A rectangular prismatic wood specimen, measuring 2 cm x 2 cm x 36 cm, was subjected to free vibration tests under free-free boundary conditions. Both longitudinal and flexural vibration modes were excited. The Fast Fourier Transform (FFT) spectra obtained from these tests are provided (Fig. 4.14). Given that the specimen's mass is 85 g, determine the longitudinal modulus of elasticity using both longitudinal and flexural resonance frequencies (E_l and E_f). Compare the results obtained from the two vibration techniques and discuss any observed discrepancies.

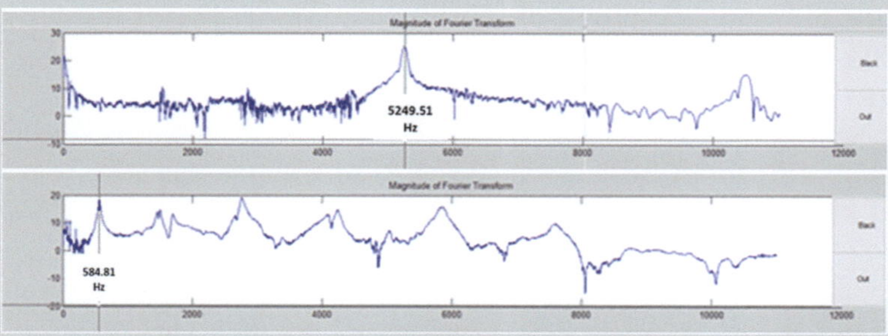

Fig. 4.14 Two graphs displaying the FFT spectrum of a rectangular beam's free vibrations. The upper graph shows the longitudinal vibration, with a peak at 5249.51 Hz representing the first mode. The lower graph shows the flexural vibration, with a peak at 551.68 Hz representing the first mode. Both graphs plot frequency (Hz) on the horizontal axis and amplitude (dB) on the vertical axis. Image from NDT-Lab® software [43]

Solution

$$D = \frac{85}{2 \times 2 \times 36} = 0.590 \times 1000 = 590\,\mathrm{kg/m^3}$$
$$\mathrm{v} = 2 \times 0.36 \times 5250 = 3780\,\mathrm{m/s}$$
$$E_l = D\mathrm{v}^2 = 590 \times 3780^2 = 8{,}430{,}156{,}000\,\mathrm{Pa}$$
$$E_l = 8.43\,\mathrm{GPa}$$

$$\alpha = \frac{I}{Al^2} = \frac{\frac{1}{12}bh^3}{Al^2} = \frac{\frac{1}{12} \times 0.02 \times 0.02^3}{0.02 \times 0.02 \times 0.36^2} \approx 2.57 \times 10^{-4}$$
$$\Rightarrow \left(\frac{E_f}{D}\right)_1 = \left(\frac{4\pi^2 l^2 f_1^2}{\alpha \cdot m_1^4}\right) \Rightarrow \left(\frac{E_f}{590}\right) = \left(\frac{4 \times \pi^2 \times 0.36^2 \times 585^2}{2.57 \times 10^{-4} \times 4.73^4}\right)$$
$$\Rightarrow E_f = 8{,}030{,}445{,}121\,\mathrm{Pa}$$
$$E_f = 8.03\,\mathrm{GPa}$$

The flexural vibration test yielded a lower modulus, a result consistent with the anticipated influence of shear deformation, which inherently reduces the apparent stiffness. Conversely, the longitudinal vibration test, produced a higher value (about 5%) for dynamic modulus. The stress wave test, which primarily induces longitudinal waves, has a minimal shear deformation.

In a free flexural vibration test, conducted on a beam with free-free ends, both radial-longitudinal (LR) and tangential-longitudinal (LT) vibrations can be utilized to estimate longitudinal elastic moduli. However, a slight discrepancy between the moduli derived from LR and LT vibrations is observed, primarily due to variations in shear deformation errors inherent to these respective planes. Given that radial planes are commonly employed in musical instrument resonators, particularly the radial (R) plane, the beam is typically excited by striking the tangential (T) direction on the radial (R) plane.

In laboratory research, a significant divergence between the dynamic moduli estimated from LR and LT flexural vibration tests serves as an indicator of potential internal hidden defect within the wood sample. Such samples are subsequently classified as structurally unsound [46].

Practical Exercise 4.7

A rectangular plate, measuring 4 mm x 218 mm x 572 mm, was subjected to free flexural vibration test under free-free boundary conditions (Fig. 4.15). The Fast Fourier Transform (FFT) spectra obtained from this test is provided. Given that the specimen's mass is 196 g, determine the

modulus of elasticity parallel to grain direction using flexural resonance fundamental frequency.

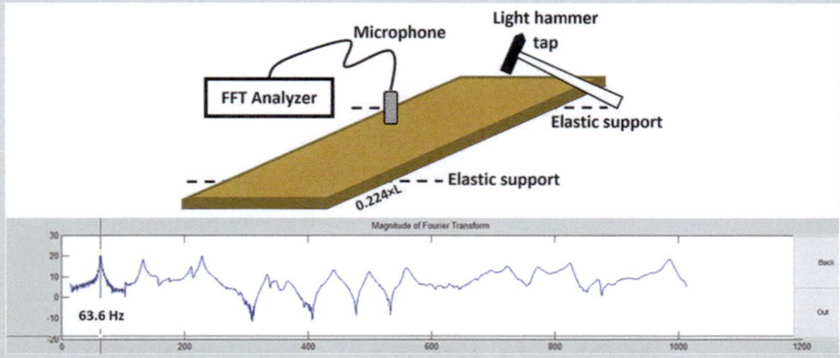

Fig. 4.15 Schematic of a free-free wooden plate undergoing flexural vibration testing along the grain direction. The plate is supported at 0.224 × L from each end with elastic supports, and excited by a hammer impact on one end. A microphone, placed at the mid-span, captures the sound, which is processed by an FFT analyzer to obtain the frequency spectrum. Graph displaying the FFT spectrum of a rectangular plates flexural vibration, with a peak at 63.6 Hz representing the first mode. Graph plots frequency (Hz) on the horizontal axis and amplitude (dB) on the vertical axis. Image from NDT-Lab® software [43]

Solution

$$D = \frac{196}{0.4 \times 21.8 \times 57.2} = 0.393 \times 1000 = 393\,\text{kg/m}^3$$

$$\alpha = \frac{I}{Al^2} = \frac{\frac{1}{12}bh^3}{Al^2} = \frac{\frac{1}{12} \times 0.218 \times 0.004^3}{0.004 \times 0.218 \times 0.572^2} \approx 4.08 \times 10^{-6}$$

$$\Rightarrow \left(\frac{E}{D}\right)_1 = \left(\frac{4\pi^2 l^2 f_1^2}{\alpha \cdot m_1^4}\right) \Rightarrow \left(\frac{E}{393}\right) = \left(\frac{4 \times \pi^2 \times 0.572^2 \times 63.6^2}{4.08 \times 10^{-6} \times 4.73^4}\right)$$

$$\Rightarrow E = 10{,}054{,}346{,}542\,\text{Pa}$$

$$E \approx 10\,\text{GPa}$$

While frequencies of 500 Hz or 60 Hz in these exercises may appear relatively low, the inherent viscoelastic nature of wood still introduces errors in stiffness estimations. Although the magnitude of these errors is reduced at these frequencies, they remain significant and acceptable. Therefore, it is suggested to adjust stiffness values obtained through various methods against those derived from modal vibration testing. Specifically:

- **Static flexural modulus adjustment**: Increase the static flexural modulus by approximately 10% to account for viscoelastic effects.

- **Dynamic modulus from longitudinal resonance**: Retain the dynamic modulus of elasticity obtained from longitudinal resonance testing of long beams with low cross section dimensions without modification.
- **Dynamic modulus from flexural resonance**: Retain the dynamic modulus of elasticity obtained from flexural resonance testing of long beams or long plates without modification or an optional increase upto 5%.
- **Ultrasonic stress wave velocity adjustment**: Reduce the dynamic modulus derived from direct ultrasonic stress wave velocity testing by approximately 15%, provided it has not been previously calibrated and reduced by the manufacturer to align with modal vibration dynamic modulus values.

These adjustments aim to achieve relative equivalence with standardized dynamic stiffness values obtained through beam modal vibration testing, such as those specified by ASTM standards, C1548-02 [1] or D6874 [2]. This ensures that results from various testing methodologies and instruments can be reliably compared.

During a free-free beam flexural vibration test, the longitudinal acoustic velocity of the specimen can be indirectly estimated using the dynamic stiffness and bulk density, as described by Eq. 4.5. This method leverages the relationship between these properties to approximate the speed of sound propagation along the longitudinal axis.

To accurately assess wood vibration in free beams and plates, luthiers can utilize reliable systems, beyond the NDT-Lab® system I previously presented in above practical exercises, which also operate under free vibration of beams or plates with free boundary conditions (Fig. 4.16).

- **Acoustical anisotropy**:

  During my early studies in wood science, Tsoumis [58] published a book introducing the concept of acoustic anisotropy, defined as the anisotropy of sound velocity along orthotropic axes. Tsoumis theorized that wood exhibiting a higher ratio of sound velocity parallel to the grain (longitudinal direction) to sound velocity perpendicular to the grain (transverse direction) is more suitable for soundboards in musical instruments. This rationale stems from the observation that longitudinal elastic stiffness parallel to growth direction, at equal density, significantly exceeds transverse stiffness perpendicular to the growth. Consequently, the ratio of longitudinal stiffness to density is substantially greater than the transverse stiffness to density (i.e. $v_L > v_R$). This implies that for optimal acoustic performance, wood should preferentially vibrate along the grain, with minimal or weaker transverse vibration contributing to unwanted noise.

Following my Ph.D. studies, I conducted independent research to empirically validate Tsoumis's theory, which yielded confirming results [47]. Based on this evidence, Tsoumis's theory, now partially validated, holds significant promise as a guiding principle for wood selection in musical instrument manufacturing. Further research and confirmation are anticipated to solidify its status as a fundamental standard. Specifically, wood species demonstrating a greater ratio of longitudinal to transverse sound velocity (radial direction for soundboards) exhibit enhanced sound

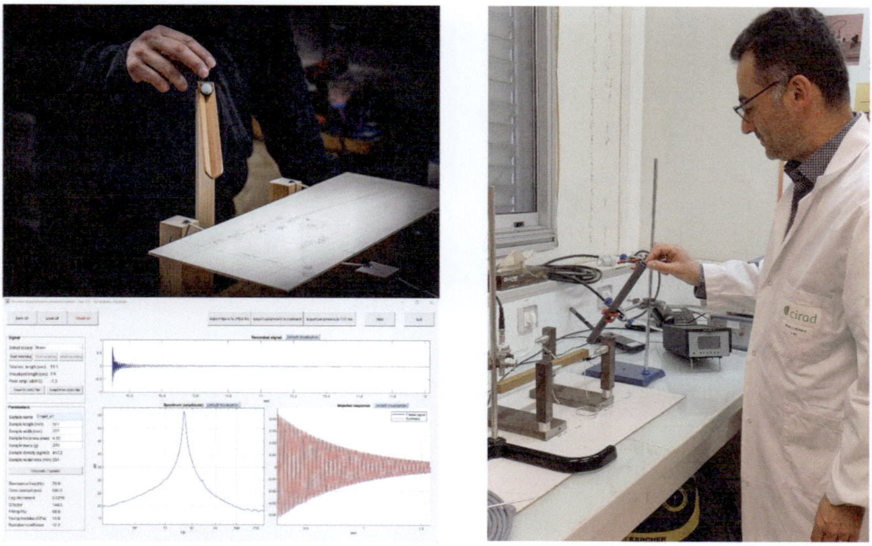

Fig. 4.16 Images showing two systems for accurately assessing wood vibration in free beams and plates. The left side shows the TPC system and software, dedicated by Mr. Giuliano Nicoletti, the right side shows the BING system and software, dedicated by Dr. Loic Brancheriau (CIRAD) to this book. Both systems operate under free vibration with free boundary conditions. The left image includes a photograph of setup and a software interface showing graphs of vibration data. The right image shows a laboratory setup with a wood rectangular beam

radiation (higher sound radiation coefficient) and/or prolonged vibrational decay (lower damping capacity).

$$Anisotropy = \frac{v_L}{v_R} \tag{4.12}$$

where, v_L is the longitudinal sound velocity in growth direction and v_R is the sound velocity in radial direction perpendicular to the growth direction.

Direct measurement of acoustic anisotropy is readily achieved using stress wave instruments equipped with piezoelectric sensors (Fig. 4.17, left). These instruments offer a straightforward approach for determining sound velocity, requiring minimal operator expertise and training. Measurements can be performed on a variety of wood forms, including plates, beams, and poles, standing trees and irregularly shaped samples.

Alternatively, experienced operators can employ modal analysis of plate bending vibrations to estimate an indirect sound velocity. By identifying and analyzing the resonant frequencies of bending modes, the sound velocity along the grain (longitudinal direction) can be determined from vibrations along the plate's length (Fig. 4.15). Similarly, the sound velocity in the radial direction (perpendicular to the grain) can be estimated from bending vibrations along the plate's width (Fig. 4.17, right). This

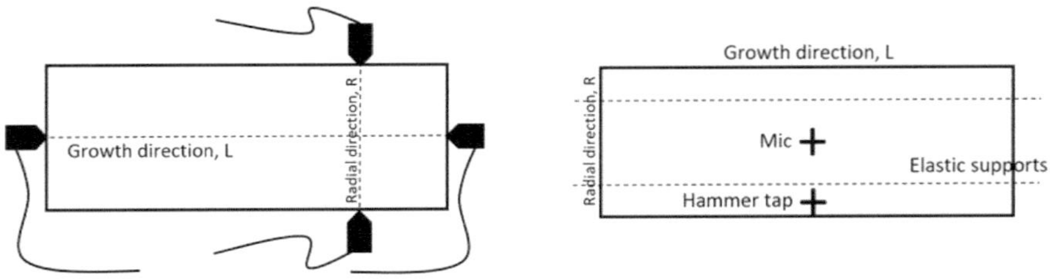

Fig. 4.17 Diagram comparing two methods for measuring acoustic anisotropy in wood. The left panel shows the direct measurement method using piezoelectric transducers to determine sound velocity along the grain (longitudinal direction, L) and perpendicular to the grain (radial direction, R). The right panel shows the indirect method using modal analysis of bending vibrations, with labels indicating microphone placement, hammer tap location, and elastic supports for measuring sound velocity along the R directions

method, however, requires significant expertise in modal identification and analysis experience (Training is essential).

4.2.3 Damping Capacity

When a piece of wood is impacted, it emits a transient sound known as a tap tone, which decays over time [50]. The damping capacity of vibrating wood and wood products quantifies the energy dissipated through internal friction. However, the presence of external resistance during vibration obscures the distinction between internal and external friction contributions to damping. Therefore, accurate measurement of internal damping capacity necessitates testing on a specimen with minimal external constraints. This approach ensures that the measured damping primarily reflects the material's inherent internal friction.

Achieving truly unconstrained vibration is challenging. Consequently, it is necessary to employ soft, thin elastic supports strategically positioned at the theoretical nodal points of vibration. At these nodal points, oscillations are theoretically minimal, effectively minimizing external damping influences. Accurate identification of these nodal points is paramount, particularly in damping capacity evaluations, as it directly impacts the reliability of the measured internal friction.

Analysis of the vibrational response of the wood plate sample from Exercise 4.7 in time domain reveals that the initial decay of the complex sound, prior to modal decomposition, exhibits a highly irregular and non-linear pattern (Fig. 4.18a). By employing the Fast Fourier Transform (FFT), a complex sound signal can be decomposed into its constituent sinusoidal components, each representing a distinct mode (Fig. 4.18b). This decomposition yields the time-varying intensity of each modal component, enabling a detailed examination of the sound decay process for individual modes. See the modal decay diagram in the time–amplitude domain, revealing a non-linear decay characteristic for the 1st mode of the complex vibration (Fig. 4.18c).

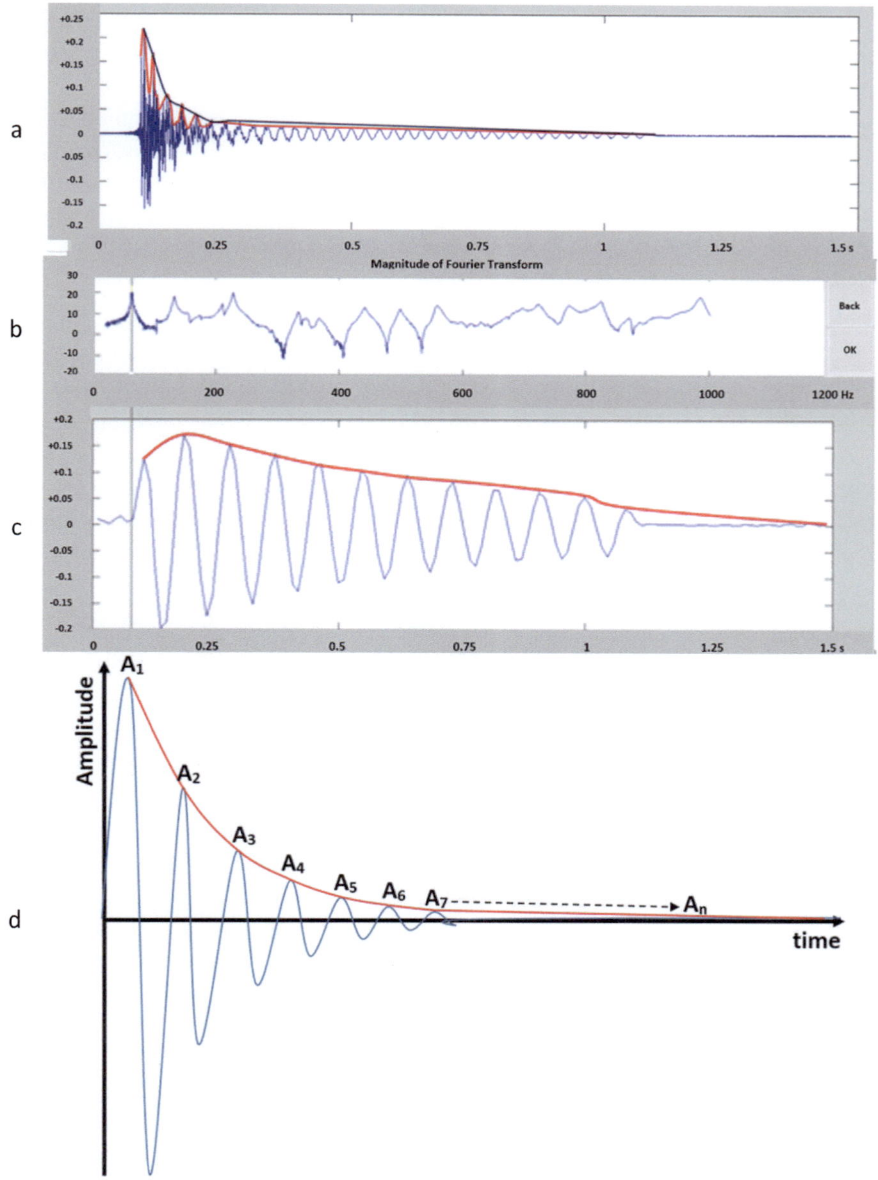

Fig. 4.18 Illustration of the analysis process for a wood plate's vibrational response. **a** The time-domain signal with irregular decay. **b** The FFT spectrum with identified modes. **c** Highlights the non-linear decay of the 1st mode. **d** An idealized decay schematic. Images from NDT-Lab® software [43]

Initially, the sound intensity (amplitude) exhibits a rapid decay rate, which subsequently decreases, indicating a time-dependent decay constant. This variation persists until the sound signal reaches complete silence.

To contextualize this empirical observation, consider the decay diagram obtained from a natural wood board, exhibiting inherent inhomogeneities due to variations in growth ring density (early and latewood), growth rate, and radial composition (from pith to bark). Even though this real-world example is simpler than the initial complex tap tone, it is still somewhat confusing when compared to the idealized schematic of regular sound decay (Fig. 4.18d).

- **The logarithmic decrement method**: While a simple definition of damping as the temporal decay of vibration intensity might seem intuitive, the process is inherently nonlinear. The rate of sound intensity attenuation is initially high and subsequently decreases asymptotically. Consequently, a linear representation of damping is more accurately achieved through a logarithmic transformation of the intensity decay. This approach yields the logarithmic decrement, a metric that effectively characterizes the rate of energy dissipation. The logarithmic decrement (δ) is defined as:

$$\delta = Ln\frac{A_i}{A_{i+1}} \tag{4.13}$$

In this equation, A, with any numerical index, i, denotes the peak height of the sine wave of vibration, as illustrated in Fig. 4.18d. The units of the peak height are not important, as the resulting fractional ratio is dimensionless. Assuming a linear trend for the Neperian logarithmic intensity decay, the attenuation rate would appear constant regardless of the peak interval chosen (e.g., first to second, second to third, etc.). However, to accurately determine the uniform linear region of the transient intensity decay, it is essential to calculate the decay rate point-by-point, peak-by-peak, across the entire decay curve (Fig. 4.18d). This detailed analysis allows for precise identification of the deviation point where the decay transitions from linear to non-linear.

Furthermore, this peak-by-peak analysis facilitates the accurate delineation of the four segments of an ADSR (Attack, Decay, Sustain, and Release) envelope (building on its introduction in Chap. 3, the ADSR envelope remains a key element in this and consequent chapters). By precisely tracking the intensity decay, the transitions between these segments can be identified, providing a more reliable characterization of the sound's temporal evolution.

Now, examine the peak-by-peak logarithmic decrement display for the same plate sample as in the previous figure (Fig. 4.19).

Recall the ADSR (Attack, Decay, Sustain and Release) envelope discussed in Chap. 3. This model provides a framework for understanding sound amplification and attenuation, particularly when viewed through Neperian logarithmic representation. However, it's necessary to note that Sometimes, especially with bowed instruments, the decay is very short or even non-existent—the sound goes

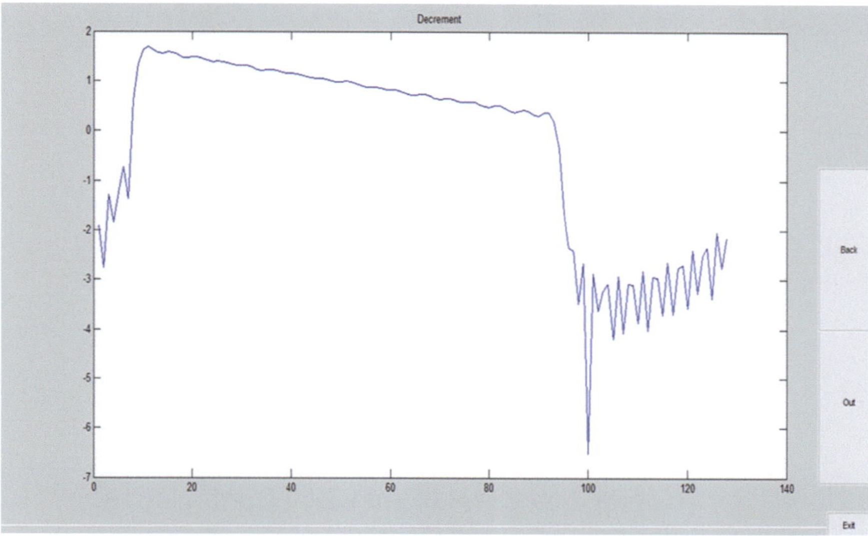

Fig. 4.19 Graph illustrating the logarithmic decrement of a vibrating system, demonstrating the decay of oscillations over time. The image highlights the use of logarithmic decrement as a measure of damping in a system. Image from NDT-Lab® software [43]

straight to the Sustain [32]. Think of a plucked string. It reaches its peak volume (the attack), then immediately starts to fade (the decay). In transiently excited vibrations, such as those produced by a single impact or stroke, the musician lacks continuous control over the sound's evolution. Consequently, the decay and sustain phases may merge, rendering them indistinguishable (Fig. 4.20). In the case of a tap tone applied to a free plate, the excitation is a singular impulse, resulting in free vibrations across the plate's natural modes. These vibrations subsequently decay and cease due to internal and external damping mechanisms. While the ADSR envelope offers a conceptual analogy, our primary focus here is on the damping capacity and logarithmic decrement of the wood e.g. a free plate.

The prior reference to the ADSR envelope is intended for illustrative purposes only. A more comprehensive analysis of ADSR envelopes, particularly in the context of complete instrument testing, will be presented in the final chapter of this book.

Returning to the peak-by-peak logarithmic decrement analysis outlined in Eq. 4.13, we observe distinct characteristics in the tone profile of a plate's first bending mode tap-tone. In the initial phase, prior to the onset of the tap-tone excitation, low-amplitude noises are evident. The low amplitude of the pre-excitation signal renders it inherently vulnerable to interference from ambient noise.

Upon initiation of the tap-tone, the signal amplitude rapidly increases, effectively masking the pre-existing noise. Subsequently, as the signal reaches its peak amplitude, the decay phase starts. This decay, occurring under uncontrolled conditions, exhibits a linear trend and transitions into the sustain region. It is again note

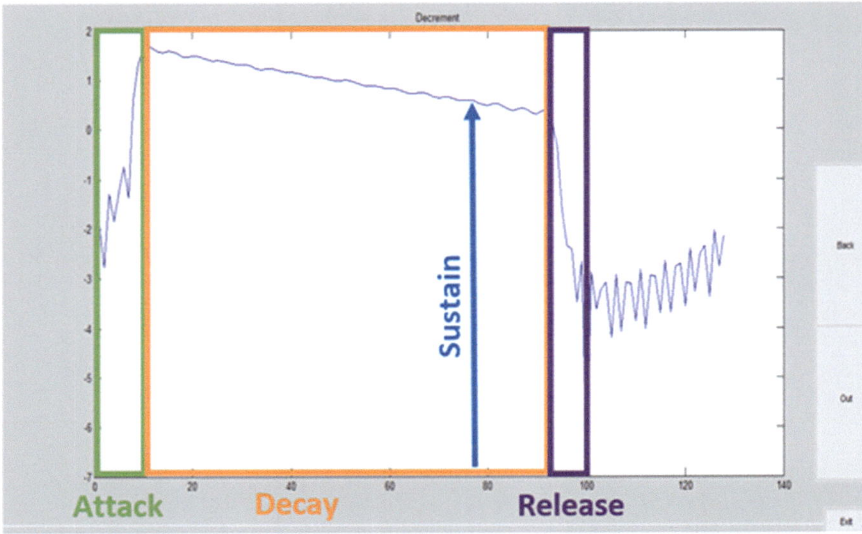

Fig. 4.20 Diagram illustrating the application of an ADSR envelope to model the logarithmic decay of sound, demonstrating how the envelope stages can represent the different phases of a decaying vibrational signal. Image from NDT-Lab® software [43]

that tap-tones, by nature, do not possess a constant sustain. Rather, the slope of this linear decay segment inversely correlates with the perceived sustain: a lower slope indicates a longer perceived sustain and a lower logarithmic decrement rate.

For accurate logarithmic decrement measurements, we focus solely on this linear decay segment. Beyond this region, the signal undergoes a rapid amplitude drop, signifying the onset of the release phase. Logarithmic decrement measurements are not conducted within the release region. Finally, as the tap-tone signal diminishes below the ambient noise floor during the release phase, extraneous environmental weaker noises once again become discernible in the measurement data.

The comprehensive analysis of vibration decay and damping for characterizing the acoustic properties of wood in luthiery, necessitates computational methods. Manual execution of these operations is impractical due to the complexity and volume of data involved. While the simplified logarithmic decrement equation is often presented as a visual representation, depicting the reduction of n sinusoidal oscillation peaks simultaneously, this approach is primarily didactic and does not reflect the actual computational procedures employed:

$$\delta = \frac{1}{n} Ln \frac{A_i}{A_{i+n}} \tag{4.14}$$

For precise logarithmic decrement measurement, the peak-to-peak method outlined in Formula 4.13 is recommended. This approach facilitates the isolation of the linear decay phase, thereby mitigating interference from other ADSR envelope components. In contrast, the direct use of Eq. 4.14, often presented in educational texts, which incorporates numerous peaks concurrently, risks introducing errors due to the inclusion of non-linear decay segments.

In practice, sophisticated software algorithms are required to accurately determine the logarithmic decrement and damping capacity from measured vibration signals. Several commercially available software packages are specifically designed for luthiers to perform these analyses very fast and accurate.

Equations 4.13 and 4.14 demonstrate that the logarithmic decrement is a dimensionless quantity. For practical reporting, the logarithmic decrement is commonly normalized by dividing it by π. This normalized value is then referred to as the damping capacity or loss tangent ($\tan\delta$).

$$\tan\delta = \frac{\delta}{\pi} \tag{4.15}$$

Damping, a frequency-dependent material property, significantly influences the vibrational behavior of wood. In the context of wooden musical instrument acoustics, where flexural vibrations are paramount, the industrial method for evaluating wood's damping capacity involves first-mode flexural vibration analysis using dedicated software. This approach minimizes the influence of higher-order modes and provides a relevant measure of damping under conditions closely resembling those encountered during musical performance.

- **Half-power bandwidth method**: The damping capacity of a material is also assessed by analyzing the bandwidth of modal resonance peaks in its amplitude-frequency response curve (Fig. 4.21).

$$\tan\delta = \frac{f_2 - f_1}{f_0} \tag{4.16}$$

To measure damping capacity, we first identify the resonance frequency (f_0), where the vibrational system has its peak amplitude [50]. Then, we locate the "half-power" points. Two criteria are used to determine the half-power point:

- A 3 dB reduction from the maximum logarithmic sound intensity (dB): This method, employed in logarithmic sound amplitude measurements, directly corresponds to halving the sound power and is used to calculate the half-power bandwidth in frequency analyses.
- Dividing the maximum sound power (W/m^2) by 2: This method is used in linear sound intensity measurements. In this context, dividing by 2 equates to halving the sound power.

The selection of the appropriate method hinges on whether the sound intensity is measured on a logarithmic (dB) or linear (W/m^2) scale. To illustrate the conversion between these linear and logarithmic sound intensity measures, this equation can be used (Eq. 4.17).

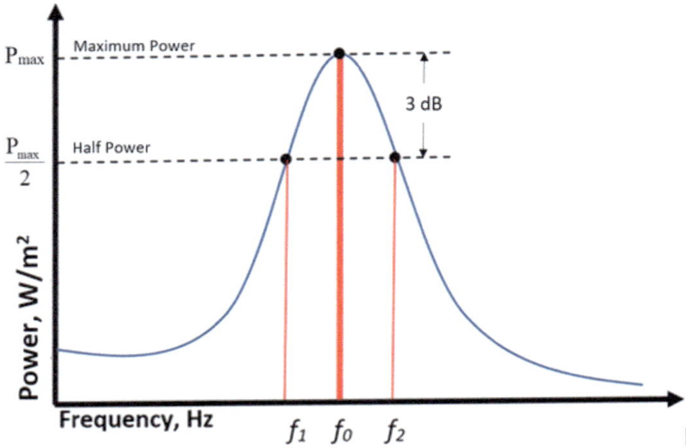

Fig. 4.21 Diagram of a resonance curve demonstrating the bandwidth method for measuring damping. The graph plots intensity (dB or W/m^2) against frequency (Hz). The peak amplitude is labeled "P_{max}" and the frequencies $f1$ and $f2$ are marked where the half power amplitude is "$P_{max}/2$." The difference between $f2$ and $f1$ represents the bandwidth, which is used to determine the damping capacity

$$dB = 10 \times \log \frac{P}{P_0} \tag{4.17}$$

where, P is the Sound power in W/m^2 and P_0 is the reference sound power, i.e. the human hearing threshold defined as 10^{-12} W/m^2. For instance, if the peak power of the first mode of a tap-tone is 0.01 W/m^2, halving this power results in a 3 dB decrease in sound intensity.

$$dB_{max} = 10 \times \log \frac{P_{max}}{P_0} = 10 \times \log \frac{0.01}{10^{-12}} = 100\,dB$$
$$dB_{halfpower} = 10 \times \log \frac{P_{halfpower}}{P_0} = 10 \times \log \frac{0.005}{10^{-12}} \approx 97\,dB$$
$$dB_{max} - dB_{halfpower} = 100 - 97 = 3\,dB$$

The bandwidth is the frequency difference between these points. The 3 dB standard arises from the logarithmic nature of decibels, which express power ratios, not absolute values. A 3 dB drop always means half the power, crucial for accurately determining energy dissipation.

As illustrated in Fig. 4.22, vibrational modes characterized by narrow peaks and low bandwidths. A narrow bandwidth indicates low damping, while a wide bandwidth signifies high damping.

Accurate determination of Bandwidth frequencies is a delicate process, fraught with potential for error. Instrumentation limitations, the proximity of other resonant modes, inconsistencies in test protocols, and inherent system nonlinearities

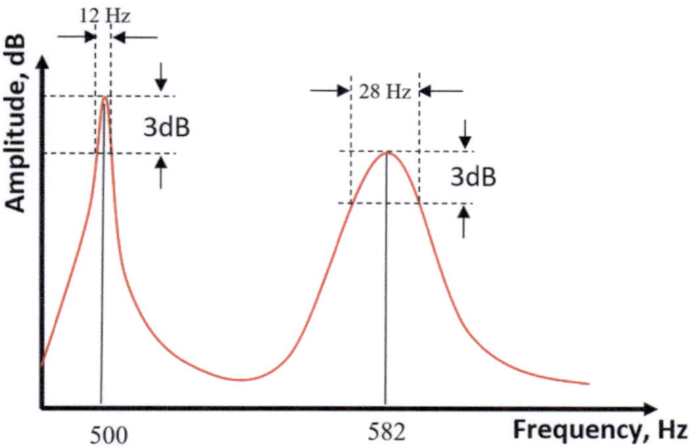

Fig. 4.22 Graph illustrating the half-power bandwidth method for comparing damping in two consecutive vibration modes. The x-axis represents frequency (Hz), and the y-axis represents amplitude (dB). Two resonance peaks are shown, one at 500 Hz with a bandwidth of 12 Hz, and the other at 582 Hz with a bandwidth of 28 Hz. Both peaks have a 3 dB drop indicating the half-power points. As defined by Eq. 4.15, a 12 Hz bandwidth implies a lower damping of 0.024, and a 28 Hz bandwidth implies a higher damping capacity of 0.048

all contribute to measurement variability. However, it is the rate of frequency increment that poses a particularly significant challenge. Exceeding a critical increment rate inevitably leads to an overestimation of damping, rendering the results unreliable [57]. Therefore, the bandwidth method is recommended solely for accurate devices and operators or for qualitative observations and relative comparisons, only.

Continuing our discussion of damping capacity, it's important to remind that the above-mentioned techniques necessitate precise computer software. Furthermore, the relationships derived from peak-by-peak logarithmic decrement or the FFT spectrum in the bandwidth method are not amenable to manual or non-computational analysis. But acquiring commercial tools and software is not always the optimal solution, particularly when cost-effectiveness is a priority. For alternative approaches, one can implement a custom damping measurement application using either the logarithmic decrement or bandwidth method. These methods, grounded in mathematical principles, ensure repeatable and accurate calculations. In such applications, errors are primarily attributed to operator input. By constructing stable support structures, accurately identifying vibration nodes for elastic supports, and meticulously collecting data for input, the software will reliably perform its intended function. Therefore, careful data acquisition is paramount for accurate results.

- **Ono and Norimoto empirical relationship**: An alternative method exists for estimating damping capacity, particularly useful for those possessing the means to measure stiffness and density of their tonewood but seeking to avoid the complex logarithmic peak-by-peak sine wave formulas. This approach leverages the empirical relationship (Eq. 4.18) between specific stiffness and damping capacity, initially proposed by Ono and Norimoto [38, 39] based on several data and subsequently refined by Brémaud et al. [8] and validated by Beauchêne et al. [3] using more data again.

$$\tan \delta = 10^{-1.23} \times \left(\frac{E}{D \times 10^6} \right)^{-0.68} \tag{4.18}$$

The stiffness, E, must be entered in pascals and the density in kilograms per cubic meter. Ono and Norimoto's regression equation, which utilized stiffness in gigapascals and density in grams per cubic centimeter, required the inclusion of a 10^6 numerical coefficient to convert the results to SI base units.

Despite the moderate coefficient of determination ($R^2 = 0.5 \sim 0.6$) for this regression relationship, the data provides evidence of a relationship between these two parameters, though not a perfect correlation. This experimental regression relationship provides an approximate estimation of damping capacity. While it is inherently an approximation, numerous studies across a wide range of wood species have consistently reported similar correlation trends, establishing it as a statistically meaningful and valid relationship. The empirical curve derived by Ono and Norimoto is recognized as a standard reference for the comparative analysis of novel experimental data [3, 8] (Fig. 4.23).

While this method for estimating damping capacity may not achieve the precision of peak-by-peak logarithmic decrement or bandwidth approach, it offers a practical method for determining wood damping capacity through manual calculation, given dynamic stiffness and density values. Although research laboratories have found this technique to yield sufficiently reasonable results, it is particularly valuable for instrument-making workshops, which often lack comprehensive laboratory controls. Studies indicate that minor discrepancies between damping capacity values obtained using the Ono and Norimoto relationship and actual measurements may be attributed to variations in the chemical composition of the wood cell wall [3, 8, 9]. Consequently, for typical, healthy woods used in musical instrument construction, the results obtained through this method are generally reliable.

However, woods with abnormal cell wall structures or chemical composition, such as compression and tension reaction wood, juvenile wood, and those exhibiting significant deviations in cellulose, hemicellulose, and lignin content, or microfibrils angles, will likely exhibit damping capacities that differ substantially from predictions based on the Ono and Norimoto relationship. Fortunately, such abnormal woods are readily identifiable and are not utilized in musical instrument construction.

Therefore, in the absence of specialized software for directly estimating damping capacity from free-free beam flexural vibration, the Ono and Norimoto relationship provides a viable and practical means for estimating the damping capacity of wood.

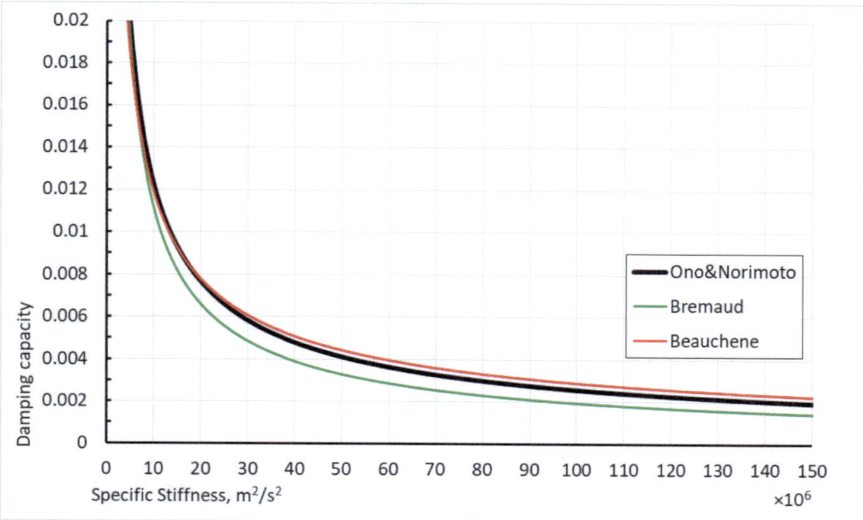

Fig. 4.23 Plot comparing experimental regression relationships for damping capacity and specific stiffness in wood. The graph shows data from three studies: Ono & Norimoto [38], Brémaud et al. [8], and Beauchêne et al. [3]. All three studies show a consistent trend of decreasing damping capacity with increasing specific stiffness, supporting the validity of this relationship for estimating damping capacity in wood

Practical Exercise 4.8

Calculate the damping capacity of the tested wood plate, using the Ono and Norimoto equation, utilizing the data and results from Exercise 4.7.

(D = 393 kg/m³ and E = 10 GPa or 10 x 10⁹ Pa).

Solution

$$\tan\delta = 10^{-1.23} \times \left(\frac{10{,}054{,}346{,}542}{D \times 10^6}\right)^{-0.68} = 0.0065 \approx 0.007$$

The damping capacity of this wood plate, determined via the peak-by-peak logarithmic decrement method of flexural vibration using NDT-Lab® software, was measured as 0.007, closely aligning with the result predicted by the Ono and Norimoto relationship, with a minor deviation. Results from both tests are seemingly identical.

Damping capacity is sometimes expressed using the quality factor, Q, which is defined as the inverse of $\tan\delta$.

4.2.4 Mechanical Impedance

In stringed instruments, mechanical impedance of the wood acts like a barrier to vibrations. Imagine when a string is plucked; it vibrates. These vibrations need to transfer efficiently to the instrument's body for a pleasant sound. Mechanical impedance is the resistance the wood offers to this transfer.

Different woods have varying impedances. High impedance wood transmits vibrations less, producing a weaker sound. Conversely, low impedance wood transmits better, resulting in a louder, more resonant sound [50].

Instrument makers select woods with suitable impedance, harmonizing it with other components for optimal sound. Factors like wood type, density, and grain direction all influence mechanical impedance.

Essentially, the wood's mechanical impedance is a key factor in determining the sound quality of stringed instruments.

In Chap. 3, we initially introduced the mechanical impedance in the context of wave reflection, using the example of two ropes with differing mechanical properties (Fig. 3.21). We noted that a wave encountering an obstacle with higher mechanical impedance experiences reflection, akin to hitting a barrier. Extending this concept, and building upon the fundamental principles of density and stiffness discussed in present chapter, we can explain wave transmission between media. When a wave on a first medium encounters a second medium, it must overcome the second medium's inertia and engage its elastic stiffness. If the impedance mismatch is not too great, a portion of the wave will propagate into the second medium, causing it to vibrate. This leads us to a more detailed discussion of mechanical impedance in acoustic systems, specifically focusing on:

- The string as a wave initiator.
- The bridge as a transmitter.
- The top-plate as a resonator.
- The back plate (resonator box) as a reflector.

Traditional explanations of resonance often emphasize the necessity of matching natural frequencies between materials. However, we propose that both frequency matching and impedance matching are essential for efficient resonance. Without impedance matching, the wave cannot effectively enter the second medium and amplify its vibrations. This implies that the transmitted intensity approaches zero when there is a significant mismatch between the mechanical impedances, z_1 and z_2. This occurs when either z_1 is substantially lower than z_2, or conversely, when z_2 is substantially lower than z_1 [14, 61]. In essence, a significant impedance mismatch, even with frequency matching, inhibits resonance.

In acoustic instruments, the vibration wave originating from the string must achieve impedance matching with the "bridge-plate" system to facilitate efficient transmission. The bridge acts as an impedance transformer, mediating between the string and the top-plate [59]. The top-plate, designed with low density and high elasticity, minimizes inertial resistance and maximizes vibrational reversibility. Finally, the back plate, with high density and elasticity, maximizes sound reflection within the resonator box (Fig. 4.24).

The bridge's role is critical. It must possess specific elasticity and density characteristics to effectively receive the string's vibration and provide impedance matching between the string and the top-plate, which are inherently different materials. This impedance matching, in effect, creates a gradual transition for the wave, enabling efficient energy transfer. These interactions of stiffness and density result in mechanical impedance. The acoustic energy transfer circuit in a stringed instrument can be summarized as follows [10, 53].

In stringed instruments, precise control of mechanical impedance across components is crucial for optimal sound production. The string, possessing its intrinsic input impedance, must efficiently transfer vibrational energy to the bridge and soundboard (top-plate), while minimizing energy dissipation at the nut and pegbox. To achieve this, the nut and pegbox are designed with high mechanical impedance, effectively reflecting the majority of the string's vibrational energy back into the string itself.

Subsequently, the bridge acts as an impedance transformer. This impedance matching normalizes reflection and optimizes transmission of vibrational energy. Additionally, airborne coupling contributes to soundboard excitation. The soundboard, driven by the string's forced frequency, resonates in its natural tap-tone modes (forced with frequency match of the input frequency), amplifying the sound.

A portion of this amplified sound radiates outward, reaching the listener. Another portion propagates within the instrument's resonant cavity towards the back plate. For effective amplification, the back plate should ideally possess high mechanical impedance, reflecting the sound waves back towards the soundboard. This superposition of reflected waves reinforces the soundboard's vibrational modes, further enhancing sound output. (However, some designs intentionally allow slight back plate vibration, necessitating a lower impedance than usual.)

The transfer of vibrational energy is governed by the mechanical impedances of the materials involved, which are determined by their internal friction, mass inertia, and stiffness. When the incoming vibrational energy surpasses these resistive forces, transmission occurs. Conversely, if the energy is insufficient, reflection predominates.

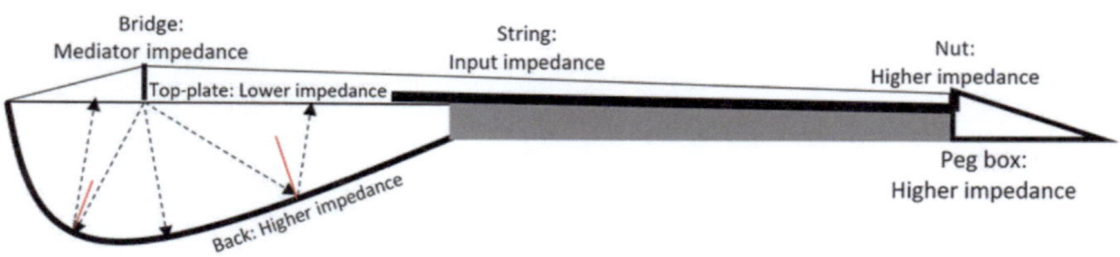

Fig. 4.24 Diagram of a hypothetical stringed instrument illustrating the flow of vibrational energy and impedance matching. The string is labeled with "input impedance." The bridge is labeled with "mediator impedance." The top-plate is labeled with "lower impedance." The back plate, nut, and pegbox are labeled with "higher impedance." Arrows indicate the direction of energy flow, with dotted lines representing internal reflections. Impedance matching facilitates efficient energy transfer and sound amplification

Therefore, understanding the functional roles of each component in the vibrational pathway is essential for optimizing instrument design.

Recall that the elastic stiffness of wood was introduced as a parameter K, the spring constant, with units of Newton per meter (N/m), or as a modulus of elasticity, parameter E, with units of Newton per square meter (Pa). Stiffness, represented by K, varies with the cross-sectional dimensions and length of the beam or plate. Specifically, thinner and longer beams or plates exhibit lower springy constants. Conversely, stiffness, represented by E, is a material property and is independent of the material's dimensions. We previously established the relationship between E and K in Eq. 2.9 of Chap. 2.

Turning to mechanical impedance, we encounter two related parameters. The first, denoted as z, represents the complex mechanical impedance that varies with the thickness and other dimensions of the wood or the string, measured in Newton-second per meter (N s/m). The second, denoted as z_0, is the acoustic characteristic impedance, a material property independent of dimensions, with units of Pascal-second per meter (Pa s/m). Characteristic impedance assumes a material of infinite bars or thin plates, simplifying the analysis. However, in complex mechanical impedance, we deal with finite dimensions and reflections, leading to a more complex system [14].

Mechanical impedance is complex, possessing both real and imaginary components, whereas acoustic characteristic mechanical impedance is purely real:

- **Complex mechanical impedance**[1] **z***: When an oscillator, such as the bridge-plate system, is driven by an external force, like a vibrating string, determining the mechanical impedance becomes crucial for assessing potential impedance matching or mismatch. Given the complex nature of mechanical impedance, consider the following:

$$z* = z' + iz'' = R + iX \tag{4.19}$$

where, R is the mechanical resistance, the real part and X is the mechanical reactance, the imaginary part of the complex mechanical impedance.

The simplified mass-spring model, even with effective mass and stiffness modifications, is inadequate for comprehensively modeling the complex behavior of continuous systems like beams or plates undergoing flexural vibration. This model inherently considers only a single vibration mode, neglecting the distributed nature of mechanical impedance, internal friction (damping), and reactance, which are dependent on spatial coordinates, mode shapes, and boundary conditions.

To achieve accurate modeling, more sophisticated approaches, such as mechanical impedance relations derived from wave equations and dynamic analyses of beams, plates, or volumes, as well as finite element calculations, are necessary.

[1] The complex mechanical impedance equations, derived for idealized mass-spring systems, exhibit limitations when applied to beams or plates with distributed loads and anisotropic timber. Consequently, while these equations are presented here, their primary purpose is to introduce the concept of impedance and its constituent components: mechanical resistance (attributed to internal friction), mass reactance, and stiffness reactance.

However, these methods are often impractical for a luthier's workshop and for luthiers without a mechanical engineering background. Therefore, the discussion here will utilize the simplified mass-spring model to introduce the concept of mechanical impedance and the various resistive forces that oppose vibration.

For those with a mechanical engineering background, finite element analysis offers a powerful tool. This method involves discretizing a volume, surface, or line into a large number of finite mass and spring elements. Solving the equations governing these elements yields the complex mechanical impedance, accounting for spatial distribution and boundary conditions (free, simply supported, fixed and etc.).

To introduce the concept of complex impedance, we will model beams and wooden plates as simplified mass-spring systems. While mass and stiffness correction methods will be applied, the primary focus remains on the pedagogical value and conceptual understanding of complex impedance. When calculating mechanical impedance, it's essential to use the effective mass (m_{eff}) and effective spring stiffness (K_{eff}), which vary depending on the vibration system.

- **Mechanical resistance, *R***: The real component (z′) of the impedance represents mechanical resistance, which is equivalent to internal friction. We have already established damping capacity as an indicator for internal friction, allowing us to calculate this real component of mechanical impedance [23].

$$R = \tan\delta\sqrt{m_{eff}K_{eff}} \tag{4.20}$$

Substituting the modulus of elasticity, E, for the spring constant, K (from Eq. 2.9):

$$R = \tan\delta\sqrt{m_{eff}\frac{E_{eff}A}{L}} \tag{4.21}$$

Given a springy constant in N/m and mass in kilograms, the mechanical resistance, which is the real part of the mechanical impedance, has the dimensions of N s/m, as previously defined.

- **Mechanical reactance, *X***: The imaginary part (z″) of mechanical impedance, represents the opposition to motion in a mechanical system caused by energy storage in components such as mass and springs, rather than energy dissipation [23]:

$$X = X_m - X_K = 2\pi f m_{eff} - \frac{K_{eff}}{2\pi f} \tag{4.22}$$

highlighting that it depends on frequency and system parameters. At resonance, where the mass reactance, X_m and stiffness reactance, X_K become equal, the reactance can be zero, but off-resonance, it determines how the system stores and releases energy, influencing motion dynamics.

By setting mass reactance equal to stiffness reactance, and examining the relationship with frequency, it demonstrates that at constant mass, increasing stiffness increases the resonant frequency. And it shows that at constant stiffness, increasing mass decreases the resonant frequency.

$$2\pi fm = \frac{K}{2\pi f} \Rightarrow f = \sqrt{\frac{K}{4\pi^2 m}} \tag{4.23}$$

Mass reactance arises from the inertia of a mechanical system, which is the resistance to changes in motion. Consequently, a greater system mass results in higher mass reactance. Furthermore, mass reactance increases with vibration frequency, signifying a greater resistance to velocity changes at higher frequencies. Mass reactance stores and releases vibrational energy as kinetic energy.

Stiffness reactance is determined by the system's elasticity, or resistance to deformation. Higher system stiffness leads to greater stiffness reactance. Conversely, stiffness reactance decreases with increasing vibration frequency, indicating reduced resistance to deformation at higher frequencies. Stiffness reactance stores and releases vibrational energy as potential energy.

As previously stated, when calculating mass reactance and stiffness, it's essential to use the effective mass (m_{eff}) and effective springy stiffness (K_{eff}), which vary depending on the vibration system. While a detailed discussion of boundary conditions and vibration types is beyond the scope of this book, for practical applications in luthiery workshops, it's important to understand that for a free-free, uniform beam undergoing in the first mode of longitudinal vibration, the effective mass is half the total mass and 0.4857 multiplying to mass for the flexural vibration [17, 56]. This result arises from the mass distribution along the beam and the first mode shape (a half-sine wave). The simplified model, which neglects effective mass and stiffness correction factors, is applicable only to a simple mass-spring system, not a distributed beam. Without accounting for the complex force distribution within the beam, mass reactance and stiffness reactance will not be equal at the natural frequency. We wish to make clear again that the mass reactance increases with vibration frequency, signifying a greater resistance to velocity changes at higher frequencies. Conversely, stiffness reactance decreases with increasing vibration frequency, indicating reduced resistance to deformation at higher frequencies. Consequently, treating such systems as simple point mass-spring systems leads to the calculation of a distinct resonance frequency where mass and stiffness reactance are indeed equal. While this calculated resonance frequency has not, to my knowledge, been utilized in wood acoustics research, I suggest its potential value warrants further investigation. It is conceivable that future applications may be identified.

A practical exercise will further illustrate this concept. It's important to note again that mechanical impedance is a complex quantity (Eq. 4.19), as represented by the following formula (Eq. 4.24).

$$z = \sqrt{R^2 + X^2} \tag{4.24}$$

Practical Exercise 4.9

A rectangular prismatic wood specimen, introduced in Exercise 4.6 **(2 cm × 2 cm × 36 cm, 85 g), was subjected to both the longitudinal and the flexural free vibration tests under free-free boundary conditions. Its 1st fundamental frequency was measured as 5249 Hz for longitudinal and 585 Hz for flexural vibration in FFT spectrum. The longitudinal modulus of elasticity obtained through longitudinal vibration and flexural as 8.43 and 8.03 GPa, respectively. Damping capacity was obtained as 0.007.**

Calculate the mechanical resistance and mass reactance as well as the stiffness reactance to build the mechanical impedance of these vibration systems, (1) with and (2) without considering the effective mass and stiffness. (3) Then calculate the simple resonance frequency without considering the effective mass and stiffness.

Solution

1. *With effective mass and stiffness*:

Flexural:
$m_{eff} = 0.4857 \times \text{m} = 0.4857 \times 0.085 = 0.0412845\ \text{kg}$
$K_{eff} = m_{eff} \times (2\pi \text{f})^2 = 0.0412845 \times (2\pi \times 585)^2 = 557{,}774\ \text{N/m}$

$$R = \tan\delta\sqrt{m_{eff}K_{eff}} = 0.007 \times \sqrt{0.0412845 \times 557{,}774} = 1.062\,\text{N s/m}$$

$$\begin{aligned} X &= 2\pi f m_{eff} - \frac{K_{eff}}{2\pi f} = 2\pi \times 585 \times 0.0412845 - \frac{557{,}774}{2\pi \times 585} \\ &= 151.75 - 151.75 = 0 \end{aligned}$$

$$z = \sqrt{R^2 + X^2} = \sqrt{1.062^2 + 0^2} = 1.062\,\text{N s/m}$$

Longitudinal:
$m_{eff} = 0.5 \times \text{m} = 0.5 \times 0.085 = 0.0425\ \text{kg}$
$K_{eff} = m_{eff} \times (2\pi \text{f})^2 = 0.0425 \times (2\pi \times 5249)^2 = 46{,}227{,}650\ \text{N/m}$

$$\begin{aligned} R &= \tan\delta\sqrt{m_{eff}K_{eff}} = 0.007 \times \sqrt{0.0425 \times 46{,}227{,}650} \\ &= 9.812\,\text{N s/m} \end{aligned}$$

$$X = 2\pi f m_{eff} - \frac{K_{eff}}{2\pi f} = 2\pi \times 5249 \times 0.0425 - \frac{46{,}227{,}650}{2\pi \times 5249}$$

$$= 1401.67 - 1401.67 = 0$$

$$z = \sqrt{R^2 + X^2} = \sqrt{9.812^2 + 0^2} = 9.812\,\mathrm{N\,s/m}$$

2. *Without effective mass and stiffness*:

m = 0.085 kg
Flexural:

$$K = \frac{EA}{l} = \frac{8.03 \times 10^9 \times 0.0004}{0.36} = 8{,}922{,}222\,\mathrm{N/m}$$

$$R = \tan\delta\sqrt{mK} = 0.007 \times \sqrt{0.085 \times 8{,}922{,}222} = 6.096\,\mathrm{N\,s/m}$$

$$X = 2\pi fm - \frac{K}{2\pi f} = 2\pi \times 585 \times 0.085 - \frac{8{,}922{,}222}{2\pi \times 585}$$
$$= 312.4 - 2427.4 = -2115\,\mathrm{N\,s/m}$$

Surprisingly, at the resonance frequency of the first flexural mode, the mass and stiffness reactances are not equal. In this relatively lower frequency, the mass reactance is smaller while the stiffness reactance looks greater. We anticipate that with increasing frequency, the mass reactance will rise while the stiffness reactance will fall.

Longitudinal:

$$K = \frac{EA}{l} = \frac{8.43 \times 10^9 \times 0.0004}{0.36} = 9{,}366{,}667\,\mathrm{N/m}$$

$$R = \tan\delta\sqrt{mK} = 0.007 \times \sqrt{0.085 \times 9{,}366{,}667} = 6.246\,\mathrm{N\,s/m}$$

$$X = 2\pi fm - \frac{K}{2\pi f} = 2\pi \times 5249 \times 0.085 - \frac{9{,}366{,}667}{2\pi \times 5249}$$
$$= 2803 - 284 = 2519\,\mathrm{N\,s/m}$$

Surprisingly, at the resonance frequency of the first longitudinal mode, the mass and stiffness reactances are not equal. In this relatively higher frequency, the mass reactance is greater while the stiffness reactance looks smaller. We anticipate that with decreasing frequency, the mass reactance will fall while the stiffness reactance will rise, until they react to equal values!

3. *Resonance frequency without considering the effective mass and stiffness*:

Mass Reactance = Stiffness Reactance

$$2\pi fm = \frac{K}{2\pi f}$$

$$f^2 = \frac{K}{4\pi^2 m} \Rightarrow f = \sqrt{\frac{K}{4\pi^2 m}}$$

Flexural:

$$f = \sqrt{\frac{8{,}922{,}222}{4\pi^2 \times 0.085}} = 1631\,\text{Hz}$$

Longitudinal:

$$f = \sqrt{\frac{9{,}366{,}667}{4\pi^2 \times 0.085}} = 1671\,\text{Hz}$$

These frequencies are currently theoretical. If we model the vibrating system as a simplified mass-spring system, we obtain these numerically similar frequencies, within a certain tolerance. The observed difference between these two frequency values arises from variations in the included stiffness, a consequence of the inherent errors discussed earlier regarding longitudinal and flexural free vibration tests. Understanding these theoretical frequencies is beneficial for enhancing clarity and may prove valuable in future analyses.

By applying correction factors for mass and stiffness, the distributed beam system is effectively normalized, allowing the mechanical impedance of the vibrating material to be calculated across varying frequencies using simplified mass-spring relationships. These correction factors ensure that, at the natural resonance frequency, the mechanical reactance becomes zero, increasing with deviations from this frequency. Consequently, the system opposes vibration at frequencies other than resonance.

Practical Exercise 4.10

For the rectangular plate, introduced in Exercise 4.7, the damping capacity were 0.007. Calculate (a) the mechanical resistance, mass reactance and stiffness reactance. Given that a vibration with a frequency of 82 Hz, corresponding to the second octave note Mi (E_2), is transmitted from a string to the plate, (b) calculate the mechanical impedance of the plate against this frequency.

$E = 10$ **GPa,** $\tan\delta = 0.007$, $m = 196$ **g,** $D = 393$ **kg/m^3,** $f = 63.6$ **Hz**

Solution

a:

$$m_{eff} = 0.4857 \times 0.196 = 0.095\ \text{kg}$$
$$K_{eff} = 0.095 \times (2\pi \times 63.6)^2 = 15{,}170\ \text{N/m}$$

$$R = \tan\delta\sqrt{mK} = 0.007 \times \sqrt{0.095 \times 15{,}170} = 0.266\,\text{N s/m}$$

$$X = 2\pi fm - \frac{K}{2\pi f} = 2\pi \times 63.6 \times 0.095 - \frac{15{,}170}{2\pi \times 63.6}$$
$$= 37.96 - 37.96 = 0\,\text{N s/m}$$

$$z = \sqrt{R^2 + X^2} = \sqrt{0.266^2 + 0^2} = 0.266\,\text{N s/m}$$

b:

$$X = 2\pi fm - \frac{K}{2\pi f} = 2\pi \times 82 \times 0.095 - \frac{15{,}170}{2\pi \times 82}$$
$$= 48.94 - 29.44 = 19.5\,\text{N s/m}$$

$$z = \sqrt{R^2 + X^2} = \sqrt{0.266^2 + 19.5^2} = 19.5\,\text{N s/m}$$

While the complex calculations involved in determining the precise acoustic properties of anisotropic wood, often plagued by inherent computational errors and requiring extensive further research, may seem impractical for direct application in a luthier's workshop, they provide a crucial theoretical foundation. Understanding the interplay between mechanical resistance, mass reactance, and stiffness reactance, and their frequency-dependent behavior, is essential for a luthier to intuitively grasp the instrument's response.

For instance, when the frequency of a vibrating string matches a vibrational mode of the instrument's plate, minimal reactance occurs, facilitating optimal resonance. In this scenario, the internal friction of wood becomes the primary factor governing vibration damping. Therefore, pre-calculating the wood's damping capacity at resonance frequencies helps mitigate reactance-related errors.

As will be detailed in Chap. 5, this understanding enables impedance matching through strategic adjustments to plate thickness at specific nodal and antinode points. For example, the soundboard's acoustic impedance is directly proportional to the square of its thickness. Consequently, soundboards of substantial thickness, such as those found in pianos, exhibit significantly higher impedance compared to the vibrating strings. To achieve optimal sound quality, precise impedance matching between the strings and the soundboard is therefore essential [50, 61]. Modifying mass and stiffness (springy constant) allows for targeted impedance control, enhancing or inhibiting vibration at specific frequencies. This is necessary for achieving frequency and impedance matching between the string and plate via the instrument's carving.

Conversely, when the string frequency deviates from a plate mode, the plate exhibits reactance. Consequently, a plate with a greater number of vibrational modes, as assessed through tap-tone analysis, minimizes mechanical reactance issues. Notably, both increased mass and stiffness contribute to increased reactance. However, by manipulating mass and stiffness (springy constant), the plate's impedance can be tailored for specific frequencies. The combined effect of these changes is frequency-dependent, emphasizing the importance of identifying nodal and antinode patterns for each vibrational mode, a topic explored in detail under "tap-tone" in Chap. 5. Generally, mass adjustments at antinodes and stiffness modifications near nodes have a greater impact on impedance at the corresponding frequency.

Moreover, manipulating mass and stiffness in one mode can inadvertently alter other modes due to nodal and antinode interference. Therefore, practical experience complements theoretical knowledge, a concept further elaborated upon in Chap. 5.

- **Acoustic characteristic impedance, $\mathbf{z_0}$**: This parameter, denoted as z_0, is the acoustic characteristic impedance, an intrinsic material property independent of geometric dimensions and possessing only a real component. It is expressed in units of 'rayls' or Pascal-second per meter (Pa s/m). The rayl is named after Lord Rayleigh and is used to quantify acoustic characteristic impedance. Characteristic impedance is defined for a material modeled as an infinite medium, such as an infinitely long bar or an infinitely large thin plate, thereby simplifying analytical considerations.

  Analogous to the speed of sound, the acoustic characteristic impedance is directly proportional to the square root of the modulus of elasticity (E) and the density of the material. Specifically, the characteristic impedance (z_0) of a material, such as wood, is defined as the product of the material's speed of sound (v) and its density (D):

$$z_0 = \mathrm{v} \cdot D = \sqrt{E \cdot D} \tag{4.25}$$

  The acoustic impedance of wood, ranging from 20×10^5 to 37×10^5 Pa s/m, is significantly lower than that of steel (395×10^5 Pa s/m) and cast iron (258×10^5 Pa s/m) [30, 50]. This substantial impedance mismatch necessitates impedance matching strategies between the wires and the wooden plate. Techniques to achieve this include selecting an appropriate bridge design and adjusting the thickness and dimensions of both the plate and the strings.

Practical Exercise 4.11

For the rectangular plate, introduced in Exercise 4.7, calculate the acoustic characteristic impedance of the wood. E = 10 GPa, tan δ = 0.007, m = 196 g, D = 393 kg/m^3, f = 63.6 Hz.

To further explore the acoustic properties of wood, let's consider a scenario where a vibrating steel string is used in conjunction with this wood plate. The steel string has a Young's modulus of 175 GPa and a density of 8800 kg/m^3. Calculate the characteristic impedance of this steel string and compare it to that of wood.

Solution

Wood plate:

$$z_0 = \sqrt{E \cdot D} = \sqrt{10 \times 10^9 \times 393} = 1,982,423\,\text{Pa s/m}$$

String:

$$z_0 = \sqrt{E \cdot D} = \sqrt{175 \times 10^9 \times 8800} = 39,242,834\,\text{Pa s/m}$$

The significant disparity in characteristic acoustic impedance between steel string and the wood used in this exercise presents a substantial impedance mismatch. For efficient acoustic energy transfer, impedance matching techniques are required to reduce this difference. Further details on impedance matching can be found in Chap. 5.

Exercise 4.11 highlights the importance of achieving impedance matching between a string and a wooden plate. While the characteristic impedance formula suggests material changes as a primary solution, leveraging finite element analysis reveals alternative strategies through mechanical modifications to either the string or the wood.

Given that most luthiers do not possess a background in mechanical engineering, the following observations, derived from finite element analysis, outline the effects of various factors on mechanical impedance, facilitating practical applications of impedance matching:

- **Boundary conditions (from free to simply supported, then fixed)**:
  - Mechanical resistance due to the internal friction (R): Increases (due to greater constraint and higher damping due to supports friction).
  - Mass reactance (X_m): Decreases (due to reduced effective displacement).
  - Stiffness reactance (X_K): Increases (due to increased system springy constant).
  - **Total impedance (z)**: Increases (due to higher springy constant and damping).

- **Moving away from a vibration node and towards a vibration antinode**:
  - Mechanical resistance due to the internal friction (R): Increases (due to greater motion and higher energy dissipation).
  - Mass reactance (X_m): Increases (due to greater mass displacement).
  - Stiffness reactance (X_K): Decreases (due to reduced stiffness effect at the antinode).
  - **Total impedance (z)**: Varies depending on the mode, generally decreases at the antinode.
- **Increasing the length of the vibrating material**:
  - Mechanical resistance due to the internal friction (R): Increases (due to a larger surface area for energy dissipation).
  - Mass reactance (X_m): Increases (due to greater mass).
  - Stiffness reactance (X_K): Decreases (due to a decrease in natural frequency and lower springy constant).
  - **Total impedance (z)**: Decreases (due to lower frequency).
- **Increasing the thickness of the vibrating material**:
  - Mechanical resistance due to the internal friction (R): Increases (due to greater mass and surface area for damping).
  - Mass reactance (X_m): Increases (due to greater mass).
  - Stiffness reactance (X_K): Increases (due to greater bending stiffness).
  - **Total impedance (z)**: Increases (due to higher stiffness and mass).
- **Increasing the width of the vibrating material**:
  - Mechanical resistance due to the internal friction (R): Increases (due to a larger surface area for dissipation).
  - Mass reactance (X_m): Increases (due to greater mass).
  - Stiffness reactance (X_K): Increases (due to greater bending springy constant).
  - **Total impedance (z)**: Increases (due to greater mass and springy constant).

These changes are summarized in Table 4.3.

Mechanical impedance, z, is sometimes expressed using the mechanical admittance, Y or Y_0, which is defined as the inverse of z or z_0.

Furthermore, based on the following equations, when two media in contact (z_1 and z_2) possess equal mechanical impedance, wave transmission occurs from z_1 to z_2 without reflection, and the wave propagation continues uninterrupted. Conversely, if z_2 is significantly greater than z_1, or vice versa, almost all of the input vibration intensity from z_1 is reflected back, with only a small fraction entering z_2. This phenomenon, previously illustrated in Chap. 3 using a rope analogy, is now mathematically described by Fletcher and Rossing's [14] equations.

Table 4.3 Detailing how boundary conditions, vibration location, and material dimensions affect mechanical resistance, mass reactance, stiffness reactance, and total mechanical impedance*

Factor		Mechanical resistance (R)	Mass reactance (X_m)	Stiffness reactance (X_K)	Total impedance (z)
Boundary conditions	Free to simply supported	↑**	↓***	↑	↑
	Simply supported to fixed	↑	↓	↑	↑
Vibration location	Node to antinode	↑	↑	↓	Varies (usually ↓)
Material dimensions	Increase length	↑	↑	↓	↓
	Increase thickness	↑	↑	↑	↑
	Increase width	↑	↑	↑	↑

* These effects are general and directional, and their precise magnitude depends on the mode shape, frequency, and material properties. Finite element analysis can show these changes with greater accuracy, but the overall trend remains the same

** Increase ↑

*** Decrease ↓

$$\begin{aligned}\frac{I_r}{I_1} &= \left(\frac{z_2 - z_1}{z_2 + z_1}\right)^2 \\ \frac{I_t}{I_1} &= \frac{4z_2 z_1}{(z_2 + z_1)^2}\end{aligned} \tag{4.26}$$

where, I_1 is the incident wave intensity, I_r is the reflected wave intensity, I_t is the transmitted wave intensity in W/m.

It is important to clarify that the concept of impedance matching between two adjacent vibrating media pertains to mechanical impedance. However, in 2006, Wegst presented this concept under the term 'acoustic characteristic impedance.' While these two impedance parameters are related, the rope example in Chap. 3 demonstrates the significant influence of media dimensions and weight on vibration transmission. Notably, the primary reference for the vibration transmission equations refers to mechanical impedance, which is dependent on the dimensions of the medium, not acoustic characteristic impedance, which is independent of these dimensions [14].

4.2.5 Acoustic Radiation Coefficient

The concept of acoustic or sound radiation, specifically concerning vibrating surfaces like wooden soundboards in musical instruments, involves a complex interplay of material properties and vibrational behavior. At its core, the sound radiation coefficient ('*rc*', while also denoted as 'σ', we use '*rc*' to avoid confusion with stress), quantifies how efficiently a vibrating surface converts its mechanical energy into sound waves that propagate through the surrounding medium, typically air [52].

Unlike the simple material properties, i.e. density (D) and modulus of elasticity (E), the sound radiation coefficient *rc* is a measure, highly dependent on factors beyond just the material itself. Frequency of vibration, the geometry of the vibrating surface, and the specific mode of vibration all play significant roles. Consequently, a single, universally applicable formula that directly incorporates wood density (D) and modulus of elasticity (E) to calculate *rc* could be inappropriate.

However, D and E profoundly influence *rc* indirectly by shaping the vibration characteristics of the soundboard. Wood stiffness, as defined by the modulus of elasticity (E), dictates its capacity for potential energy storage and, consequently, its vibrational behavior across different frequencies. Higher E translates to greater stiffness, influencing the resonant frequencies and vibration modes. Density (D) determines the mass of the piece of wood, creates the inertia of a mechanical system, which is the resistance to changes in motion and the interplay between mass and stiffness governs the natural frequencies of the board. These natural frequencies are crucial because they determine which frequencies are amplified and radiated most effectively.

The speed of sound within the wood, calculated as the square root of the E to D ratio, further influences wave propagation through a wood plate. How vibrational waves travel and interact with the edges of the board significantly affects the board's radiation efficiency. Optimal radiation occurs when the vibration wavelength in the air is comparable to or smaller than the dimensions of the radiating surface. The resonant frequencies and vibration modes, determined by the energy storage from E and the inertia from D, influence how well this condition is met.

While a precise calculation of *rc* necessitates intricate analysis, simplified models offer valuable insights. For instance, the simplified radiation coefficient for a piston in an infinite baffle provides a basic understanding of how surface size and vibration frequency impact radiation. However, real-world wood plates e.g. soundboards are complex shapes, and their vibration patterns deviate significantly from simple piston-like motions.

In musical instrument acoustics, a simplified factor, K (not to be confused with springy constant), is often used to compare the relative radiation efficiency of different materials. One might wonder why this factor is commonly referred to as the Sound Radiation Coefficient by luthiers, tonewood suppliers, and even wood scientists.

$$K = \sqrt{\frac{E}{D^3}} = \frac{\mathrm{v}}{D} \tag{4.27}$$

This formula provides a quick estimate of how effectively a material radiates sound based on its stiffness and density. Its unit is m^4/kg s, and can almost be called as the '*specific sound velocity*' in wood. While not directly related to the scalar sound radiation coefficient, it remains useful. Materials with higher K values are generally considered to have better sound radiation. However, it's crucial to remember that this is a simplification and doesn't account for the intricate dynamics of sound radiation.

Ultimately, understanding sound radiation involves appreciating the complex interplay of material properties, vibrational behavior, and wave propagation. While direct formulas for rc are complex, the concepts of stiffness, density, resonant frequencies, and wave propagation provide a solid foundation for comprehending how vibrating surfaces transform mechanical energy into sound.

Practical Exercise 4.12

For the rectangular plate, introduced in Exercise 4.7, calculate the sound radiation coefficient, K.

$E = 10$ GPa, $\tan\delta = 0.007$, $m = 196$ g, $D = 393$ kg/m³, $f = 63.6$ Hz.

Solution

$$K = \sqrt{\frac{E}{D^3}} = \sqrt{\frac{10 \times 10^9}{393^3}} = 12.8\,\mathrm{m^4/kg\,s}$$

It would be acceptable to be utilized as the top-plate (based on Table 4.1).

Soundboards in some of the most famous Stradivari violins achieve K-values of 16 and higher, a metric that surpasses the maximum of 12–15 typically observed in even the highest quality modern violins [36]. As the wooden plate in Exercise 4.12 falls within this range, it is suitable for a soundboard.

Notably, some references, such as Wegst [61], consider the sound radiation coefficient as a damping mechanism related to sound radiation. This is distinct from damping caused by internal friction. The coefficient's potential connection to mass and stiffness reactances might explain its classification as a form of damping.

4.2.6 *Acoustic Conversion Efficiency*

Notably, some references, such as Wegst [61], consider the sound radiation coefficient as a damping mechanism related to sound radiation. This is distinct from damping caused by internal friction. The coefficient's potential connection to mass and stiffness reactances might explain its classification as a form of damping.

Now, whatever the reason for calling it a form of damping, when considering sound radiation as a mechanism for vibration attenuation, alongside damping due to internal friction, it's essential to acknowledge the coupled interaction of these two damping types within the wood used in musical instrument components. Modifying one form of damping can influence the other. Consequently, focusing solely on either sound radiation or internal friction damping while neglecting the other can lead to inaccurate predictions of vibration behavior.

The acoustic performance of wood in musical instruments hinges on a delicate balance between energy dissipation mechanisms. Ideally, a sounding board should efficiently convert vibrational energy into radiated sound, minimizing internal energy losses. This principle can be quantified through the acoustic conversion efficiency (*ACE*), a metric that reflects the material's ability to transform mechanical vibration into acoustic output.

The *ACE* is influenced by the interplay between radiation damping and internal friction. High radiation damping facilitates the transfer of vibrational energy into sound waves, while low internal friction minimizes energy dissipation within the wood's microstructure. This relationship is often expressed as:

$$ACE = \frac{K}{\tan\delta} \tag{4.28}$$

where, *ACE* is Acoustic conversion efficiency in m^4/kg s, and *K* is radiation efficiency in m^4/kg s, reflecting the material's ability to radiate sound. $\tan\delta$ is the loss tangent, representing internal friction within the material.

High-quality tonewoods, suitable for resonators, exhibit *ACE* values exceeding 2000 m^4/kg s [50, 61].

However, for idiophones (e.g., xylophones), where sound generation relies on the direct vibration of the instrument body, *ACE* is less critical. Instead, optimal performance depends on a combination of high modulus of elasticity, high density, high hardness (defined in Chap. 2), and low internal friction.

Furthermore, it was already stated that the anisotropy of wood, characterized by varying sound velocities along longitudinal, radial, and tangential directions, significantly impacts acoustic performance. Sound velocity ratios between these directions serve as indicators of acoustical efficiency, which are directly related to the radiation coefficient, internal friction and acoustic conversion efficiency.

Material defects and environmental degradation negatively affect *ACE*. Knots, while increasing density, disrupt grain alignment, reducing the radiation coefficient and consequently lowering *ACE*. Discontinuities such as checks or breaks compromise the modulus of elasticity, impacting vibrational modes and reducing *ACE*. Fungal decay weakens the wood's structural integrity by degrading cellulose and lignin, leading to decreased modulus of elasticity and increased internal friction, thereby diminishing *ACE*.

Environmental conditions significantly influence acoustic properties. Under typical indoor conditions (e.g., room temperature and 8% moisture content), wave dissipation due to sound radiation is maximized, while internal friction is minimized, optimizing the acoustic performance of wooden musical instruments. This optimal environment promotes a high acoustic conversion efficiency.

4.2.7 Timbre and Sound Color

The concepts of timbre and sound color, representing the unique character and signature of a sound, were thoroughly discussed for vibrating strings in Chap. 3. In this chapter, we will extend that discussion to examine these features in wood, specifically in planar and volumetric forms.

The parameters previously presented regarding wood, while sometimes simplified, should be consistently considered when selecting wood for any stringed musical instrument, regardless of whether it's a guitar, violin, Persian setar, or Chinese pipa. However, the specific manner in which these parameters influence the final sound composition of each instrument is the essence of timbre or sound color. This can vary significantly depending on the particular piece of wood.

As previously stated, the soundboard is after the string, the primary sound-producing component of an instrument. Among bending, longitudinal, and torsional vibrations, bending vibrations are the most dominant for sound generation [14]. Therefore, we will primarily focus on bending vibrations. However, we will also consider other vibration modes, as they contribute to the overall tonal characteristics of the soundboard. Understanding the sound spectrum and vibration modes is essential for shaping the soundboard's unique tonal color and character.

In a rectangular beam undergoing bending vibration, the primary particle oscillation is constrained to the direction of excitation. For instance, if a beam is struck on its radial surface, precisely along the LT (longitudinal-tangential) plane, the unique geometry of the beam dictates that wood particle oscillation will occur predominantly in the tangential direction, with the resulting wave propagating along the beam's length. This aligns with the fundamental principle of transverse (bending) vibrations: particle oscillation is perpendicular to the direction of wave propagation. Due to the beam's form factor, alternative wave propagation modes are effectively suppressed (Fig. 4.25).

Conversely, modal analysis of a plate subjected to an impact reveals a more complex scenario. Excitation of the plate's radial surface can induce vibrations along

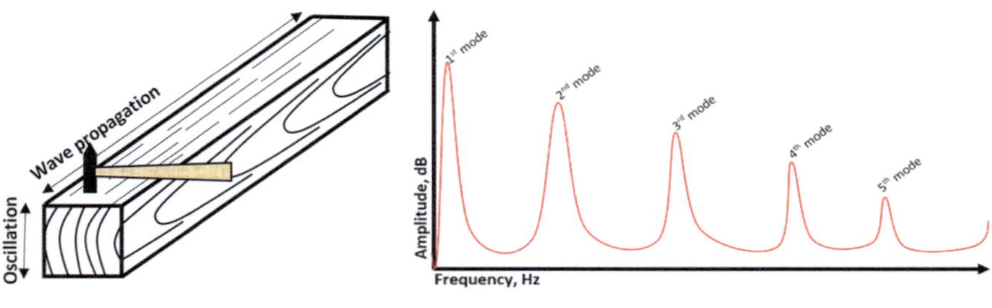

Fig. 4.25 Illustration of a tap-tone experiment on a wood beam. The diagram shows the hammer tap exciting oscillation in the tangential direction and wave propagation in the longitudinal direction. The FFT spectrum shows the first five flexural vibration modes, which contribute to the timbre of the sound

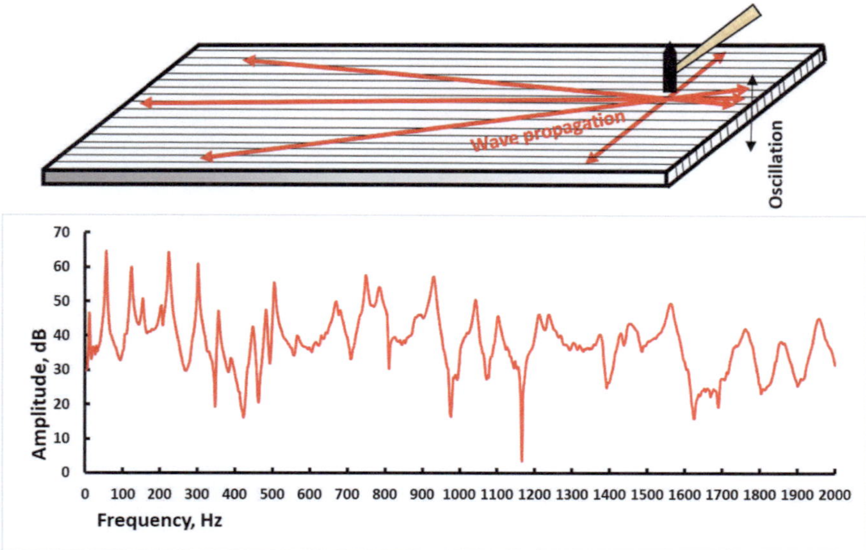

Fig. 4.26 Diagram depicting a tap-tone test on the radial surface of a spruce free plate (572 × 216 × 4.07 mm, 210 g). The image shows a hammer tap exciting tangential particle oscillation, with arrows indicating complex wave propagation pathways of varying lengths. An accompanying FFT spectrum displays a mixed pattern of vibration modes, contributing to the timbre of the tap-tone. The axes of the FFT spectrum are labeled "frequency, Hz" and "amplitude, dB"

its length, width, and diameter, generating a multitude of wavelengths and frequencies. The superposition of these waves, governed by linear superposition principles, results in a complex sound signature. This signature can be effectively characterized and analyzed in the frequency domain using Fast Fourier Transform (FFT) techniques (Fig. 4.26).

In each vibrational mode, the mechanical reactance is minimized and the mechanical impedance of the vibrating medium is solely determined by its internal friction. Provided external constraints, such as support friction or other inhibiting factors, do not impede the vibration at the nodal points, the medium can sustain prolonged oscillations in that specific mode. Resonance occurs when the frequency of an externally applied force matches one of these natural modes. A vibrating plate with a greater number of natural frequency modes closely aligned with external driving frequencies exhibits a higher likelihood of resonance. However, the presence of undesired resonances, arising from frequency interference, can disrupt the desired tonal qualities. This phenomenon is analogous to attempting to maintain a consistent swing rhythm when an external force, applied at an incompatible frequency, interferes. In instrument making, it is essential to selectively dampen unwanted resonant frequencies while enhancing those contributing to the desired tonal range. The soundboard after it mounted efficiently must effectively resonate the fundamental frequencies of the strings and their overtones, minimizing extraneous noise. This aspect will be further explored in Chap. 5, focusing on wood treatment, modification, and tap-tuning.

Prior to any treatment, it is essential to select a wood sample possessing inherent acoustic properties suitable for resonating the desired frequencies. This selection should be based on the wood's stiffness, density, and initial timbre, which exhibits limited modifiability. Referencing Table 4.3 will aid in planning effective modifications (discussed in Chap. 5).

In tap-tone analysis, the dominant frequency, characterized by the highest sound pressure level (decibels), is typically considered the plate's fundamental resonance (some luthiers refer to this as the 'plate pitch' and attempt to tune it to a desired note). This fundamental resonance often corresponds to the first flexural vibration mode in growth direction, which generally produces a louder sound than other modes due to its greater elastic stiffness. To accurately identify this resonance, careful consideration must be given to support placement, impact location, and listening position.

To ensure the accurate observation of the first flexural vibration mode during tap-tone analysis, supports (or fingers) should be positioned to minimize external friction. For a rectangular plate, this typically involves placing supports at approximately 0.224 times the plate's dimension from each edge, as detailed in Table 4.2 and illustrated in Fig. 4.27.

In other words, to accurately observe and analyze the fundamental (first) mode of flexural vibration during a tap-tone test, proper support placement is crucial. For flexural vibrations (in L-direction), the plate should be supported along the blue dashed line. This minimizes constraints on the L-direction mode. Conversely, for flexural vibrations in the width (b-direction) direction, supports should not extend beyond the red dashed line, preventing undue damping of the b-direction mode.

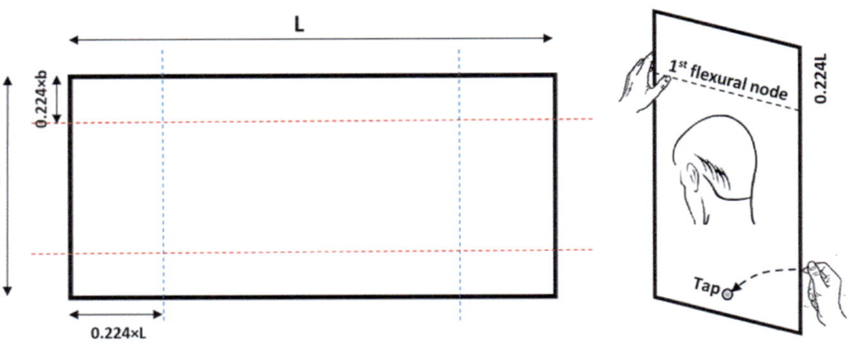

Fig. 4.27 Diagram illustrating proper support placement for tap-tone testing of a rectangular plate. The plate is labeled with length (L) and width (b). Blue dashed lines indicate support locations for minimizing constraints on the L-direction mode (0.224L from each end). Red dashed lines indicate support limits for minimizing damping of the b-direction mode (0.224b from each end). Supporting the plate at the intersection of these lines allows for simultaneous observation of fundamental modes in both directions

Ideally, supporting the plate with two fingers soft tip at the intersection of the blue and red dashed lines allows for the simultaneous excitation and observation of the fundamental modes in both the growth and perpendicular to growth directions.

For the analysis of higher-order vibrational modes in a free plate tap-tone, the nodal lines (represented by the dashed lines) specific to each mode must be identified. Refer to Table 4.2 to determine the nodal line locations for the desired mode in both the L and b directions.

Imagine tapping a wooden plate. If it makes a rich, clear sound that lasts a long time, it's considered to have a beautiful tone. This happens when the plate vibrates in many different ways (resonant modes), those vibrations don't cancel each other out (low interference), and the main vibrations fade away slowly (prolonged decay).

Natural frequencies, also known as resonant modes of vibration, are the frequencies at which a vibrating medium experiences minimal impedance. This occurs when the stiffness reactance and mass reactance are equal and opposite, resulting in their cancellation and a net reactance of zero. In essence, the distribution of stiffness and density throughout the material, in various directions, dictates its natural frequencies and associated mode shapes.

Wood, being an anisotropic and heterogeneous material, possesses a complex array of natural frequencies. This characteristic renders it suitable for producing resonant sounds with rich tonal complexity, provided its stiffness and density distribution facilitates the creation of a multitude of frequencies within the operating range of the instrument's strings. These frequencies should align with the fundamental frequencies and overtones generated by the strings. If the natural frequencies of the wood do not correspond to the string frequencies, either incomplete resonance will occur, or unwanted frequencies will be excited, leading to a dissonant or noisy sound.

This discussion here, focused on a rectangular wood plate to isolate and examine the intrinsic timbre of wood within a standardized geometric form. It is important to recognize that subsequent modifications, such as:

- Dimensional alterations (cuts, shaping, etc.)
- Boundary condition changes (transitioning from free to constrained states)
- The introduction of soundholes (size and shape)
- The implementation of a soundboard (vaulted shape) and string tension
- The integration of internal bracing (braces, bass bars, soundposts).

All significantly influence the resulting tap-tone frequency composition of the wooden instrument. A skilled luthier will effectively manipulate these variables to achieve final desired tonal characteristics.

Chapter 6 will expand upon these concepts within the context of a completed musical instruments, building upon the principles of wood treatment and modification established in Chap. 5.

When considering the timbre of woods used in musical instrument resonance boxes, several factors beyond simple homogeneous and harmonic frequency variation and response are crucial. These include:

- **Relative modal loudness and muteness**: The interplay between the loudness and the degree of weakness or muteness of each vibrational mode relative to other modes.
- **Level of sensitivity to excitability and sustain**: The process of initiating, maintaining, and extending vibrations, which can be characterized by an Amplitude-Decay-Sustain-Release (ADSR) envelope for each individual mode.
- **Intermodal comparison**: The comparative analysis of these ADSR envelopes across different vibrational modes within a single wood sample and between different wood species or samples of similar dimensions.

Consequently, precise measurement and comparative analysis of modal loudness, damping, and ADSR envelopes are crucial for accurately characterizing and understanding the timbre of wood species and resonator samples. However, comprehensive investigations in this area remain limited, highlighting a significant research gap.

It is empirically observed that two distinct musical instruments, even when playing the same fundamental notes in tune, produce perceptibly different sounds. This phenomenon arises from the unique harmonic content that constitutes each instrument's sonic signature.

As Helmholtz [24] articulated, the timbre of a musical note is determined by the number, relative frequencies, and intensities of the constituent simple tones within its complex vibration [10]. This statement effectively corroborates the aforementioned factors that influence timbre.

Musicians often describe an instrument's timbre using subjective terms such as "bright" (typically associated with pleasing sounds) or "dark," "hard," "soft," "open," "closed," "sharp," or "dull" (often associated with less desirable sounds). Acoustically, a timbre characterized by a preponderance of high-frequency components is perceived as "sharp," whereas a timbre dominated by low-frequency components is perceived as "dull" [10].

These subjective descriptors must be correlated, with varying degrees of success, to the objective properties of the sound. Specifically, the sound spectrum, which represents the frequency distribution of the sound's energy, and the ADSR envelope function, which delineates the temporal evolution of the sound's amplitude, provide acoustical correlates of timbre. To illustrate the diversity of timbre, consider the A4 note (220 Hz) played on two distinct instruments: a conventional classical guitar (Soroush brand, spruce top) at the second fret of the third string, and a conventional violin (Hofner brand) bowed on the fingered fourth string. Although both instruments produce the same pitch, they exhibit unique spectral signatures. These signatures, which repeat with each period, consist of distinct combinations of harmonic frequencies (and some noise frequencies) with varying amplitudes and durations, enabling immediate differentiation between the two instruments. Figure 4.28 presents the time-domain sound signatures along with their corresponding Fast Fourier Transform (FFT) frequency responses and waveforms.

Both recordings exhibit two noise peaks within a frequency range below the first mode. These peaks, common to both recordings, likely originate from the

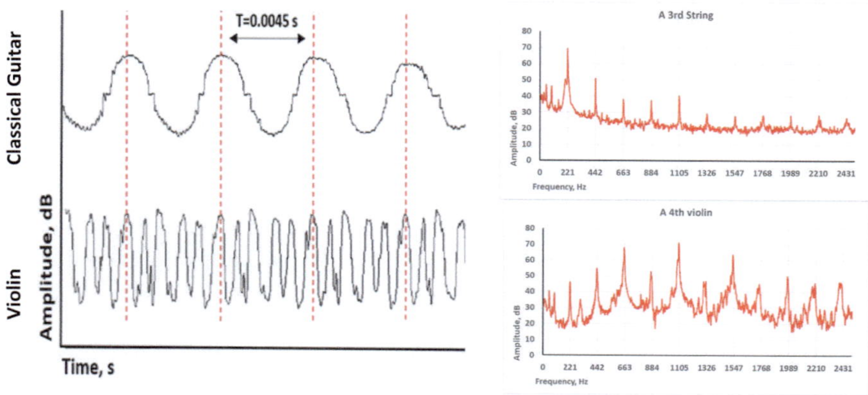

Fig. 4.28 Diagram illustrating the concept of timbre by comparing the time-domain and frequency-domain representations of the same pitch played on different instruments. The image highlights how the unique spectral signatures of each instrument, consisting of different harmonic combinations, contribute to their distinct timbres. $f_1 = 221$ Hz $\rightarrow$ T $= 0.0045$ s

recording environment, as the test was not conducted in an anechoic chamber. Potential sources include the computer's CPU cooling fan. These environmental noise peaks are temporarily disregarded for the current analysis of musical instrument timbre. Notably, the harmonic amplitude order differs between the violin and guitar recordings. Furthermore, an additional peak is observed between the first and second modes of the violin, absent in the guitar recording. This peak, therefore, is attributed to the violin's structure, not the environment. This non-harmonic frequency is classified as a noise frequency. Although resonance from other strings was mitigated during recording, the laboratory environment was not completely silent. However, luthiers routinely encounter environmental noise. As demonstrated in the FFT spectrum, environmental noise is readily distinguishable from instrument-specific noise. Given our focus on timbre, we will proceed with the analysis accordingly. While we have illustrated approximately eleven primary harmonics, both the FFT spectrum and the violin's time-domain waveform indicate a richer tonal quality for the instrument under test. In contrast, the guitar's harmonics and the specific note examined exhibit a significant attenuation in the higher frequency overtones.

The instruments presented here are solely for demonstrating timbral differences and are not of ideal quality. Therefore, these spectra should not serve as reference charts for high-quality musical instruments. A more comprehensive discussion of wood treatment, tuning, and complete instrument construction will follow in subsequent chapters. If you wish to understand how subjective descriptions like 'sonorous sound with good projection,' 'nasal character,' 'agreeable brightness,' or 'soft purity' correlate with scientifically measurable spectral and timbral parameters, please continue reading until the next two chapters.

4.3 Components of Wooden Instruments and Their Required Acoustic Characteristics

Having explored the physics of sound and vibrating strings in Chap. 3, and the acoustic parameters of wood in Chap. 2 and this chapter, we now present a more systematic categorization of musical instrument components. This section will detail the characteristics of each component.

Let us review once again Schelleng's [53] vibrational energy transfer circuit e.g. for a violin that considered by Bucur [10] to be the most intuitive theory.

The bow functions as a miniature motor, imparting both pushing and pulling forces to the string. The player's hand regulates the applied force. The segment of the string between the nut and the bow resists displacement, while the segment from the bow to the bridge contributes to the vibration. The string's vibration is not a smooth oscillation; it exhibits a 'slip-stick' dynamic. The string adheres to the bow, moving synchronously with it, and then abruptly slips, reversing direction. This rapid slip-stick motion generates the violin's characteristic sound. The slip-stick action results in the formation of a sharp discontinuity, or 'corner,' that propagates along the string in both directions (Fig. 4.29). This traveling corner is known as Helmholtz motion. The frictional interaction between the bow and the string is the driving force behind the slip-stick phenomenon. Without this friction, the string would not sustain this distinctive mode of vibration. We now recognize the significance of proper bow weight and balance. These factors are crucial for maintaining the necessary frictional force between the bow hair and the string, allowing the musician to avoid applying excessive pressure or counteracting excessive weight, which leads to hand fatigue. Similarly, bow stiffness, in conjunction with its mass, is vital for maintaining high mechanical impedance. This ensures that string vibrations are directed towards the bridge, rather than being absorbed by the bow.

We used the violin as an example due to its unique bowing mechanism, which differs significantly from the playing techniques used in other stringed instruments. Regardless of how the string is excited, the resulting vibrational wave must be effectively transmitted towards the bridge.

The bridge transmits vibrational energy from the strings to the resonator box. In instruments like the violin, which incorporate a soundpost, this component establishes a direct connection between the back plate and the primary vibration source, while also reinforcing the rigidity of the resonator box. The resonator box itself consists of a bowl-shaped structure formed by the back plate and sides, and a top-plate, or soundboard, effectively functioning as a vibrating-walled enclosure. The vibration of a resonance box's body, particularly the top-plate, causes the enclosed air chamber to act as a pump, expelling and drawing in air through the sound holes. This air movement generates Helmholtz resonance, contributing to the instrument's tonal output [10]. The back plate (or back wall of the bowl), while primarily functioning as a reflector, also vibrates to a degree, influencing the overall sound. The extent of this vibration varies with the wood's mechanical impedance, lower impedance back

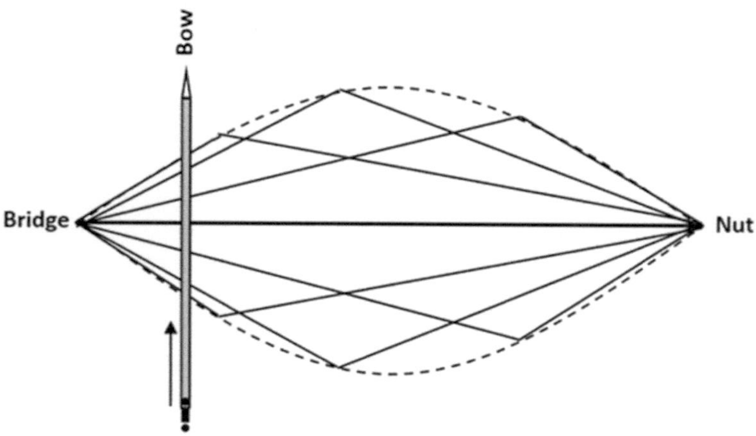

Fig. 4.29 The Helmholtz motion in violin bowing. The image shows a violin string between the bridge and nut, with a bow moving across it. The string's motion is represented by a series of lines showing the "stick–slip" behavior, with a dotted line indicating the overall envelope of the string's vibration. The bow's direction is indicated by an arrow. The image is the handart of author

plates exhibit greater activation and contribute more significantly to sound production. This balance between top and back plate vibration is a matter of tonal preference and musical style, and it can be adjusted through careful selection of wood, dimensions, and geometry. The sides of the resonance box influence the distance between the top and back plates, thereby affecting Helmholtz resonance according to acoustic tube principles. They also provide structural reinforcement.

Sound waves reflected from the back plate return to the top-plate, and the curvature of the back plate affects the focusing of sound under the bridge or other designated areas. This focusing, along with the superposition of reflected sound waves, amplifies the top-plate's vibration. This amplification, in turn, influences the bridge's movement and transmits vibrations back to the strings, sustaining their oscillation. The string vibrations then travel to the nut, and due to the nut's high impedance, reflect back to the bridge. This cyclical process continues until the sound dissipates due to internal and external friction or is overridden by subsequent notes. These events occur rapidly, potentially repeating multiple times before the next note is played.

A skilled luthier utilizes this understanding to optimize sound transmission, radiation, and reflection. By carefully selecting materials, boundary conditions, and geometric parameters, they match mechanical impedances and adjust the sound circuit. This includes:

- Choosing appropriate linings,
- Selecting support conditions e.g. free, simply supported with various soft or hard adhesives, or entirely fixed supports,
- Designing bridges with suitable dimensions, stiffness, and density, and modifying their mass and springiness through carving,
- Selecting appropriate nut and bow materials,

- Installing and adjusting sound posts, bass bars, and braces, optimizing brace designs to reduce mass without compromising stiffness, considering the intersection of braces with nodal and antinode regions of the plate, and
- …

All these adjustments aim to optimize mechanical impedance and the instrument's sound circuit while preserving the vibrational properties and reversibility of particle oscillation. This balancing act, ensuring optimal sound without compromising the instrument's function, is central to a luthier's craft, and will be further explored in Chap. 5 on treatment and tuning.

Having established a general understanding of the sound circuit in stringed musical instruments, we will now examine the characteristics of each component.

4.3.1 Top-Plate or the Soundboard

The top plate, the primary sound-producing element of the resonance box, should be quartersawn, exhibiting a radial surface with straight and parallel grains in the longitudinal direction, and possess minimal inertia to facilitate easy oscillation. This translates to a low mass reactance. To achieve this, wood with low density is preferred, minimizing the need for excessive thinning. Excessive thinning reduces the plate's springy stiffness, which is essential for reversible oscillations and vibrations. Maintaining sufficient springiness and potential energy is crucial. Therefore, the soundboard material should possess high elasticity, ensuring that necessary thinning does not compromise springy stiffness beyond acceptable limits.

However, increasing stiffness reactance elevates the plate's mechanical impedance. To counteract this and maintain a low mechanical impedance, a low-mass material is selected. These factors are subject to limitations, and the optimal balance for a soundboard, considering acoustic radiation and damping requirements, is determined through empirical experience with numerous musical instruments. Lower density coupled with higher modulus of elasticity promotes greater sound radiation and reduced damping. (Refer to the mathematical relationships between sound radiation parameters, damping capacity, density, and stiffness.)

Considering these factors, Table 4.1 presents recommended minimum and maximum values for key parameters related to the top-plate. While the values appear reasonable for traditional wood soundboards, certain entries may exceed typical wood properties. For instance, the table permits a sound velocity of up to 7000 m/s along the grain. Such wood is exceptionally rare. Similarly, the table allows a sound radiation coefficient of up to 20 m^4/kg s, a value not typically observed even in the finest spruces used in Stradivari violins.

These discrepancies may be reconciled in the future through advancements in wood treatment methods or the development of composite materials that can replace traditional top-plate wood. Alternatively, when carbon-fiber reinforced plastic composites are utilized, higher densities may be permissible if the resulting combination of density and high elastic stiffness achieves comparable radiation, damping, and efficiency values, and if mechanical impedance matching can be achieved for string-induced vibrations. Therefore, these general parameter ranges are subject to ongoing discussion and revision, contingent upon specific conditions and future developments.

4.3.2 Back-Plate, Back-Wall of the Resonance Box

The back-plate, an important component of the resonance box, contributes significantly to the instrument's overall sound quality and projection. Ideally, the back plate should also be quartersawn, with a radial surface and straight, parallel grains in the longitudinal direction. While it must possess sufficient stiffness to support the structure and withstand string tension. Furthermore, it must exhibit vibrational properties to contribute to sound production, but primarily after its initial role as a sound reflector within the instrument's body. This balance translates to a specific range for its mass reactance and stiffness reactance.

To achieve this balance, the wood for the back-wall should ideally have a higher density. Excessive density would hinder its ability to vibrate freely and reduce sound radiation. However, insufficient density could compromise its structural integrity and ability to reflect sound waves internally, which is crucial for enriching the tonal quality. The back-wall needs to possess high springy stiffness to support reversible oscillations and vibrations, contributing to the instrument's sustain and projection.

Therefore, the back-plate or back-wall material should exhibit high modulus of elasticity, ensuring that it can effectively store and release potential energy during vibration. This elasticity contributes to the vibrant and responsive nature of the instrument. However, similar to the top-plate, increasing stiffness reactance can elevate the plate's mechanical impedance. To address this, the back plate's mass must be carefully managed, taking into account the desired musical style. Should one favor an active, responsive back, characterized by lower mechanical impedance? Or, conversely, a rigid reflector with higher mechanical impedance? The optimal balance for the back-plate, considering its structural role, acoustic radiation, and damping requirements, is often determined through empirical experience and careful experimentation.

Considering these factors, the recommended parameters for the back-plate would be similar to those for the top plate, with some key differences. The acceptable range for density might be slightly higher to ensure sufficient strength and support (Table 4.1). The target values for sound velocity and radiation coefficient might be moderately lower, reflecting the back-plate's initial role in sound radiation compared.

As with the top-plate, advancements in wood treatment or the development of composite materials may lead to revisions in these parameters. For instance, carbon-fiber reinforced plastic composites could offer new possibilities for back-wall construction, potentially allowing for higher densities if the resulting combination of density and stiffness achieves desirable acoustic properties.

Ultimately, the optimal parameters for the back-wall are subject to ongoing discussion and refinement, depending on the specific design goals, materials used, and the desired sound characteristics of the instrument.

When determining the desired responsiveness of the back-plate, it's essential to recognize that its tap-tone frequency modes may differ from those of the top-plate. This discrepancy can result in either unwanted or desired frequency interference with the top-plate. The extent to which the back-plate's resonant noise contributes to a ringing sound should be carefully considered, as it directly influences the instrument's tonal character and is dependent on the musical style.

It was stated that the wood for the bowl's back should ideally exhibit a radial surface and straight grain. However, in some integrated bowl designs, tangential surfaces are unavoidable. In such cases, it's preferable for these tangential surfaces to be located on the sides rather than the back. Ultimately, if the wood's moisture-related shrinkage and swelling (as will be discussed in Chap. 5) remain within acceptable limits and do not negatively impact the key parameters of the back-wall outlined in Table 4.1, then the tangential surfaces are permissible. While a lack of straightness can reduce the wood's elastic stiffness, aesthetic considerations may sometimes have a negative effect on key parameters. Therefore, meticulous measurement and control of the wood's parameters are essential. The instrument's sound quality must take precedence, with aesthetic beauty considered secondary.

4.3.3 Side-Wall of the Resonance Box

Specifications and parameters for the wood used in the sides of resonance boxes are often presented alongside those for the back plate [10, 50, 61], and are frequently identical (Table 4.1). Consequently, the same wood species employed for the back-plate can also be used for the sides. Like the back-plate, the side walls must exhibit high stiffness and strength to provide structural reinforcement. Furthermore, they serve as an important boundary condition for the top-plate, and their high impedance facilitates the reflection of sound waves originating from the top-plate back into the top-plate again. Yet, as before, some wave absorption occurs within the sides or the plate-side connection (the binding), that must be concerned.

However, if a musical instrument is intended to possess a more responsive back and sides, wood species with lower stiffness or density can be selected. While the reflective role of the back-plate within the air cavity is significant, it is not as critical for the side walls. This difference allows for the placement of soundholes on the side walls in some instruments, rather than solely on the top-plate. Additionally, the width of the side walls determines the distance between the back and top-plates, which in turn influences the Helmholtz frequencies of the air within the body cavity.

As this book primarily focuses on wood science, it does not delve into the intricacies of bowl curvature, soundhole dimensions and placement, or precise sound tube length calculations. Nevertheless, some of their effects will be discussed in Chap. 5.

4.3.4 Bracings

Braces are wooden components adhered to the interior surfaces of the instruments's top and back plates. Their primary functions are to reinforce the instrument's structure and to shape its tonal characteristics [10, 13, 18, 34, 37, 41].

The tension exerted by the strings places significant stress on the top and back plates. Braces counteract this force, preventing deformation and damage.

By influencing the vibrational behavior of the sound plates, braces directly impact the instrument's tone and timbre. The design and placement of braces determine the instrument's resonant frequencies, thereby shaping its sonic output.

Braces for top-plate e.g. for Spanish guitars, are typically fabricated from lightweight, flexible woods such as spruce. But heavier species are also sometimes used for backs. The selection of wood species is determined by factors including the guitar's style, desired tonal properties, and required structural integrity. It is recommended that the brace material for the top plate match the top plate material [10].

Various bracing designs exist, each yielding distinct sonic characteristics (Fig. 4.30). The placement and dimensions of braces exert a significant influence on the guitar's sound. The stiffness, flexibility, and mass of the braces affect the vibrational behavior of the top plate. Luthiers can manipulate these properties through carving techniques, reducing weight and/or stiffness [13]. Specialized carving methods can reduce weight without compromising stiffness (Chap. 5).

The positioning of braces can interact with the vibrational antinodes of specific frequency modes, dampening or modifying them. This can alter the mode's vibrational shape, nodal patterns, and resonant frequency. Guitar makers carefully select bracing patterns based on the instrument's design and the desired tonal characteristics.

4.3.5 Bass Bar/Tone Bar

The bass bar is a longitudinal strip of wood, typically spruce with the same key properties of the top-plate, positioned along the bass side of the violin's soundboard, approximately under the G string. Unlike guitars, which utilize a network of braces, the violin employs a single bass bar. The presence of the bass bar disrupts the symmetry of the top plate [10]. Like bracing in other instruments, it reinforces the top-plate, preventing damage from the pressure of the bass strings. However, any additional mass or stiffness can alter the resonant frequencies and amplitudes of specific vibrational modes of the top-plate, depending on which vibrational modes of the top-plate intersect. Concurrently, the bass bar enhances the acoustic coupling between the central region of the soundboard (the 'central island' bounded by the f-holes) and the larger upper and lower plate surfaces. This improved coupling facilitates more efficient vibration transfer across the entire soundboard, resulting in increased sound radiation [20].

Fig. 4.30 Image depicting various bracing ideas for guitar construction. This image is a handart by the author, enhanced using AI

While the mass of the bass bar is relatively small compared to the total mass of the violin's top-plate, its influence on the lower-order modal frequencies is generally considered to be minor. This has been supported by empirical studies, such as those conducted by Gough [20], which demonstrated a minimal impact of the bass bar on the first four tap-tone modes of a violin soundboard. However, it is reported that applying tensions to the bass bar (increasing its springy constant and applied force to the soundboard) for the first and fourth vibrational modes resulted in an increase in resonant frequency. This technique could be considered for plate tuning [55]. The effect of stiffness changes of the bass bar on the violin tone quality was also reported [10].

Given that the bass bar and soundboard are typically constructed from the same wood species, the key material parameters determined for the soundboard (Table 4.1) can serve as a reasonable approximation for the bass bar. However, it's important to acknowledge that variations in grain orientation, density, and local stiffness may exist between the two components, potentially influencing their acoustic behavior.

4.3.6 Soundpost

Let me guide you through the intricate world of the violin, where a small wooden component known as the soundpost plays a starring role in a scientific saga of sound and structure. Imagine a slender wooden cylinder, carved from spruce, with a diameter ranging from 5 to 6.4 mm and a length of 54 mm. Though it appears unassuming, its influence is profound, particularly in high-quality violins, while its effect is more subdued in mass-produced models [10].

Our exploration starts with its precise placement—positioned just 5 mm behind the treble-side foot of the bridge, the critical junction where the strings' vibrations meet the violin's body. Skilled violin makers fine-tune the bridge and soundpost to unlock the instrument's fullest tonal potential and enhance its responsiveness, a craft that extends across the entire violin family.

The soundpost's influence first came to light through pioneering experiments [10], which revealed its ability to reshape the violin's vibrational patterns. Researchers observed that it eliminates the first vibrational mode of the top plate and introduces asymmetry to the second, enriching the sound's character [10, 54]. Scientific measurements showed that installing the soundpost raises the Helmholtz air resonance by four semitones, a subtle shift that stiffens the violin's acoustic framework [10].

Picture a theoretically simplified violin body, other than its conventional body, a rigid structure resonating with high-frequency modes [20]. Here, the soundpost acts as a disruptor, imposing an asymmetric constraint. Simulations, showcased in Fig. 4.31, illustrate this effect in frequency of 1656 Hz for example. The details are illustrated in the literature [20].

As illustrated in the figure, the placement of the soundpost relative to the nodal and antinodal positions of the violin's vibrational modes significantly influences the mode shapes, resulting in frequency shifts, either upward or downward. While Gough [20] and Bucur [10] have provided valuable insights into the interpretation of these mode shapes, it is crucial to emphasize a key aspect: the longitudinal modulus of elasticity in wood, aligned with the grain direction, is the highest modulus. Consequently, acoustic parameters such as the sound radiation coefficient, damping capacity (as studied by Ono and Norimoto), and acoustic conversion efficiency are enhanced with increasing modulus. Therefore, a primary objective in the assembly of musical instrument resonance boxes should be to direct vibrational energy along the wood's grain direction. This figure, though depicting a particular vibrational mode for this simplified violin model, clearly demonstrate the sound post's effect. Specifically, a frequency shift from 1656 to 1243 Hz, is seen, effectively illustrates how the soundpost channels vibrations along the grain direction, while constraining vibrations

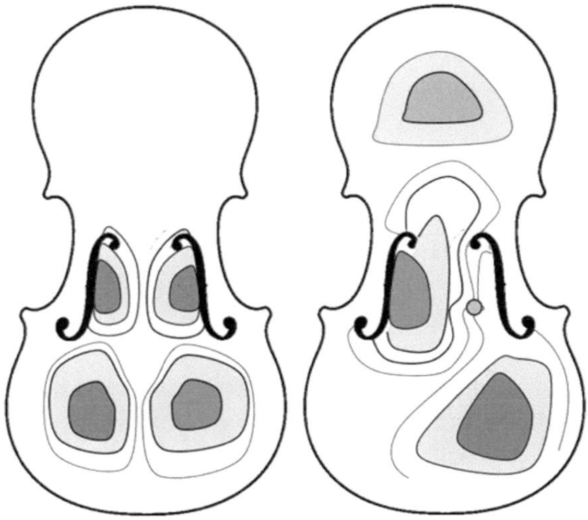

Fig. 4.31 This image demonstrates the significant influence of soundpost placement on a simplified violin's vibrational modes. As shown, the soundpost alters the modes, evidenced for an example by a frequency shift from 1656 Hz (before soundpost insertion) to 1243 Hz (after soundpost mounting). This critical component also helps direct vibrational energy efficiently along the wood grain, thereby enhancing the instrument's acoustic parameters and contributing to a more open and resonant sound. Additionally, note how the f-hole stops the extension of the mode shapes in a direction perpendicular to the wood grain

perpendicular to it. This redirection of vibrational energy contributes to a more open, sustained, and radiant sound in the violin.

The soundpost generally raises the resonant frequencies of most vibrational modes, particularly the fundamental mode. This increase is likely due to the soundpost's contribution to the soundboard's springy constant (stiffness constant). By providing partial tension, the soundpost effectively increases the soundboard's resistance to deformation, resulting in higher stiffness reactance and, consequently, elevated resonant frequencies.

Noteworthy, orienting the soundhole along the grain direction, beside the soundpost, not only facilitates the propagation of longitudinal vibrations but also minimizes transverse stress induced by the bridge and soundpost. By aligning the soundhole with the grain, the tension generated by these components are directed primarily along the longitudinal direction of the wood, reducing the risk of transverse deformation.

Beyond disruption, the soundpost is a partner in sound production. Held in place by friction within the violin's body, it enables the bridge's asymmetrical rocking motion to couple with the plate's lowest modes. This synergy generates robust monopole radiation, amplifying the sound's reach, a phenomenon [20].

In this section, it is important to revisit and consider two key points.

First, the frequencies and modes presented as examples are specific simplifications for the violin and may exhibit some numerical differences compared to similar values in a conventional violin. However, the underlying logic of the changes is effectively demonstrated.

Second, we previously noted that soundpost, by slightly pressing the plate, induces an upward tension. Consequently, by increasing the plate's stiffness constant (stiffness reactance), we would anticipate an increase in its vibration frequencies. Yet, we now observe that the placement of the soundpost, while indeed tensing the plate, simultaneously directs the mode shapes along the longer, longitudinal axis of the plate (grain direction). This redirection causes a frequency shift towards lower frequencies (e.g. Fig. 4.31). The profound effect of the soundpost should not be overlooked: it enhances stiffness, thereby improving the acoustic radiation coefficient, while concurrently lowering a number of the plate's frequencies.

Thus, the soundpost—a modest spruce cylinder—emerges as a scientific marvel, weaving asymmetry, resonance, and efficiency into the violin's essence.

4.3.7 Bridge and Saddle

In the realm of stringed instruments, the terms "bridge" and "saddle" describe key components to support the string boundary condition at one of its ends, on the soundboard. Though their usage and relationship can sometimes be nuanced. Generally speaking, the bridge serves as the primary structural element. It's the part that supports the strings and, critically, transmits their vibrations to the soundboard or body of the instrument, a function essential for sound production. Instruments like violins, mandolins, and banjos all feature a component referred to as a bridge. However, on instruments like guitars, the terminology becomes more specific. While guitars also have a bridge, they often incorporate a separate component known as the saddle. This saddle is the bearing surface where the strings make contact, and it often plays a role in fine-tuning aspects of the instrument's setup, such as string height (or action) and intonation (the accuracy of pitch along the fretboard). So, while the bridge provides the main support for the strings, the saddle is often a part of the bridge that offers a more refined point of contact. It's important to note that the terminology can vary slightly across different instruments, and some bridges may be simple, single pieces, while others are more complex assemblies. Consistent and precise terminology is crucial when discussing musical instrument components. Terminology discrepancies can lead to confusion, particularly when comparing instruments from different families. For instance, the 'saddle' in a guitar and related instruments typically refers to the component located on the bridge that directly supports the strings. However, in violin terminology, 'saddle' often designates the points where the strings connect to the body plate and the tailpiece.

To maintain clarity in this section, we will adopt the guitar-related definition of 'saddle,' referring to the string support on the bridge. The violin-specific usage of 'saddle' will not be employed here, as our focus is on the bridge and its constituent parts. Ultimately, understanding the distinction lies in recognizing that the bridge is the core structural element, while the saddle, when present, is a more specific component that often aids in precision and adjustment.

The instrument bridge, particularly in the context of the violin, is a complex component warranting extensive study. However, for the purpose of this discussion, we will focus on the fundamental principles applicable to various instruments.

The bridge's primary function is to transform the predominantly horizontal vibrations of the strings into vertical vibrations, perpendicular to the soundboard [10]. This transformation facilitates efficient energy transfer to the soundboard, enabling sound production. As an impedance matching element between the high characteristic impedance string and the lower characteristic impedance soundboard, the bridge must exhibit a carefully balanced combination of mass reactance and stiffness reactance. This intermediate impedance ensures effective vibration transmission in both directions, from the string to the soundboard and vice versa. The selection of bridge material and design hinges on the manipulation of mass and stiffness. These parameters significantly influence the bridge's interaction with the soundboard's vibrational modes. Increasing the bridge's mass, while keeping stiffness constant, generally reduces the transmission of high-frequency vibrations and lowers the resonant frequencies of soundboard modes influenced by the bridge's support points. Conversely, increasing stiffness while maintaining constant mass raises resonant frequencies and can also decrease vibration transmission. The combined effect of mass and stiffness alterations depends on the dominant parameter. These effects have been studied extensively, particularly in relation to the violin's bridge heel frequency [10]. Chapter 5 will delve into practical techniques for adjusting the bridge's mass and stiffness. For now, it's essential to understand that an ideal bridge material should possess low density and moderate stiffness. A bridge with both low density and very low stiffness will exhibit excessively low mechanical and characteristic impedance, failing to effectively couple the string and soundboard.

Maple, specifically Field maple (*Acer campestre*) from Vosges Mountains, France, with a radial grain orientation, is widely recognized as the optimal wood for violin bridges. The strings rest on the tangential edge of the bridge, leveraging the maple's radial grain to enhance sound velocity in that direction. This radial grain orientation is crucial, as the natural stiffness and sound velocity along the radial axis are significantly lower than along the growth rings. To achieve optimal acoustic performance for each instrument, the selection of components such as strings, soundboards, bridges, and associated masses must be carefully matched. This matching process considers factors including string type, soundboard material and dimensions, string and soundboard mass, bridge wood species, and the dimensions and stiffness of strings and other mechanical elements, all to ensure the desired mechanical impedance.

Placing the strings on the cross-sectional edge of the wood is undesirable due to the differential hardness of the annual growth rings, leading to uneven string penetration. Similarly, using sparsely veined maple can result in inconsistent string seating due to variations in spring and summer wood hardness. To enhance bridge stability and potentially improve sound transmission, in some instruments, one might consider incorporating a thin but stiff bearing, similar to those found in guitars, as a support for the strings. This approach can effectively increase the bridge's rigidity and minimize unwanted vibrations.

Otherwise, the tangential orientation provides a more uniform surface hardness profile, minimizing differential string penetration. Furthermore, the maple's inherent radial stiffness is sufficient for bridge applications. Engineering techniques can be employed to further refine the bridge's mass and stiffness, achieving optimal performance.

To produce sound, the string vibrations clearly have to transfer energy to the radiating shell modes via the asymmetrically rocking bridge [21].

4.3.8 Neck and Fingerboard

The harmonious marriage of material science and musical artistry finds a compelling expression in the selection of woods for the necks and fingerboards of stringed instruments. These components, far from being mere structural supports, are integral to the instrument's playability, tonal character, and long-term stability. The discerning luthier, therefore, approaches wood selection with a profound understanding of the material's inherent properties and their impact on the instrument's performance.

The neck, acting as the foundation upon which the fingerboard rests, must exhibit exceptional stiffness and stability. Under the considerable tension exerted by the strings, any degree of flexure or torsion would compromise intonation and diminish the instrument's responsiveness. To counter these forces, woods such as maple, mahogany, and ash are favored, their high modulus of elasticity providing the necessary rigidity. Furthermore, the neck's dimensional stability is paramount, preventing warping and distortions that would arise from fluctuations in temperature and humidity. Quarter-sawn wood, with its inherent resistance to warping, is often preferred, alongside meticulous drying and seasoning processes (Chap. 5). The grain direction, running longitudinally along the neck, further reinforces its strength and stiffness. While the neck's contribution to resonance is secondary to that of the soundboard, its material composition nonetheless influences the instrument's overall tonal palette.

The fingerboard, the intimate interface between the musician and the instrument, demands a different set of material characteristics. Hardness and density are paramount, ensuring a durable surface that can withstand the relentless pressure of the strings and the player's fingers. Ebony and rosewood, renowned for their exceptional hardness and high density, are thus the materials of choice. These woods, with their fine grain and natural oil content, provide a smooth and comfortable playing surface, while their density contributes to a bright and clear tone, enhancing sustain. Elevated mechanical impedance of the fingerboard at the fingering point, significantly increases the reflection of string vibrations, particularly in fretless instruments where direct finger-string-fingerboard contact occurs. In fretted instruments, the mechanical impedance of the fret material also contributes to the reflection characteristics.

Stability remains a critical consideration, preventing warping and maintaining consistent playability. The fingerboard's contribution to the instrument's tone is undeniable, adding a layer of brightness or warmth that complements the overall sonic character.

The interplay between the neck and fingerboard is a testament to the synergistic relationship between form and function. These components, working in concert, provide a stable and responsive playing surface, allowing the musician to translate their artistic vision into sound. The selection of woods for these components is not arbitrary; it is a meticulous process, guided by an understanding of the material's properties and their impact on the instrument's performance. The luthier, therefore, acts as both artist and scientist, carefully selecting and shaping the materials to create an instrument that is both beautiful and functional.

Totally, optimal wood selection for instrument necks prioritizes high stiffness, and, provided it doesn't excessively increase instrument weight or player fatigue, high density. Additionally, minimal inhomogeneity in moisture-induced shrinkage and swelling is essential (as further discussed in Chap. 5). Conventionally suitable neck woods exhibit stiffness exceeding 10 GPa, densities around 700 kg/m^3, and a tangential-to-radial shrinkage ratio (TR ratio, detailed in Chap. 5 regarding wood-water interactions) below 2.

Similar principles apply to fingerboard selection, with high stiffness and density being recommended. Due to the fingerboard's thin profile, the increased density does not significantly contribute to overall instrument weight. Concurrently, maximizing hardness and minimizing the TR ratio (tangential-to-radial moisture shrinkage ratio) are crucial. Suitable fingerboard woods typically display stiffness above 10 GPa, densities around 950 kg/m^3, a TR ratio below 2, and a Janka hardness of approximately 10,000 N.

4.3.9 Bow

The bow's wood material characteristics, particularly density and stiffness, are carefully selected to provide the necessary control and responsiveness. The interaction between the wood, hair, and strings involves a complex interplay of mechanical impedance and energy transfer, resulting in the rich and nuanced sound produced by stringed instruments.

The ideal wood for a violin bow exhibits high stiffness and density, resulting in high acoustic impedance. However, to minimize player fatigue, the bow should also be lightweight, achievable through a thin octagonal profile while keeping its acceptable springy stiffness [10]. For instance, Pernambuco wood, renowned as the premier bow material, typically possesses a Young's modulus of approximately 22 GPa and a density of around 950 kg/m^3 [10]. Acoustic velocity has become a common metric in the musical instrument industry for assessing bow wood quality [60]. Commercial advertisements often cite sound velocities ranging from 4500 to over 6000 m/s for Pernambuco. Luthiers should independently verify these claims using the methods detailed in this chapter, as commercial information can be dramatically manipulated for marketing purposes that are unverifiable or unbelievable.

The creation of a fine bow, an essential tool for stringed instrument musicians, hinges upon the selection of a material that harmonizes strength, flexibility, and acoustic properties. For nearly two centuries, Pernambuco (*Caesalpinia echinata*), or Brazilwood, has reigned supreme in this domain.

François Xavier Tourte's meticulous selection of Pernambuco, originating from the Atlantic coastal regions of Brazil, was no accident. Its properties, meticulously documented, reveal a material ideally suited for bow construction. Pernambuco exhibits remarkable tensile strength. Its ability to be shaped through dry heat bending and retain its form, coupled with a high stiffness-to-mass ratio, ensures optimal responsiveness. Moreover, its lustrous surface and resistance to transverse fracture contribute to its aesthetic and structural integrity. Tourte's observation that wood with "twists, knots, and fissures" possessed the desired characteristics underscores the importance of inherent material properties in achieving acoustic excellence. The ability of Pernambuco to maintain the bow's curvature (camber) further enhances sound control, allowing musicians to express nuanced articulations.

However, the popularity of Pernambuco has come at a cost. Deforestation has led to its classification as an endangered species, prompting the use of reclaimed wood. This scarcity has spurred the search for alternatives. Other Brazilwoods, such as *Haemotoxylum brasiletto* and *Caesalpinia sappan*, have been erroneously marketed as substitutes, but their inferior vibration transmission and reduced durability render them unsuitable for professional-grade bows. Modern materials, such as carbon fiber, offer a lower-cost alternative for student-level instruments. While carbon fiber bows exhibit clear resonances, their greater impedance levels and lack of the natural feel and tonal qualities of Pernambuco have limited their adoption by professional musicians [10].

A deeper understanding of Pernambuco's mechanical and acoustic properties reveals the reasons for its enduring appeal. Its elastic properties, including a longitudinal wave velocity of about 5000 m/s and shear wave velocities between 1066 and 1380 m/s, indicate a material with high longitudinal stiffness and transverse flexibility, ideal for bow dynamics. Fracture behavior, typically occurring in the LR plane due to ray weakness, rather than the longitudinally strong direction, highlights its structural characteristics.

The selection of bow materials is guided by several key criteria. Efficient vibration transmission, flexural formability, durability, and aesthetic appeal are paramount. Pernambuco excels in all these areas, offering a balanced combination of density and stiffness, the ability to be heat-bent and maintain camber, resistance to repetitive stress and abrasion, and a rich aesthetic tradition. While alternatives exist, they do not replicate Pernambuco's overall performance.

So, Pernambuco remains the primary material for high-quality bows, due to its unique combination of mechanical and acoustic properties. While alternative materials are used for lower-cost instruments, professional players favor Pernambuco. The conservation of this valuable resource is crucial, and the ongoing research into sustainable alternatives is essential to ensure the future of fine bow making.

However, these above knowledge are firmly rooted in the Western violin family. What happens when we shift our gaze to other bowed string traditions, such as the fiddles of Turkey and Iran, where the bow diverges significantly in design and material?

In the musical traditions of Turkey, Iran, and neighboring regions, instruments like the kamancheh—a spiked fiddle with a long history—rely on a bow known locally as "kaman" or "kamaneh". Unlike the violin bow, which maintains a fixed hair tension adjusted by a screw, the kamancheh bow allows the player to manually control tension during performance, typically by pressing the hair with the thumb or fingers. This fundamental difference influences the choice of materials for the bow stick. While the recommending Pernambuco is recommended for its ability to retain a concave camber under heat and sustain a tension of 60–75 N [10], the kamancheh bow often employs more readily available woods such as walnut (*Juglans regia*) or mulberry (*Morus alba*), reflecting both practicality and regional availability.

Walnut, a common choice in Persian and Turkish bow-making, has a density of approximately 600–700 kg/m^3 and a Young's modulus around 10–12 GPa—significantly lower than Pernambuco's 22 GPa [10, 25]. Mulberry, another favored wood, is even lighter, with a density of 500–600 kg/m^3 and a modulus of roughly 8–10 GPa [15]. These woods lack the exceptional stiffness-to-mass ratio of Pernambuco (noted as 0.2% higher in octagonal sections), and they are rarely subjected to the meticulous heat-bending process that defines the violin bow's camber curvature. Instead, kamancheh bows are often straighter or possess a natural, minimal curvature, relying on the wood's inherent flexibility rather than engineered shaping. This simplicity in construction aligns with their shorter length—typically 50–60 cm compared to the violin bow's 74–75 cm [10].

Why, then, do these traditions avoid the violin bow's design? The answer lies in a blend of technical functionality, musical style, and cultural heritage. The manual tension control of the kamancheh bow, a feature absent in the violin bow's rigid structure, allows the player to dynamically adjust the hair's tautness mid-performance. This flexibility is important for the expressive, improvisational nature of Middle Eastern music, where rapid shifts in dynamics and microtonal inflections are common [27]. In contrast, the kamancheh's variable tension enables a warmer, more fluid tone, often across multiple strings simultaneously due to the instrument's flatter bridge—a stark contrast to the violin's curved bridge designed for single-string articulation.

4.4 Final Words

The study and identification of tonewood are integral to the craft of instrument making, bridging the gap between science, tradition, and innovation. Understanding the acoustic properties, anatomical structures, and historical significance of various wood species allows luthiers to create instruments with exceptional tonal quality. Through centuries of experimentation and refinement, makers have identified and perfected the use of species such as spruce, maple, and mahogany, each contributing unique characteristics to the sound and playability of stringed instruments.

This chapter has underscored the importance of scientific analysis in tonewood selection. Traditional identification techniques, such as visual assessment and tap-tone evaluation by ear, remain invaluable, yet they are now complemented by modern methodologies, including microscopy, spectroscopy, and X-ray densitometry. These advancements allow for a more precise understanding of wood properties, providing deeper insight into density, elasticity, and internal structure. As a result, instrument makers can make informed decisions, ensuring that each piece of wood selected contributes positively to the overall tonal characteristics of an instrument.

The influence of environmental factors and geographic origins on tonewood properties cannot be overlooked. Wood sourced from different regions varies in grain structure, growth rings, and internal composition, all of which affect its resonance and durability. Historical analysis has revealed how master luthiers of the past carefully selected their materials based on availability and quality, leading to the legendary sound of instruments from the Stradivari and Guarneri workshops. Understanding these regional differences is crucial for modern luthiers who seek to replicate or innovate upon historical designs while maintaining the acoustic integrity of their instruments.

Sustainability is another pressing issue in the field of tonewood selection. With traditional wood sources under increasing threat from deforestation and climate change, the need for responsible harvesting and sustainable forestry practices is more urgent than ever. Certification programs, reforestation initiatives, and the exploration of alternative materials, such as thermally modified woods and synthetic composites, offer viable solutions to ensure the continued availability of high-quality materials for instrument making. The balance between tradition and sustainability is delicate, requiring a collaborative effort between musicians, luthiers, and conservationists to safeguard tonewood resources for future generations.

The future of tonewood research holds exciting possibilities. Emerging technologies, including machine learning and computational modeling, are revolutionizing the way wood properties are analyzed and utilized. These advancements have the potential to enhance consistency in wood selection, reduce waste, and expand the range of available materials while maintaining high acoustic standards. As digital tools become more integrated into the craft, they provide invaluable assistance in both preserving traditional techniques and pushing the boundaries of innovation.

Ultimately, the art of tonewood selection remains a harmonious blend of science, craftsmanship, and musical artistry. By deepening our understanding of wood properties and embracing sustainable practices, we can continue to build instruments that resonate with clarity, warmth, and richness. The legacy of great luthiers is carried forward not only through the instruments they created but also through the knowledge they have imparted to future generations. As we move forward, maintaining a balance between honoring tradition and embracing innovation will be key to preserving the essence of fine instrument making.

In conclusion, the study of tonewood identification extends beyond mere classification; it is a journey into the heart of musical expression. By combining empirical wisdom with modern scientific insights, we refine our ability to select and utilize woods that bring music to life. Whether in the hands of a skilled luthier or a passionate musician, the right tonewood is not just material—it is the foundation of artistry, carrying forward a tradition of sound that has shaped cultures and captivated audiences for centuries. The ongoing pursuit of knowledge in this field ensures that future generations will continue to craft instruments of beauty and resonance, preserving the profound connection between nature and music.

References

1. ASTM C1548-02, *Standard Test Method for Dynamic Young's Modulus, Shear Modulus, and Poisson's Ratio of Refractory Materials by Impulse Excitation of Vibration* (ASTM International, 2020)
2. ASTM D6874-22a, *Standard Test Methods for Nondestructive Evaluation of the Stiffness of Wood and Wood-Based Materials Using Transverse Vibration or Stress Wave Propagation* (ASTM International, 2022)
3. J. Beauchêne, N. Amusant, J. Cigna, S. Koese, B. Thibaut, Using specimens from the CIRAD Kourou wood collection to build a database of properties. Bois Forêts Trop. **352**, 61–70 (2022). https://doi.org/10.19182/bft2022.352.a36936
4. E.R. Bell, E.C. Peck, N.T. Krueger, Modulus of elasticity of wood determined by dynamic methods. Forest Products Laboratory Report, pp. 1–20 (1954)
5. J. Bodig, B.A. Jayne, *Mechanics of Wood and Wood Composites* (Van Nostrand Reinhold, 1982)
6. P.A. Bordonné, Module dynamique et frottement intérieur dans le bois. Mesures sur poutres flottantes en vibrations naturelles. Doctoral dissertation, Institut National Polytechnique de Lorraine [INPL], 1989
7. L. Brancheriau, Corrections for poisson effect in longitudinal vibrations and shearing deformations in transverse vibrations applied to a prismatic orthotropic body, in *Mechanical Vibrations: Types, Testing and Analysis*, ed. by A.L. Galloway (Nova Science Publishers, 2010), pp. 205–223. https://agritrop.cirad.fr/561113/
8. I. Brémaud, Y. El Kaïm, D. Guibal et al., Characterisation and categorisation of the diversity in viscoelastic vibrational properties between 98 wood types. Ann. For. Sci. **69**, 373–386 (2012). https://doi.org/10.1007/s13595-011-0166-z
9. I. Brémaud, J. Ruelle, A. Thibaut, B. Thibaut, Changes in viscoelastic vibrational properties between compression and normal wood: roles of microfibril angle and of lignin. Holzforschung **67**(1), 75–85 (2013). https://doi.org/10.1515/hf-2011-0186
10. V. Bucur, *Handbook of Materials for String Musical Instruments* (Springer, 2016). https://doi.org/10.1007/978-3-319-32080-9
11. V. Bucur, A review on acoustics of wood as a tool for quality assessment. Forests **14**(8), 1545 (2023). https://doi.org/10.3390/f14081545
12. G. Caldersmith, E. Freeman, Wood properties from sample plate measurements I. J. Catgut Acoust. Soc. **1**(series II), 8–12 (1990)
13. P. Dumond, N. Baddour, Can a brace be used to control the frequencies of a plate? SpringerPlus **2**, 558 (2013). https://doi.org/10.1186/2193-1801-2-558
14. N.H. Fletcher, T.D. Rossing, *The Physics of Musical Instruments* (Springer, 2012)

15. Forest Products Laboratory, *Wood Handbook: Wood as an Engineering Material* (U.S. Department of Agriculture, Forest Service, Forest Products Laboratory, 2010)
16. A. Formato, D. Ianniello, A. Pellegrino, F. Villecco, Vibration-based experimental identification of the elastic moduli using plate specimens of the olive tree. Machines **7**(2), 46 (2019). https://doi.org/10.3390/machines7020046
17. B. Garven, Creating and validating finite element models of stiffness measures and failure loads in cadaveric ulnae under static and harmonic loading. Master's thesis, Ohio University. OhioLINK Electronic Theses and Dissertations Center, 2014. http://rave.ohiolink.edu/etdc/view?acc_num=ohiou1399041718
18. Y. Giro, J.L. Le Carrou, A. Vincenti, S. Dartois, R. Viala, B. Navarret, Predicting the effects of bracing pattern modifications on acoustic guitar soundboards, in *Forum Acusticum* (2023), pp. 3419–3422. https://doi.org/10.48465/fa.2023.1090
19. R. Gonçalves, A.J. Trinca, G.C. dos Santos Ferreira, Effect of coupling media on velocity and attenuation of ultrasonic waves in Brazilian wood. J. Wood Sci. **57**(4), 282–287 (2011). https://doi.org/10.1007/s10086-011-1177-y
20. C. Gough, The violin: Chladni patterns, plates, shells and sounds. Eur. Phys. J. Spec. Top. **145**, 77–101 (2007). https://doi.org/10.1140/epjst/e2007-00149-0
21. C.E. Gough, Violin acoustics. Acoust. Today **12**(2), 22–30 (2016)
22. C. Guan, J. Liu, H. Zhang, X. Wang, L. Zhou, Evaluation of modulus of elasticity and modulus of rupture of full-size wood composite panels supported on two nodal-lines using a vibration technique. Constr. Build. Mater. **218**, 64–72 (2019). https://doi.org/10.1016/j.conbuildmat.2019.05.086
23. C.M. Harris, *Shock and Vibration Handbook*, 6th edn. (McGraw-Hill, 2002)
24. H. Helmholtz von, On the sensation of tone. Reprint 1954, Dover New York (1863).
25. R.B. Hoadley, *Understanding Wood* (Taunton Press, Newtown, 2000), p.211
26. ISO 13061-4:2014, *Physical and Mechanical Properties of Wood—Test Methods for Small Clear Wood Specimens Part 4: Determination of Modulus of Elasticity in Static Bending* (International Organization for Standardization, 2014)
27. R. Khaleqi, *Sargozasht-e Musiqi-ye Iran* (Enbe-Sina Publication, 1974)
28. E. Khosravi, M. Roohnia, A. Lashgari, A. Jahanlatibari, A. Tajdini, Evaluation of pin penetration probing technique for the assessment of basic density in air-dried wood. BioResources **16**(4), 6577–6586 (2021). https://doi.org/10.15376/biores
29. M. Kohantorabi, A.H. Hemmasi, M. Talaei Poor, M. Roohnia, B. Baziyar, The study of joints in Oak wood beam compared to the location of node and anti-node of longitudinal vibration. Iran. J. Wood Pap. Sci. Res. **36**(3), 256–263 (2021). https://doi.org/10.22092/ijwpr.2021.352061.1639
30. F.F.P. Kollmann, W.A. Coté, Jr., *Principles of Wood Science and Technology: Solid Wood. I* (Springer, 1968)
31. Y. Kubojima, S. Sonoda, H. Kato, Application of a bending vibration method without weighing specimens to the practical wooden members conditions. J. Wood Sci. **68**(39) (2022). https://doi.org/10.1186/s10086-022-02046-1
32. F. Leccese, G. Salvadori, G. Bernardini, P. Bernardini, The bowed string instruments: acoustic characterization of unique pieces from the Italian lutherie. IOP Conf. Ser. Mater. Sci. Eng. **364**(1), 012022 (2018). https://doi.org/10.1088/1757-899X/364/1/012022
33. Z. Liu, Y. Liu, J. Shen, Influence of the vibrational properties of the resonance board on the acoustical quality of a piano. BioResources **8**(2), 2135–2144 (2013). https://doi.org/10.15376/biores.8.2.2135-2144
34. M. Mihălcică, M.D. Stanciu, S. Vlase, Frequency response evaluation of guitar bodies with different bracing systems. Symmetry **12**(5), 795 (2020). https://doi.org/10.3390/sym12050795
35. P. Mirbolouk, M. Roohnia, Evaluation of dynamic modulus of elasticity of medium density fiberboard panel from longitudinal vibration tests on specimens. BioResources **10**(1), 613–621 (2015). https://doi.org/10.15376/biores.10.1.613-621

36. K.C. Nefdt, Improvement of the acoustic properties of tonewood. Doctoral dissertation, Stellenbosch University, 2024
37. A. Okuda, T. Ono, Bracing effect in a guitar top board by vibration experiment and modal analysis. Acoust. Lett. **29**(1), 103–105 (2008). https://doi.org/10.1250/ast.29.103
38. T. Ono, M. Norimoto, Study on Young's modulus and internal friction of wood in relation to the evaluation of wood for musical instruments. Jpn. J. Appl. Phys. **22**(4R), 611 (1983). https://doi.org/10.1143/JJAP.22.611
39. T. Ono, M. Norimoto, On physical criteria for the selection of wood for soundboards of musical instruments. Rheol. Acta **23**, 652–656 (1984). https://doi.org/10.1007/BF01438805
40. D. Ouis, On the frequency dependence of the modulus of elasticity of wood. Wood Sci. Technol. **36**(4), 335–346 (2002). https://doi.org/10.1007/s00226-002-0145-5
41. A. Pilch, Modification of the guitar bracing using optimization. Vibr. Phys. Syst. **34**(2), 1–7 (2023). https://doi.org/10.21008/j.0860-6897.2023.2.21
42. Rayleigh, The theory of sound. Dover Publication, New York, 1945 of the 2nd second edition 1894 (1877).
43. M. Roohnia, NDT-LAB; system to evaluate the mechanical properties of wood. Iranian patent, IR-patent no. 44032/22-08-1386, 2007
44. M. Roohnia, A. Yavari, A. Tajdini, Elastic parameters of poplar wood with end-cracks. Ann. For. Sci. **67**(4), 409 (2010). https://doi.org/10.1051/forest/2009129
45. M. Roohnia, S.E. Alavi-Tabar, M.A. Hossein, L. Brancheriau, A. Tajdini, Dynamic modulus of elasticity of drilled wooden beams. Nondestr. Test. Eval. **26**(2), 141–153 (2011). https://doi.org/10.1080/10589759.2010.533175
46. M. Roohnia, N. Manouchehri, A. Tajdini, A. Yaghmaeipour, V. Bayramzadeh, Modal frequencies to estimate the defect position in a flexural wooden beam. BioResources **6**(4), 3676–3686 (2011). https://doi.org/10.15376/biores.6.4.3676-3686
47. M. Roohnia, A. Tajdini, N. Manouchehri, Assessing wood in sounding boards considering the ratio of acoustical anisotropy. NDT E Int. **44**(1), 13–20 (2011). https://doi.org/10.1016/j.ndteint.2010.09.001
48. M. Roohnia, A. Tajdini, Identification of the severity and position of a single defect in a wooden beam. BioResources **9**(2), 3428–3438 (2014). https://doi.org/10.15376/biores.9.2.3428-3438
49. M. Roohnia, L. Brancheriau, Orientation and position effects of a local heterogeneity on flexural vibration frequencies in wooden beams. Cerne **21**(2), 339–344 (2015). https://doi.org/10.1590/01047760201521021674
50. M. Roohnia, Wood: vibration and acoustic properties, in *Reference Module in Materials Science and Materials Engineering* (Elsevier, Oxford, 2019), pp. 1–13. https://doi.org/10.1016/B978-0-12-803581-8.01996-2
51. S. Shamaiirani, M. Roohnia, Dynamic modulus of wood containing water-resistant glue finger joint after severe steaming. BioResources **11**(4), 8900–8911 (2016). https://doi.org/10.15376/biores.11.4.8900-8913
52. A. Santoni, Sound radiation and sound transmission in building structures: numerical modelling and experimental validation. Ph.D. dissertation, 2017. Retrieved from https://sfera.unife.it/handle/11392/2488167
53. J.C. Schelleng, The violin as a circuit. J. Acoust. Soc. Am. **35**, 326–338 (1963). https://doi.org/10.1121/1.1918462
54. J.C. Schelleng, The action of the soundpost. Catgut Acoust. Soc. Newslett. **16**, 11–15 (1971)
55. E.B. Skrodzka, B.B. Linde, A. Krupa, Effect of bass bar tension on modal parameters of a violin's top plate. Arch. Acoust. **39**(1), 145–149 (2014). https://doi.org/10.2478/aoa-2014-0017
56. W.T. Thomson, *Theory of Vibration with Applications*, 4th edn. (Springer US, 1998)
57. P.J. Torvik, On estimating system damping from frequency response bandwidths. J. Sound Vib. **330**(25), 6088–6097 (2011). https://doi.org/10.1016/j.jsv.2011.06.027
58. G. Tsoumis, *Science and Technology of Wood: Structure, Properties, Utilization*, vol. 115 (Van Nostrand Reinhold, New York, 1991)

59. M. Van Walstijn, S. Mehes, An explorative string-bridge-plate model with tunable parameters, in *Proceedings of the 20th International Conference on Digital Audio Effects (DAFx-17)* (Edinburgh, 2017), pp. 291–298
60. Vermont Violins, Pernambuco vs. carbon fiber bows. Retrieved from https://www.vermontviolins.com/faqs/2018/7/26/pernambuco-vs-carbon-fiber-bows (2018)
61. U.G.K. Wegst, Wood for sound. Am. J. Bot. **93**, 1439–1448 (2006). https://doi.org/10.3732/ajb.93.10.1439
62. A. Yavari, A. Hemmasi, M. Roohnia, R. Marušák, Dynamic Young's modulus of scarf-and finger-jointed beams using longitudinal vibration method. BioResources **10**(4), 6886–6895 (2015). https://doi.org/10.15376/biores.10.4.6886-6895

Chapter 5
Tonewood: Treatment and Tuning

Abstract This chapter explores the critical role of wood treatment and tuning in the crafting of musical instruments, particularly highlighting the necessity of a scientific approach to tonewood selection. It underscores that relying solely on wood species for quality assessment is insufficient due to significant variations in acoustic properties within the same species. These variations arise from diverse factors such as the tree's origin, the specific section used, drying and storage methods, and prevailing environmental conditions. Consequently, the chapter argues against the adequacy of visual inspection and advocates for the adoption of precise acoustic parameter measurements in modern lutherie. The transformation of raw wood into a resonant musical instrument is presented as a complex process significantly influenced by the inherent variability of wood as a biological material. To overcome these inconsistencies and ensure the suitability of tonewood, the chapter emphasizes the importance of wood treatment. This crucial step allows luthiers to modify and optimize the fundamental acoustic properties of the wood to achieve desired tonal qualities. Specifically, the chapter mentions the manipulation of density, the enhancement of stiffness, and the careful control of damping characteristics as key objectives of wood treatment. It references the work of Ono and Norimoto (Jpn J Appl Phys 22:611, 1983), who demonstrated the dependence of damping on density and elastic stiffness, further highlighting the scientific basis for these treatment approaches. The ultimate aim of wood treatment is to mitigate the natural variability inherent in wood, which is influenced by geographical differences and varying growth conditions. The chapter further elaborates on the important acoustic properties that directly determine an instrument's sound. These include density, stiffness, sound velocity, damping capacity, and acoustic impedance. Understanding and manipulating these parameters through appropriate treatment methods are presented as essential for achieving the desired sonic characteristics of a musical instrument. By carefully controlling these properties, luthiers can fine-tune the wood to resonate in a specific way, ultimately influencing the instrument's tonal balance, projection, and sustain. In essence, this chapter positions wood treatment and tuning as indispensable processes in modern lutherie. By moving beyond subjective visual assessments and embracing scientific evaluation methods focused on precise acoustic parameters, luthiers can effectively address the inherent variability of wood. This controlled modification of the material's properties

M. Roohnia, *Properties and Treatment of Tonewood for String Instrument Construction*, Springer Series in Materials Science 355, https://doi.org/10.1007/978-3-032-00824-4_5

through targeted treatments enables the creation of musical instruments with optimized tonal qualities, ensuring a higher level of consistency and sonic excellence. The chapter sets the stage for a more detailed exploration of specific wood treatment techniques and their impact on the acoustic behavior of tonewood.

5.1 Review of Acoustic and Vibrational Parameters of Wood

Chapter 4 underscored the complexity of wood selection for musical instruments, emphasizing that species alone are insufficient indicators of quality. Significant property variations exist within species, influenced by factors like tree section, drying, storage, and environmental conditions. Consequently, visual inspection is inadequate, necessitating a scientific approach grounded in precise acoustic parameter measurements.

The intricate transformation from tree to instrument is further complicated by these inherent wood variations. To ensure suitability, modern luthiers are urged to employ scientific evaluation methods, recognizing that visual assessments are unreliable. Crucial acoustic properties, including density, stiffness, sound velocity, damping capacity, and acoustic impedance, directly impact an instrument's sound.

5.1.1 The Motivation for the Wood Treatment

Wood treatment is crucial for modifying and optimizing the inherent acoustic properties of tonewood, enabling luthiers to achieve desired tonal qualities. By carefully altering density, enhancing stiffness, and controlling damping characteristics (Ono and Norimoto [74] also demonstrated that it depends on density and elastic stiffness), these treatments address the natural variability present in wood, a biological material influenced by regional differences and growth conditions. The aim is to standardize key acoustic parameters within acceptable ranges for instrument making. To understand the potential for modification, it's essential to consider the fundamental composition of wood, as described in Chap. 2 and summarized again here for a better understanding.

As a complex chemical compound, wood primarily consists of oxygen (approximately 44%), carbon (approximately 49%), hydrogen (approximately 6%), and a small amount of nitrogen [76]. These elements form the structural polymers of cellulose (40–50%), hemicellulose (10–30%), and lignin (20–30%). Lignin, a phenolic polymer, acts as a structural cement within the wood matrix.

Beyond these structural components, trees absorb and store various compounds, including nutrients, natural fungicidal and insecticidal toxins, water, and minerals, in the form of extractives—commonly referred to as wood sap—within their tissues, cavities, and resinous channels. While these extractives do not play a primary structural role, they contribute to the overall density of the wood.

Cellulose, hemicellulose, and lignin are the key structural elements, and any degradation or loss of their elastic stiffness will negatively impact the musical quality of the wood.

The inherent heterogeneity of wood also necessitates treatment considerations. Its spongy, porous structure comprises various cell types, including fibers, parenchyma (wood blades), vessels, and tracheids. Each cell type has a specific function, and their shape and size vary throughout the tree's growth cycle, resulting in the distinct characteristics of springwood and summerwood.

Furthermore, wood exhibits anisotropy, meaning its properties differ depending on the direction of measurement relative to the grain. Approximately 90% of wood cells are oriented along the longitudinal axis of the tree, with the remaining cells oriented perpendicularly. This longitudinal orientation (tracheids and fibers) and the presence of transverse elements (wood rays or parenchyma) contribute to stiffness, particularly in the radial direction.

Macroscopic examination reveals the layered structure of a tree:

- **Outer bark (dead layer)**: A protective, corky layer composed of dead tissues and cells, varying in thickness with age and species.
- **Inner bark (living layer)**: A soft, moist layer of living tissues responsible for transporting nutrients throughout the tree.
- **Cambium (generative layer)**: Located between the bark and the wood, this layer produces new phloem (outer) and xylem (wood-forming) cells.
- **Wood (xylem)**: Situated immediately after the cambium, the wood transports water, stores nutrients, and provides mechanical support. It is further divided into:
  - *Outer wood (sapwood)*: The younger, lighter-colored layer located after the cambium, containing living cells responsible for sap transport.
  - *Heartwood*: The central, often darker-colored core of the trunk, formed when sapwood cells become inactive. Heartwood primarily stores extractives and provides mechanical resistance.
  - *Pith*: A soft, central region of the heartwood with limited industrial value.
- **Rays (medullary rays)**: Radial strips of cells extending from the bark to the pith, perpendicular to the annual rings. Their primary function is nutrient transport and storage, but they also contribute to radial stiffness. An abundance of rays and pith flecks can be less desirable for instrument faceplates.
- **Springwood and summerwood (earlywood and latewood)**: Within each annual growth ring, springwood (formed early in the growing season with ample water) features large cells with thin walls and a lighter color. Summerwood (formed later with less water) has narrower cells with thicker walls and a darker color. The balance between springwood and summerwood characteristics significantly influences the wood's acoustic properties. Wider springwood typically results in

lighter wood with lower elastic stiffness. Achieving a proper balance between springwood and summerwood thickness is crucial for optimal tonewood.

Given the inherent variability of wood as a biological material, sourced from diverse tree origins and growth conditions, treatment processes are employed to maximize uniformity and ensure that key acoustic parameters fall within the desired ranges for crafting high-quality musical instruments.

5.1.2 The Relationship Between Moisture and Wood

The fundamental understanding of wood treatment for acoustic and musical applications begins with the relationship between water and moisture content within the wood structure. Therefore, this section will first explore the interaction between water and wood, followed by a discussion of proper wood drying techniques and methods for evaluating their success. These constitute the initial and most crucial acoustic treatment of wood.

As discussed in detail in the second chapter, concepts such as shrinkage and swelling, equilibrium moisture content, and the fiber saturation point are foundational, and here, we present practical methods and approaches based on this prior knowledge. To consolidate our understanding, water is a natural component of all living parts of a tree. Typically, in the trunk, water comprises more than half the weight of the wood tissue [78]. When a tree is harvested and its wood is processed into lumber, veneer, or other forms, the material immediately begins to release moisture into the surrounding environment. As drying progresses, the wood's dimensions and physical properties start to change. However, even after processing, some water remains within the cell wall structure.

The physical and mechanical properties, resistance to biodegradation, and dimensional stability of any wood product are significantly influenced by its moisture content. Given that water affects nearly all characteristics of wood and its derived products, comprehending the nature of water within wood and its relationship to the material's microstructure and properties is essential. The accompanying figure illustrates the location of water within a wood cell. As long as free water is present in the cell lumen (cavity), the cell wall remains saturated, having physically absorbed the maximum possible amount of water. Importantly, most acoustic properties of wood, excluding weight and density, are not significantly affected by the quantity of water in the cell lumen.

As drying commences, the free water within the cell lumen is the first to be expelled. Once this free water is removed, the bound water within the cell wall begins to dissipate.

- **Different states of water in wood**:

  Generally, water exists in three states within wood:

  - Free water: Located in liquid or vapor form within the cell lumens. Held by capillary forces, it is readily removed from wood in warm, dry conditions and is the first type of water lost during the drying process. Hence, it is termed "free water."
  - Bound water (ımpregnated water): Situated within the cell walls, bound to cellulose chain molecules through surface adsorption forces. Compared to free water, its removal requires more energy. Consequently, wood loses free water before bound water during drying.
  - Structural water: Integrated into the molecular structure of the cell membrane. It is not easily separated from the wood, requiring decomposition of the material for its removal. This water does not influence standard moisture content measurements. The temperature required to remove structural water would result in the wood turning to ash. Therefore, this moisture is not included in wood moisture calculations.

- **Wet wood or green wood**:

  Wood containing both free water and bound water, such as freshly cut timber, is termed "wet wood" or "green wood." In this state, the moisture content is high, and its specific amount varies depending on factors such as tree species, the specific part of the tree (trunk, branch, heartwood, sapwood), tree age, specific gravity, the amount of extractives, soil type, and other environmental variables.
- **Fiber saturation point (FSP)**:

  As already illustrated, the fiber saturation point is the critical moisture content at which all free water has been removed from the cell lumens, but the cell walls remain fully saturated with bound water. Below this point, changes in moisture content directly cause alterations in the wood's properties, including dimensional stability and strength. The FSP is approximately 30% moisture content, based on dry weight, though it can range between 25 and 30% depending on the wood species. Wood used in environments where it does not come into contact with liquid water will typically have a moisture content below the FSP. For instance, outdoor wood typically has a moisture content of 10–12%, while indoor wood ranges from 6 to 7%.

The standard method for measuring wood moisture content relies on Eq. 2.4 (Chap. 2), where a wood sample is weighed and subsequently dehydrated to a zero moisture content. For musical applications, it is crucial that this dehumidification process is performed at a slow rate to avoid detrimental effects at the cellular level. Consequently, wood intended for moisture measurement should be placed in an oven chamber at 65 °C for 72 h. This extended duration ensures the removal of all free water, resulting in a completely dry, or anhydrous, sample. While higher temperatures

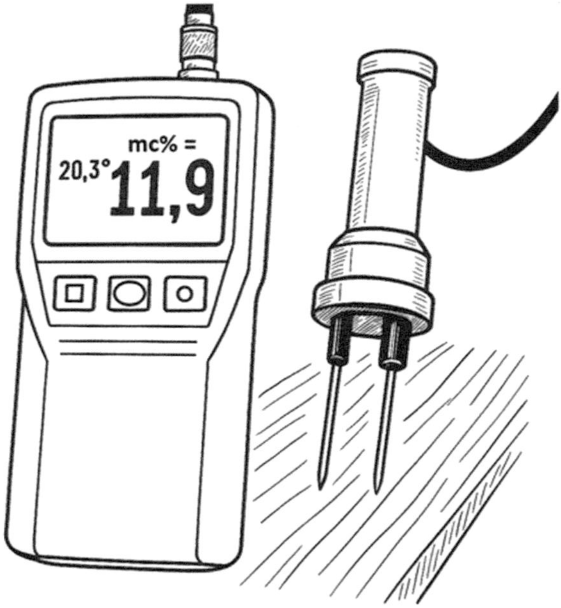

Fig. 5.1 Pin-type timber moisture meter displaying a reading of 11.9 on a piece of wood. Handart by author, enhanced using AI

and shorter drying times may be practical for moisture measurements in other wood industries, the specific requirements of musical instrument construction necessitate a more nuanced and controlled approach.

For those who find this waiting time too long and are willing to prioritize speed over absolute accuracy, acceptable electronic hygrometers offer an alternative. Selecting the correct device hinges on identifying your specific needs and the degree of accuracy required for your particular use (Fig. 5.1).

Electrical moisture meters for wood are devices used to measure the amount of moisture present in wood. These devices operate based on the changes in the electrical properties of wood as its moisture content varies.

Most electrical wood moisture meters utilize the principle of electrical conductivity or electrical resistance. Dry wood is a good electrical insulator, but as its moisture content increases, its electrical conductivity increases, and its electrical resistance decreases. Electrical moisture meters typically have two pins or electrodes that are inserted into the surface or depth of the wood (Fig. 5.1). A small electrical current is then applied through these pins into the wood. The device determines the moisture content by measuring the electrical resistance or conductivity between these two pins. Higher moisture levels result in lower resistance and higher conductivity.

More advanced models employ other methods such as capacitance or radio frequency waves to measure moisture without the need for pin insertion. These types of moisture meters are suitable for measuring surface or near-surface moisture and avoid damaging the wood.

The accuracy of electrical wood moisture meters depends on various factors, including, Device Quality, Wood Species, Temperature, Moisture Range and Usage Method.

Generally, pin-type electrical moisture meters typically have an accuracy of around ± 1% to ± 3% within the common moisture range of wood (approximately 6%–30%). Pinless moisture meters may have lower accuracy but are very useful for their specific noninvasive applications.

To ensure measurement accuracy, it is important to use a calibrated moisture meter and carefully follow the manufacturer's instructions. Additionally, considering the wood species and environmental conditions can help in interpreting the results more accurately.

A reasonable approximation of the wood's equilibrium moisture content can be estimated by knowing the relative humidity of the surrounding air. Simply wait until the wood in question stops shrinking (Table 2.1—Chap. 2).

- **Disadvantages of accelerated wood drying**:

  As the moisture content of wood decreases from its green state to the fiber saturation point (approximately 30%), the wood's density diminishes, while its stiffness remains relatively unchanged. Consequently, specific stiffness, sound radiation coefficient, and acoustic conversion efficiency generally increase, and damping capacity decreases.

  As drying progresses below the fiber saturation point, the elastic stiffness begins to increase, intensifying the aforementioned beneficial changes.

  Proper drying is the essential first acoustic treatment for wood, preventing the negative consequences of improper seasoning. However, accelerating water removal through high temperatures or neglecting proper drying techniques can lead to detrimental effects. Rapid moisture loss can cause the wood cell walls to rupture [88, 92] and differential shrinkage in the orthotropic directions can result in a significant reduction of elastic stiffness (Fig. 5.2).

  Visible damage and cracks in your tonewood sample are a clear advantage, enabling you to readily exclude the flawed material from instrument making through direct observation. The difficulty lies in identifying defects that are too fine and subtle for the naked eye or other senses. Wood with such concealed damage can present a seemingly sound exterior and reasonable density, yet suffer from reduced elastic stiffness. This can occur when a piece, initially healthy during production and packaging, is exposed to unfavorable weather conditions during transit, treatments or tunings, leading to microscopic cracks and damage from repeated drying and wetting cycles and the stresses of orthotropic shrinkage and swelling. Consequently, the necessity and recommendation for employing accurate wood measuring tools in musical instrument making are further reinforced.

  Improper drying techniques or fluctuations in the ambient humidity during storage and transport can lead to warping in wood. It's crucial to understand that these warping cause the wood fibers' growth direction to deviate from the sample's longitudinal axis, negatively impacting its longitudinal elastic stiffness.

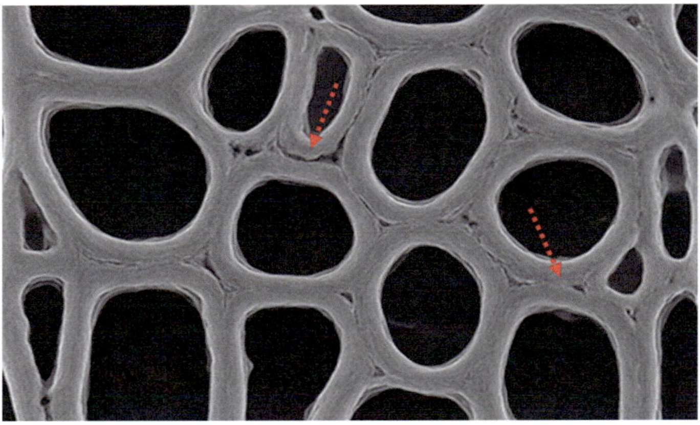

Fig. 5.2 A schematic microscopic image illustrating micro-cracks in wood at the cellular level. The image shows a cross-section of wood with a fracture and separation in the middle lamella, indicated by a red arrow. These micro-cracks, caused by rapid moisture and heat changes during drying or hydrothermal treatment, can significantly affect wood's elastic stiffness even if not visible to the naked eye. A handart by author, enhanced using AI

Attempting to rectify this deviation through carving or sanding might result in a visually straight piece of wood, but it will contain numerous cuts across the natural grain (Fig. 5.3). Consequently, these corrective methods cannot compensate for the stiffness lost due to the initial warping and can, in fact, exacerbate the problem.

Furthermore, if a seemingly sound wood sample exhibits a rough, fuzzy surface during carving or sanding that resists becoming smooth and even as expected, it indicates that the fibers' growth direction is misaligned with the sample's geometric orientation. This misalignment was likely concealed by prior sanding or carving to become invisible to the customer.

- **Dimensional Instability in wood: shrinkage and swelling**:

The various warping observed in wood arise from the inherent differences in the magnitude of shrinkage and swelling along its orthotropic axes.

Wood is classified as an inhomogeneous and anisotropic material, specifically exhibiting orthotropic anisotropy. This means its properties vary depending on the direction of measurement. A readily observable manifestation of this directional dependence is dimensional change, or shrinkage.

Unlike isotropic materials like a simple sponge, orthotropic wood does not shrink uniformly. Understanding this non-uniform behavior is crucial for preventing shrinkage-related defects, which may not become apparent until months or even years after a wood product is finished, only to suddenly emerge with fluctuations in humidity.

Shrinkage is quantified as a percentage representing the dimensional reduction of wood as it transitions from a green (fully saturated) state to an anhydrous (completely dry) state.

A handart by author, enhanced using AI, but inspired from Wikimedia Commons, free resources (it was registered before January 1, 1930): https://commons.wikimedia.org/wiki/File:Cassells_Carpentry.133_Planks_Warped_according_to_position_in_Tree.png

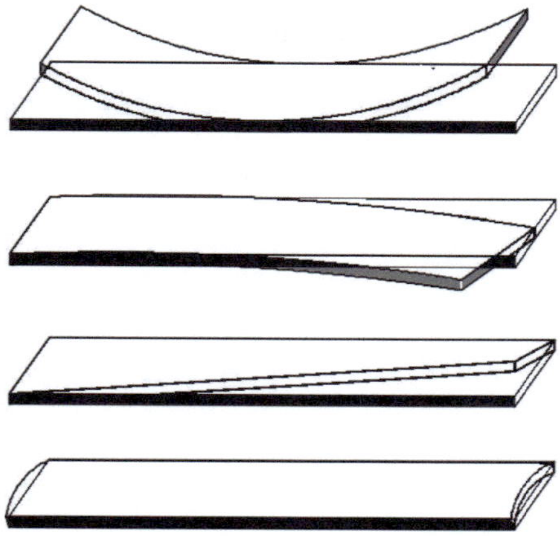

Image from Wikimedia Commons, free resources (Creative Commons Attribution licemse): https://commons.wikimedia.org/wiki/File:Houtvervorming_001.png

Fig. 5.3 The image is illustrating wood warping. The left section shows a cross-section of a log sawn into planks, depicting various forms of warping including bow, crook, cup, and twist, with lines indicating the distorted grain direction. The right section shows four individual planks exhibiting different types of warp: bow (curved along the length), crook (curved along the edge), twist (ends not lying in the same plane), and cup (curved across the width). These warpings cause the wood fibers' growth direction to deviate from the longitudinal axis, reducing elastic stiffness, and that carving or sanding cannot restore this lost stiffness

Longitudinal shrinkage, the change in length along the grain, is typically minimal (around 0.1–0.2%) and often considered negligible in most applications.

Radial shrinkage, occurring perpendicular to the growth rings, varies significantly between wood species, ranging from <2% in dimensionally stable species to approximately 8% in unstable ones. The majority of wood species exhibit radial shrinkage in the 3–5% range. Tangential shrinkage, occurring along the growth rings, also varies, from about 3–12%, with most woods falling between 6 and 10%.

The relationship between tangential and radial shrinkage is expressed by the tangential-to-radial shrinkage ratio, or T/R ratio. This ratio serves as an indicator of shrinkage uniformity and, consequently, the dimensional stability of the wood. Ideally, wood species with good stability will possess a low T/R ratio (see Chap. 2.2.2).

Average shrinkage values for normal wood across numerous species are documented in scientific literature (e.g. [62, 75, 86, 104]). Additionally, the relationships outlined in Chap. 2, based on ISO standards, can be used to independently determine the percentage of shrinkage and the T/R ratio. It is important to remember that deviations from normal wood, such as the presence of reaction wood, juvenile wood, or physical/biological damage, can significantly alter these established values. Any such observed abnormalities typically disqualify that piece of wood for use in instrument making, as abnormal wood also exhibits atypical stiffness and density properties beyond just unwanted shrinkage.

5.2 Wood Drying—A Foundational Acoustic Treatment

This extensive introduction regarding the interaction between water and wood serves as a foundation for understanding the crucial process of wood drying for the musical instrument industry. Therefore, acquiring a thorough knowledge of wood moisture content is a prerequisite before any utilization of timber intended for musical instruments.

The efficacy of wood drying relies on a comprehensive understanding of moisture dynamics and the mechanisms of water movement within the wood structure.

As previously mentioned, moisture in wood exists in three distinct states. Firstly, constituent water is integral to the wood's cell molecules and remains even in seemingly dry timber. It is critical to note that the complete removal of this moisture is not feasible. Attempting to dry wood to zero percent moisture would result in its degradation to ash. Secondly, bound water, located within the cell walls, requires careful management during the drying process. Improper removal of this moisture can lead to irreversible structural damage and compromise the wood's integrity. Thirdly, free water, present as liquid water and vapor within the lumens (cavities) of wood cells, is the most readily removed. It dissipates relatively quickly when exposed to warm or dry air.

A primary concern during wood drying is shrinkage, a phenomenon that can cause significant defects. While many wood suppliers may claim their stock is well-dried, it's essential to recognize wood's inherent hygroscopic nature—its capacity to both lose and absorb moisture. Upon introducing purchased wood into a workshop environment, it will gradually acclimatize to the ambient humidity. For instance, wood brought into a workshop with 40 percent relative humidity will exchange moisture with the surrounding air until equilibrium is reached. Furthermore, machining processes can also induce localized drying and warping. The friction generated by dull cutting tools can heat the wood, leading to uneven moisture loss and potential deformation or cell-wall fractures.

Therefore, the principal objective of wood drying is to achieve moisture equilibrium with the intended service environment. This involves reducing the wood's moisture content to match the typical humidity levels of the region where the final musical instrument will be used.

The equilibrium moisture content (EMC) of wood varies significantly across different regions of the globe due to differing climatic conditions. EMC refers to the moisture content that wood reaches and maintains after prolonged exposure to an environment with constant temperature and relative humidity, indicating a balance between moisture absorbed from the air and moisture lost to the air.

Here's an overview of the average EMC in various regions with different climates:

- Hot and Humid Regions (e.g., Tropical Rainforests) experience very high relative humidity (typically above 80%) and warm, relatively constant temperatures, resulting in a high average EMC, typically between 16 and 20% or even higher. Due to the high humidity in the air, wood has a strong tendency to absorb moisture (e.g., Manaus-Brazil, Singapore).
- Temperate and Humid Regions (e.g., Coastal Areas with Moderate Climates) with moderate to high relative humidity (60–80%), seasonal variations in temperature but generally mild, the average EMC is moderate, typically between 12 and 16%. This can vary depending on the season and proximity to water sources (e.g., Seattle-USA, London-UK).
- Temperate and Dry Regions (e.g., Inland Continental Areas with Hot Summers and Cold Winters) with moderate to low relative humidity (40–70%), significant seasonal changes in temperature, the average EMC is moderate to low, typically between 8 to 12%. During cold, dry winters, EMC can reach even lower values (e.g., Denver-USA, Ankara-Türkiye).
- Hot and Dry Regions (e.g., Deserts) with very low relative humidity (typically below 40%), very hot temperatures during the day and cold at night, the average EMC is low, typically between 5 and 9%. In these regions, wood loses moisture rapidly (e.g., Phoenix-USA, Riyadh-Saudi Arabia).
- Cold and Dry Regions (e.g., Polar or High Altitude Mountainous Areas) with low relative humidity, very cold temperatures, the average EMC is low, similar to hot and dry regions, typically between 5 to 9%. Cold air has a limited capacity to hold moisture (e.g., Fairbanks-USA, Alaska, Lhasa-Tibet).

Actual EMC can vary slightly depending on local conditions, air ventilation, and the specific wood species. In many regions, EMC will fluctuate throughout the year due to changes in temperature and relative humidity. Understanding the EMC of the intended region is crucial for various wood applications, such as construction, furniture making, and the maintenance of wooden structures. Significant changes in wood moisture content can lead to problems like warping, cracking, and dimensional instability.

So, the average equilibrium moisture content of wood is strongly influenced by the surrounding climatic conditions. Regions with high humidity tend to have higher EMC, while regions with low humidity have lower EMC. Understanding these differences is essential for the proper and sustainable use of wood.

Consider the scenario of a tree felled in a Tropical Rainforest intended for use in Ankara-Türkiye. After processing the timber, transporting it to Ankara, and drying it to the Ankara equilibrium moisture content, further considerations arise if the final instrument is destined for an even drier location like Riyadh-Saudi Arabia. Two primary solutions exist.

1. The wood can be processed into the final musical instrument in Ankara, and its entire surface meticulously sealed with a coating. This effectively blocks the wood's pores, preventing further moisture exchange with the environment.

However, this approach is often impractical for musical instruments where significant unvarnished surfaces are desired due to wood's inherent affinity for water.

2. The moisture content of the wood can be adjusted to the equilibrium level of the final destination. For instance, wood intended for use in Riyadh should be dried to approximately 7% moisture content. This can be achieved by temporarily manipulating the temperature and humidity within the workshop environment (Table 2.1, Chap. 2).

A question then arises: Does commercially purchased "dried" wood require further drying? Due to variations in ambient humidity and temperature between the wood's origin and its current location, internal stresses and moisture imbalances are likely to be present. Therefore, the answer is unequivocally yes; purchased wood typically necessitates further acclimatization and drying to reach equilibrium with the local workshop environment.

Finally, it is imperative to recognize that different wood species, possessing unique densities and pore structures, require individual drying protocols. For example, mulberry and walnut should be dried separately, as their shrinkage characteristics and equilibrium moisture points differ. Furthermore, even within the same species, such as walnut, variations in growing climate can influence drying time and necessitate separate drying schedules to account for potential differences in the wood's initial moisture content and cellular structure.

To preserve the acoustic quality of resonance wood, the musical instrument industry generally avoids kiln drying tonewood, even using gentle schedules, as it can lead to internal tension and other microcellular phenomena [95].

5.2.1 *Air-Drying Wood Outdoors (Yard and Garden)*

To effectively air-dry wood in an outdoor setting like a yard or garden, consider the following steps (Fig. 5.4):

- **Ground preparation**:

  Select a drying area with a ground slope between 5° and 10° to facilitate water runoff. Avoid swampy or poorly drained locations. To prevent moisture accumulation beneath the woodpile, it is recommended to prepare the ground by pouring a layer of cement and covering it with a non-absorbent material such as plastic sheeting.
- **Construction of a simple drying shed:**

  Erect temporary walls on the two sides of the wood stack that will face the prevailing wind direction. Install a roof supported by these walls, creating a structure similar to a gazebo but open on two sides. This design promotes consistent, unidirectional airflow through the stack. The roof should be constructed from a non-light-transmitting and non-moisture-absorbent material to shield the wood

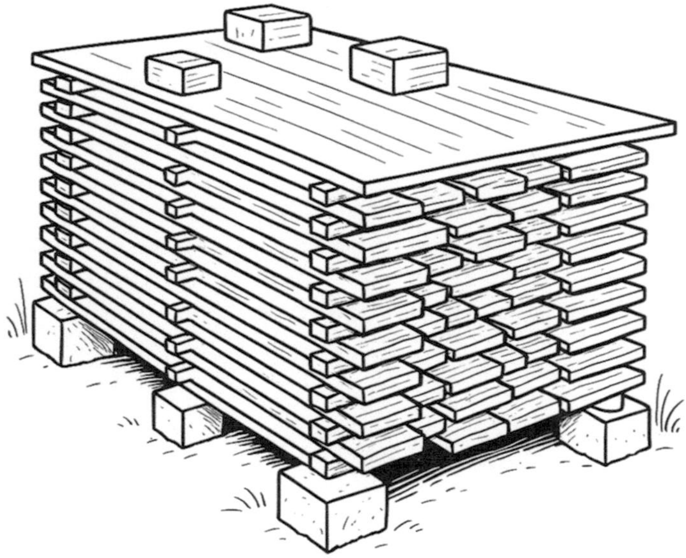

Fig. 5.4 A black-and-white schematic illustration of an open-air wood drying setup. The woodpile is elevated above the ground using concrete blocks, with planks stacked horizontally and evenly spaced by thin wooden sticks (stickers) to allow airflow. A sloped, rigid cover is placed over the pile to protect it from rain, and weights are positioned on top to secure it. The setup is designed to promote proper air circulation for effective drying. A handart by author, enhanced using AI

from direct sunlight, rain, and snow. The walls should also be made of similar materials.

- **Stacking logs**:

  Sort logs by length to ensure consistent stacking. For instance, if using 3-m logs, place cement blocks on the leveled ground at intervals of 1 m or less to provide stable support.

  Position a dowel (a rectangular piece of wood, 2–4 cm in cross-section and equal in length to the width of the logs, preferably made from a denser wood species than the logs being dried, or a non-moisture-absorbent polymer/plastic dowel; grooved or triangular dowels can aid drying) on each cement block.

  Place the first layer of logs onto these dowels. For subsequent logs in the same layer, place another dowel parallel to the first on the same cement block before positioning the next log.

The height of the stack should allow for adequate air circulation, maintaining a suitable distance from the shed's roof and approximately 50 cm of clearance from the walls. Ensure that boards within a single stack are of similar thickness. If thinner boards are present and a separate stack isn't feasible, they can be placed on top of thicker, similarly stacked logs, arranging thicknesses from greatest at the bottom to least at the top.

Orient the lumber in the stack so that the cut ends (where annual rings are visible) face a retaining wall or a temporary windbreak. This encourages cross-ventilation, allowing dry air to enter from one side and humid air to exit from the other and preventing direct airflow onto the wood sections. The walls and roof of the drying shed should extend approximately 1 m beyond the length of the stack on both ends to facilitate this airflow pattern.

To minimize common drying defects, consider sealing the cut ends of the wood with oil or paint, or applying S-irons (S-shaped metal fasteners) to prevent splitting.

The duration required for wood to dry varies depending on the season and the wood species. Warmer periods like spring and summer generally offer more favorable drying conditions.

5.2.2 *Drying Wood Indoors*

Freshly cut wood takes a significant amount of time to dry in a closed workshop. To accelerate this process slightly, an outlet fan can be strategically placed at one side of the stacked wood, directed along the middle of the boards. This creates airflow that helps expel humid air from the workshop. It's worth noting that while inlet fans are generally avoided in tonewood workshop, they are sometimes employed in carpentry. The same drying principles apply to semi-dried wood purchased from a lumber store, although the initial moisture content is lower.

To dry wood indoors effectively, designate a specific area within the workshop for stacking. Choosing a location with good natural ventilation will further reduce the drying time.

It's essential to remember that moisture evaporates from the end grain of wood 10–12 times faster than from its sides. This rapid moisture loss from the ends is a primary cause of end-checking (cracking). To prevent this, the end grain should be sealed immediately after cutting with a moisture-retarding material, as even a delay of a few hours or days can be detrimental.

Various sealants can be used, including paraffin wax, polyurethane paint, varnish, shellac, or latex paint. Apply a thick, even coat to the entire end surface to impede rapid moisture escape and promote more uniform drying throughout the board.

During the initial stages of drying, the path of water vapor escaping the wood should be managed from the core outwards. This can be facilitated by maintaining surface moisture through occasional misting with water (direct spraying should be avoided to prevent uneven moisture absorption).

However, once the wood's moisture content reaches the fiber saturation point (approximately 30%), surface misting becomes ineffective. From this point onwards, it's crucial to avoid any sudden fluctuations in wind, humidity, or temperature to prevent stress and potential defects.

Below the fiber saturation point, shrinkage begins, and the differential shrinkage between the radial and tangential directions becomes a significant concern, potentially leading to warping or cracking. Therefore, it's vital to avoid rushing the drying process.

The wood is considered to have reached its equilibrium moisture content when the shrinkage in thickness and width has relatively ceased. This indicates that the wood's moisture has balanced with the temperature and humidity conditions of your drying space. At this stage, the wood is generally ready for use.

5.2.3 *Predicting Musical Properties of Properly Dried Wood*

The tonewood you receive will likely have a moisture content that differs from that of your workshop or its intended final environment.

All the acoustic tests described in this book provide data specific to the wood's condition at the time of testing. To predict the acoustic properties of your wood after it has been properly dried to the desired moisture content, you can adjust its density and elastic stiffness accordingly.

To correct the measured density to the target moisture content, one could account for both changes in mass and volume (shrinkage and swelling) using a relationship similar to Eq. 2.3 in Chap. 2. However, considering that woods suitable for musical instruments generally exhibit small shrinkage and swelling coefficients, and given the unavoidable errors in non-laboratory settings, such a precise prediction is primarily useful for research and certain applications in the construction industry. For non-research purposes and practical calculations, ISO Standard No. 13910 [43] allows for predicting wood density at the target moisture content using the following relationship:

$$D_{target} = D_{test}\left(\frac{100 + mc_{target}}{100 + mc_{test}}\right) \tag{5.1}$$

Inwhich, D_{test} is the apparent density of the tonewood in testing condition before drying, D_{target} is the apparent density of the tonewood, predicted in target condition after drying. mc is the moisture content of tonewood before and after drying.

As previously mentioned, while factors like wood species and extractive content influence density, the ISO standard provides an acceptable method for estimating density at the target moisture content. As a result, feel free to use it for your practical needs.

As discussed in Chap. 4, all acoustic parameters of wood stem from the interplay between elastic stiffness and density. Notably, even damping capacity, an indicator of sound attenuation, exhibited a significant regression correlation with the stiffness-to-density ratio, as established by Ono and Norimoto [74]. Consequently, the ability to predict elastic stiffness based on variations in wood moisture content would allow us to effectively estimate all acoustic parameters of wood. Divos and Kiss [24] proposed the following relationship to predict the static modulus of elasticity, as required by the European standard EN 338, from the dynamic modulus, considering changes in wood moisture content. This relationship was subsequently revised and validated by Omrani-Ghahjaverestani et al. [73].

$$E_{target} = 0.87E_{test}\left(1 + \frac{mc_{test} - mc_{target}}{50}\right) + 0.6 \tag{5.2}$$

As is clear, E_{test} and E_{target} are the modulus of elastisity of the tonewood in testing condition before drying and predicted in target condition after drying, respectively.

This estimated modulus of elasticity, as defined by the European standard BS [31], represents the static value. However, for the modulus of elasticity pertinent to instrument making, a ten percent upward adjustment of this static value is necessary, based on the guidelines and sources [11] discussed in Chap. 4.

The adequacy of wood drying for maintaining wood integrity can be assessed by comparing estimated density and elastic stiffness values (based on target moisture content) with actual values measured after drying [47].

In essence, failing to achieve the predicted elastic stiffness for a given moisture content after drying, allowing for minor variations, likely indicates wood damage during the process and an unsuccessful drying outcome. The closer the measured stiffness is to the prediction, the more successful the drying; the greater the deviation, the more significant the damage.

Using predicted elastic stiffness and density values, we can calculate a piece of wood's sound velocity, sound radiation coefficient, acoustic characteristic impedance, damping capacity [74], and acoustic conversion efficiency. This analysis provides insight into the wood's suitability for musical instrument applications, potentially leading to more informed material selection.

Just as a reminder, we'll revisit these relationships from Chap. 4 now.

Practical Exercise 5.1

A piece of wood with dimensions 72 x 9 x 4 cm and a weight of 1.400 kg exhibited a fundamental natural frequency of 2810 Hz (F octave 7) in a free longitudinal vibration test of a free-free beam. Given that the wood's moisture content was 18% at the time of testing but is expected to reach 8% in service conditions after successful drying, predict the stiffness, density, and other relevant acoustic parameters of this wood under those service conditions.

Solution

$$v_{test} = 2 \times L \times f = 2 \times 0.72 \times 2810 = 4046 \text{ m/s}$$

$$D_{test} = \frac{mass}{volume} = \frac{1.4}{0.72 \times 0.09 \times 0.04} = 540 \text{ kg/m}^3$$

$$D_{target} = 540 \times \frac{100+8}{100+18} = 494\ \text{kg/m}^3$$

$$E_{test} = D_{test} \times \text{v}_{\text{test}}^2 = 540 \times 4046^2 = 8.84 \times 10^9\ \text{Pa} = 8.84\ \text{GPa}$$

$$E_{target} = 0.87 \times 8.84 \times \left(1 + \frac{18-8}{50}\right) + 0.6 = 9.52\ \text{GPa}$$

$$\text{v}_{\text{target}} = \sqrt{\frac{E_{target}}{D_{target}}} = \sqrt{\frac{9.52 \times 10^9}{494}} = 4390\ \text{m/s}$$

$$K_{target} = \sqrt{\frac{E_{target}}{D_{target}^3}} = \sqrt{\frac{9.52 \times 10^9}{494^3}} = 8.9\ \text{m}^4/\text{kg s}$$

$$z_{target} = \sqrt{E_{target} \cdot D_{target}} = \sqrt{9.52 \times 10^9 \times 494} = 2{,}168{,}612\ \text{Pa s/m}$$

$$\tan \delta_{target} = 10^{-1.23} \times \left(\frac{E_{target}}{D_{target} \times 10^6}\right)^{-0.68}$$

$$= 10^{-1.23} \times \left(\frac{9.52 \times 10^9}{494 \times 10^6}\right)^{-0.68} = 0.008$$

$$\text{ACE}_{target} = \frac{K_{target}}{\tan \delta_{target}} = \frac{8.9}{0.008} = 1112.5\ \text{m}^4/\text{kg s}$$

Comparing these results with the wood's acoustic properties under the test conditions reveals that proper drying is the primary treatment that reduces density and damping capacity while increasing elastic stiffness, sound velocity, sound radiation coefficient, and acoustic conversion efficiency. Importantly, because a successful drying process leads to a simultaneous increase in elastic stiffness and a decrease in density, the acoustic characteristic impedance of the wood remains relatively unaffected.

Note 1: Incorrect predictions after drying indicate a flawed process and potential wood damage.
Note 2: Stiffness corrections are applicable for a target moisture content difference of smaller than 18% $(mc_{test}–mc_{target}) \leq 18\%$.
Note 3: Stiffness and density corrections apply only below the saturation point.

5.3 Chemical Treatments

5.3.1 Solvent Washing

The primary goal of solvent washing is to eliminate non-structural extractives from the wood's cell cavities. These extractives do not typically contribute to the wood's stiffness or mechanical resistance.

A suitable solvent is defined as any chemical agent capable of penetrating the wood structure, dissolving the extractives within, and effectively removing them during the washing process. However, distilled water, ethanol, and acetone solutions are particularly effective in extracting sugars, resins, oils, and similar compounds from wood. Notably, the TAPPI and ASTM standards for wood, cellulose, and paper identify two main solvents as sufficient for the complete removal of wood extractives [4, 51, 100].

Solvents used for wood washing should ideally have a neutral pH of approximately 7 to avoid acidic or alkaline contamination. Both acids and bases are detrimental to the integrity of the wood cell wall. Furthermore, the use of detergents and bleaches, such as laundry detergent, dishwashing liquid, and bleach (including bleach solutions), should be avoided as they also damage the wood cell wall.

- **Washing wood with distilled water**:

This process removes water-soluble extractives, such as sugars and certain mineral salts, from the wood. Using non-distilled water as a solvent would cause its dissolved salts to precipitate within the wood, counteracting the treatment's primary goal: to lighten the wood without compromising the cell wall, stiffness, or mechanical resistance.

The risks associated with this treatment method, such as successive cycles of drying and wetting and the differential tangential and radial shrinkage, necessitate careful wood drying after the washing stage.

While this treatment has a historical precedent in traditional instrument making [15], some relevant scientific research are certifying it and are illustrated in future paragraphs.

- **Washing wood with organic solvents (ethanol and acetone)**:

This specific washing process aims to remove remaining extractives that are soluble in organic materials, including fats, proteins, waxes, and certain mineral salts. A mixture of one part acetone and three parts ethanol (96°) is recommended for the efficient extraction of these organic-soluble extractives [25].

Due to the water content in the ethanol, the wood will become wet during this treatment. The subsequent cycles of wetting and drying, along with the inherent differences in tangential and radial shrinkage, are significant risks associated with this method. Therefore, proper wood drying procedures are crucial following this treatment.

It is important to note that most of the organic solvents e.g. ethanol and acetone are highly flammable solvents. Consequently, strict adherence to safety protocols and the use of closed containers are essential during this extraction method (Fig. 5.5). While this methodology aims for accuracy, it does not encompass all potential safety concerns. Users must establish suitable safety, health, and environmental protocols and ascertain any relevant regulatory limitations prior to implementation.

Which type of washing is required to extract the wood extractives content? The answer to this question varies. Depending on the specific extractives targeted, sometimes only water washing is necessary, sometimes only organic solvents washing is

Fig. 5.5 Plastic barrels with sealed lids, designed for safely storing flammable solvent-soaked wood during organic solvents extraction. The barrels are depicted in varying sizes, with circular, clamp-secured lids. A handart by author, enhanced using AI

sufficient, and in other cases, both sequential treatments are required. Conversely, neither washing method may be necessary for certain applications.

When both water washing and organic solvents washing are deemed necessary, the water washing step should always precede the organic solvents washing. For instance, washing can sometimes decrease the density below the acceptable limit for wood used in the back and sides of the bowl. In rare cases, the removal of extractives can also reduce the elastic stiffness, as observed in Pernambuco (*Paubrasilia echinata*—formerly listed in the genera *Caesalpinia* and *Guilandina*) used for violin bows [57, 58].

Traditional Iranian instrument makers sometimes employ a pre-treatment for soundboards. Selected wood is immersed in distilled water with bread yeast for a period before washing. This pre-fermentation facilitates the release of extractive substances, simplifying the subsequent cleaning process with water, alcohol, and acetone. Below, there are laboratory experiments on solvents washing with or without pre-fermentations and total wood extractive removal. Continued research on other wood products is recommended.

Water washing increased the sound radiation coefficient and decreased the damping capacity of maple wood pellets. Subsequent alcohol washing of these water-washed maple samples further increased the sound radiation coefficient and further decreased the damping capacity [82].

In contrast, water washing reduced the sound radiation coefficient and increased the damping capacity of Persian silk wood (*Albizia julibrissin*). However, alcohol washing of these water-washed Albizia samples again resulted in a smaller sound radiation coefficient and a larger damping capacity [32]. This suggests that extractives in Albizia wood may contribute to the structural integrity of its cell walls. Similarly, in Pernambuco wood, the removal of extractives, through either water or organic solvent washing, resulted in a decreased sound radiation coefficient and an increased damping capacity [2, 58].

Water washing of pine wood (*Pinus silvestris*) increased the sound radiation coefficient. This increase was more pronounced in pre-fermented samples. Washing with an ethanol-acetone mixture also increased the sound radiation coefficient. The magnitude of this increase surpassed that observed with water washing. However, the increase resulting from ethanol-acetone washing was similar in both pre-fermented and non-pre-fermented samples. Water washing reduced vibration damping, and fermentation pre-treatment further decreased these values. Ethanol-acetone washing caused a substantial reduction in vibration damping compared to water washing. Fermentation pre-treatment amplified this decrease in damping values [21].

In ash wood, water washing decreased the sound radiation coefficient and increased the damping capacity. However, subsequent ethanol washing of these water-washed samples resulted in a higher sound radiation coefficient and a lower damping capacity. Fermentation pre-treatment did not significantly alter these effects [22].

The soaking process of walnut wood (*Juglans regia*) with water and ethanol-acetone mixture washing increased the acoustic coefficient. Pre-treatment in both methods led to a further enhancement of this factor. Water soaking without pre-treatment showed minimal impact on the damping factor. However, pre-treatment prior to water soaking resulted in a significant decrease in damping factor values. Ethanol-acetone mixture washing significantly reduced vibration damping values. Pre-treatment before ethanol-acetone washing further amplified this reduction in damping factor values, suggesting that extractives soluble in the ethanol-acetone mixture negatively influence the damping factor of walnut wood [108].

Both water washing and the removal of Arizona cypress wood extractives by organic solvents significantly reduced vibration damping. The improvement of this acoustic property (reduction of damping capacity) was more pronounced with ethanol-acetone washing compared to water washing. Fermentation pre-treatment amplified the reduction in damping observed with both water washing and ethanol-stearate washing. The sound radiation coefficient did not exhibit significant changes under water washing, ethanol-acetone washing, or fermentation pre-treatment [109].

Similarly, both water washing and the removal of spruce wood extractives using organic solvents significantly reduced vibration damping and increased the sound radiation coefficient [61].

In Pterocarpus wood (*Pterocarpus erinaceus*), a higher extractive content correlated with lower damping capacity and a slightly higher sound radiation (Traoré et al. 2010). This suggests that certain extractives in this wood likely contribute to cell wall structural integrity, influencing its acoustic properties. While this study examined wood samples with varying extractive content, Brémaud et al.'s investigation [12], into solvent washing of African padauk—another species of this genus (*Pterocarpus soyauxii*)—extractives revealed that complete removal of both lumen and cell wall extractives could improve and reduce the damping capacity of the wood. Notably, this process did not significantly alter the specific stiffness, an index related to sound radiation.

Laboratory studies on mulberry wood generally indicate that higher extractive content correlates with weaker sound radiation and higher attenuation [81]. Consequently, solvent washing of mulberry wood has become a focus of research. Water washing has been shown to reduce the damping capacity of mulberry wood [80], without a positive effect on the sound radiation coefficient [91]. However, extraction of extractives using organic solvents can effectively reduce the damping capacity and increase the sound radiation coefficient [91].

The influence of water-soluble extractives on the acoustic properties of *Arundo donax* (reed) was investigated across varying relative humidity [70]. At low relative humidity, the extractives increased both the density and the dynamic Young's modulus of the reed, potentially having a limited impact on the sound radiation coefficient. This effect diminished with increasing relative humidity due to the deliquescence of the extractives. It was proposed that the extractives enhance reed quality by reinforcing its dry state. Furthermore, the extractives significantly increased the reed's damping capacity, suggesting that their removal would decrease damping without necessarily increasing sound radiation. These alterations might affect the reed's frequency response, contributing to a richer and softer tone. The tonal qualities lost upon extractive removal were restored by impregnating the reed with glucose, the primary component of natural extractives.

Nefdt [65] highlighted the potential of extracting wood extractives with non-destructive solvents to reduce density while preserving elastic stiffness, thereby enhancing the sound radiation coefficient. However, our observations indicate that this outcome is not universally consistent. In some instances, extractive removal has paradoxically led to a loss of elastic stiffness and a subsequent decrease in the sound radiation coefficient. Therefore, we advise against the indiscriminate solvent washing of all wood species, as the efficacy of each treatment is context-dependent. Furthermore, accurate prediction of treatment effects and selection requires the measurement of acoustic parameters prior to intervention.

The impact of wood extractive removal on the sound radiation coefficient (K) is complex, stemming from their influence on both density (D) and modulus of elasticity (E), where a higher E and lower density yield a higher sound radiation coefficient.

Extractives, a diverse group of compounds within wood, can affect sound radiation coefficient through several mechanisms.

Firstly, their removal often decreases density, as extractives are typically denser than the wood's cellulose, leading to an increase in sound radiation coefficient. However, in rare instances where extractives occupy significant volume, their removal might slightly increase density, potentially decreasing sound radiation coefficient.

Secondly, the effect on modulus of elasticity varies. Removing weakly bound extractives can sometimes increase modulus of elasticity by improving cell structure integrity, thus increasing sound radiation coefficient. Conversely, extractives contributing to structural strength can, upon removal, decrease E and consequently sound radiation coefficient.

Thirdly, the specific types and amounts of extractives, which differ significantly across wood species, dictate the variability of these effects. Finally, the extraction method itself can selectively remove compounds and even induce structural changes (e.g., through improper drying), further complicating the prediction of sound radiation coefficient's behavior post-treatment. Therefore, a nuanced understanding of the extractives' role within a specific wood species and the chosen extraction technique is important for predicting and optimizing its acoustic properties.

Reducing density without altering stiffness leads to an increased acoustic radiation coefficient and decreased damping capacity. If density reduction is accompanied by a decrease in stiffness, the acoustic radiation coefficient and damping capacity can either increase, decrease, or remain unchanged, depending on the relative magnitudes of the density and stiffness reductions. When both the radiation coefficient and damping capacity decrease, evaluating the combined acoustic conversion efficiency is important for decision-making. Cellulose provides wood's mechanical resistance, lignin its stiffness, and hemicellulose its flexibility [10]. We now need to determine which other structural components of wood, besides extractives, have been affected by our extractive material removal method.

5.3.2 Enhancing Acoustic Properties of Wood with Additives

Generally, introducing any additive to wood increases its mass, consequently raising its density. Given that these additives typically don't offer specific reinforcement to the cell wall structure, a significant increase in density due to their application will likely result in a decreased acoustic radiation coefficient and, consequently, an increased damping capacity. This raises the question of whether applying such additives is necessary or appropriate. Could this treatment compromise the inherent characteristics of the wood, especially for applications similar to musical instruments? However, exceptions may exist where certain additives are predicted to integrate into the mechanical structure of the cell wall. In such cases, by enhancing stiffness relative to density, these additives could potentially reduce damping capacity and increase the acoustic radiation coefficient.

Throughout history, musical instrument makers globally have treated wood with additives using a range of traditional techniques [15, 99]. While some of these methods have proven useful, many are undoubtedly now deemed unacceptable and inapplicable based on current knowledge. Moreover, the original reasons for adding these substances often weren't acoustic, and their continued use today appears to be a repetitive and aimless practice rooted in tradition.

From antiquity, concurrent with the professionalization of musical instrument creation, historical evidence reveals human experimentation with various wood treatment methods. A notable example is the application of substances such as Roman hydraulic cement and volcanic ash to violin wood, frequently with or without varnish [6]. Intriguingly, these techniques have persisted to the present day, and their rationale—whether aimed at improving acoustics or enhancing pest resistance—deserves scholarly investigation and judicious replication, rather than uncritical adherence as tradition.

To begin with, it's essential to measure the acoustic parameters of the wood. This allows for targeted treatment selection based on whether a specific parameter needs enhancement or reduction. When wood is intended for a soundboard, any additive should either increase its stiffness or decrease its density. Ideally, if an additive inevitably increases density, its contribution to stiffness should outweigh this increase. Conversely, if the wood is for the back and sides of the instrument's body, the additive must not compromise its elastic stiffness. Consequently, alkaline and acidic substances, known to damage cellulose and lignin, should be avoided in wood treatment. Any historical or traditional use of such substances should not be replicated without thorough investigation.

Meanwhile, certain additive treatments can positively influence the acoustic properties of wood in musical instruments, while others can be detrimental. Interestingly, some additive treatments are necessary for practical reasons and thus require careful application to minimize acoustic damage. For instance, various types of varnishes used to enhance the wood's durability are applied out of necessity, a point that will be elaborated upon in the subsequent paragraphs.

Malvermi et al. [55] investigated the impact of alkaline treatments in the priming process on the mechanical properties of wood used for musical instruments. They stated that luthiers (violin makers) commonly use alkaline solutions as a pretreatment to enhance wood durability and prevent biological attacks. This study explored how alkaline primers weaken the vibrational properties of tonewood. The researchers found that pretreating wood with alkaline solutions can reduce its specific stiffness. Notably, the decrease in specific stiffness was much more severe with KOH compared to NH_3, which appeared to leave the specific stiffness largely unaffected. They also showed that alkaline pretreatments influence the penetration of subsequent treatments (coating), ultimately enhancing the vibrational properties of the wood, made this step necessary before coating.

It has been previously emphasized that most chemical treatments with additives found in old musical instruments were intended for purposes other than acoustics, such as durability and aesthetics [65]. However, their unavoidable effects on the wood's acoustic properties have misguided many instrument makers seeking to replicate historical patterns and methods.

Our initial exploration into additive treatments for musical instruments, specifically using alkalis, stems from recent studies. While we acknowledge that employing such treatments without a thorough understanding of their underlying mechanisms might seem unconventional, these studies have reported high concentrations (up to 1000 ppm) of sodium and other elements (including potassium, chlorine, calcium, copper, and zinc) within the cell walls of Stradivari and Guarneri violins [98, 99]. This presence could be attributed to various additive treatments involving alkaline salts. One plausible explanation is that logging companies might have used salt in waterways to transport timber, potentially leading to beneficial alterations in the wood's chemical structure for musical instruments [65]. Furthermore, a French researcher named Palissy also noted the positive effects of certain salts on wood tone [65]!

These findings suggest that specific minerals introduced through chemical or preservative treatments might have contributed over time to enhancing the acoustic properties of historical musical instruments.

Researches indicated that controlled, short-term, and weak alkaline treatments can increase the crystallinity index of the cell wall, potentially leading to improved acoustic characteristics [106]. Notably, studies have shown that the sodium compounds sodium hydroxide and sodium sulfite reduce wood density by degrading hemicellulose and lignin while preserving cellulose [54, 102].

In contrast to acidic solutions like H_2SO_4, alkaline solutions do not damage the S2 layer of the cell wall [46, 59]. Instead, they primarily loosen the middle lamella, which has the highest concentration of hemicellulose, especially when combined with heat treatments at 100 °C and 120 °C. As mentioned, both sodium hydroxide and sodium sulfite have demonstrated the ability to degrade hemicellulose.

The established link between increased crystallinity index and improved acoustic properties of wood implies that any substance capable of even partially enhancing wood's crystallinity could positively influence its vibroacoustic behavior. Qi et al. [77] revealed that treating Chinese pine, white oak, and poplar wood with varying concentrations of NaOH improved their crystallinity. Their findings indicated that NaOH effectively removed hemicellulose. Similarly, Barman et al. [8] found that increasing NaOH concentration from 2.5 to 10% effectively facilitated hemicellulose removal. However, concentrations exceeding 10% damaged the wood stiffness. Specifically, the crystallinity index increased with concentration up to 10%, but a decrease was observed beyond 20%. It's important to note that the apparent increase in the crystallinity index might be a relative effect. The degradation of amorphous components like hemicellulose and lignin would inherently increase the proportional representation of crystalline cellulose within the cell wall. Therefore, the results reported by these researchers warrant further investigation, as alkaline treatments are fundamentally known to be detrimental, reducing wood crystallinity at higher concentrations and prolonged exposure times. This is because sodium hydroxide essentially disrupts the OH– bonds between cellulose molecules within the crystallization zones, leading to the dissolution of crystalline cellulose.

Evidently, a higher percentage of crystalline material within the cell wall layers would facilitate the propagation of sound waves through the wood, as crystals are denser than the othe components. However, whether alkali salts directly promote cellulose crystallinity or merely make the existing crystallinity more prominent through the preferential removal of amorphous hemicellulose and lignin in weak treatments requires further research. This explanation, which aligns with previous reports [65, 99], is not solely my interpretation for the purpose of this book.

Nevertheless, based on these research observations, a higher proportion of crystalline solids within the cellular structure could potentially lead to a slight increase in the speed of sound in wood. Conversely, prolonged exposure during chemical, thermal, or degradation treatments leads to a decrease, or disappearance, of the crystallinity index, ultimately negatively impacting the acoustic properties.

Given these problems and issues, it's important to question the true value of solely resorting to chemical additives to enhance the acoustic properties of wood, especially when simpler acoustic treatment methods are available. The answer is likely no. While negative reports are also exist, such as the finding that NaOH treatment cannot be repeated on spruce due to compromised wood structure and surface cracks [65], treating wood with a sodium solution to improve acoustic properties is therefore not considered a feasible approach. However, when additives are necessary for reasons beyond acoustics, we must still monitor their impact on the wood's sound response. If the effect is beneficial, we can embrace it, if detrimental, we should avoid it if possible, or at least minimize the negative impact. Furthermore, many seemingly positive attributes observed when comparing aged, high-quality musical wood with modern wood, often attributed to chemical treatments, might instead be a result of natural aging processes and the continuous vibrations from playing. For instance, the reduction in hemicellulose and lignin content, along with the preservation of cellulose—characteristics linked to mild, controlled alkaline additive treatments—have also been found in the wood of old Stradivarius violins, suggesting that aging may be a significant contributing factor [13, 99].

Despite these concerns, there is a promising perspective that justifies research into chemical treatment with additives: the potential to accelerate wood aging. If certain chemical additive treatments can indeed expedite the aging process, and if aged wood demonstrably possesses superior acoustic qualities based on experience and research, then this scientific and practical avenue is certainly worth exploring. Indeed, several researchers have shown interest in this very topic, which we will now discuss.

5.3.3 Aging to Enhance the Acoustic Properties of Wood

Let's review the changes wood undergoes during the aging process that can enhance its musical properties.

In the short term (3–10 years), wood achieves dimensional stability and complete moisture equilibrium. This means that dimensional changes due to moisture absorption and loss are minimized.

During the medium term (up to approximately 100 years), structural changes occur within the wood. These include an increase in cellulose crystallinity, chemical changes in lignin, and a decrease in hemicelluloses. These alterations can effectively improve the wood's acoustic properties.

However, in the long term (beyond 100 years), the physical, mechanical, and acoustic properties of wood gradually decline. Consequently, the use of very old wood (e.g., wood aged several centuries) is not recommended for manufacturing musical instruments [13].

In summary, wood stabilizes in the short term, undergoes structural changes that enhance acoustic properties in the medium term, and gradually loses these properties in the long term.

To accelerate short-term stabilization, careful drying of the wood is essential. To expedite medium-term changes, excess materials like extractives or hemicellulose should be removed chemically or thermally, depending on the desired properties for the specific instrument component.

However, excessively accelerating wood aging can cause significant damage, as seen in Guarneri and Stradivari violins. This suggests that the sound quality of these instruments likely deteriorated in this aspect over extended periods, peaking at around or before 100 years.

Of course, there's another aspect in the sound quality of old instruments; the effect of playing. It is widely accepted that instruments improve with time and continued playing. A violin, for example, requires a period of playing (months or years) to reach its optimal acoustic performance [13].

Conversely, an instrument's acoustic quality diminishes when it is not played for an extended period, due to the relaxation of static and dynamic loading from string tension [13]. The instrument experiences loading from both its vibration during playing and the static pressure exerted by the strings.

Examining the rheology of the violin system reveals fatigue under the combined influence of a dynamic regime (vibration during playing) and a static regime (tension from the strings).

Continuous vibration can induce subtle structural changes in the wood that affect its sound. For instance, the breaking of hydrogen bonds between cellulose molecules through continuous vibration can reduce the wood's internal damping [13].

Another hypothesis is that continuous vibration expels workshop dust from the microscopic pores of the interior, unvarnished surfaces of the wood and could therefore possibly be useful in improving the sound of certain notes and modes that have been repeatedly resonated and vibrated. To determine the validity of this hypothesis, specific experimental studies are necessary. These studies could include:

- Particle analysis within the pores of both aged and new wood.
- Measurement of changes in wood's acoustic properties before and after continuous vibration, considering the presence or absence of dust.

- Application of microscopic imaging techniques to directly observe the movement of dust particles within wood pores during vibration.

while this hypothesis presents an intriguing possibility, it currently lacks empirical support and necessitates rigorous scientific investigation.

However, valuable research has investigated the positive impact of wood aging on the acoustic properties of tonewood, and some examples are mentioned here.

One of the earliest studies on this topic was conducted by Barducci and Pasqualini [5], who investigated the effect of natural aging on the acoustic properties of wood. They used the resonant frequency method to measure sound velocity and the quality factor in the longitudinal and radial directions of spruce and maple wood. Their findings indicated that the best acoustic properties were observed after 10 years of natural aging.

Holz [42] also examined the influence of aging on the acoustic and mechanical properties of wood. This research involved selecting wood samples aged 2–4 years, 60 years, and 180 years, and determining the longitudinal elastic modulus and internal friction using the resonance method. Bucur [13] suggested that a submicroscopic analysis of the wood material is necessary for a deeper understanding of the changes caused by wood aging.

She proposed that combining non-destructive methods like X-ray micro-densitometry, X-ray diffraction, and ultrasonic techniques could yield more accurate results.

Sobue and Okayasu [93] analyzed the effect of continuous resonant vibration on wood quality.

Cai and Tai [15] explored the knowledge surrounding wood used in the construction of the ancient Chinese guqin. They aimed to understand how the sound quality of this instrument improves over time by translating and interpreting ancient Chinese texts and comparing this traditional knowledge with modern scientific findings. In this context, they drew parallels between beliefs about sound improvement in ancient Chinese stringed instruments (guqin) and Italian violins, noting the shared belief that high-quality instruments improve with age.

It is reported that artificial aging with UV radiation can enhance the acoustic properties of tonewood [96], mimicking the natural aging process, by increasing crystallinity and altering the polymer structure. Analyses revealed that the chemical structure of polysaccharides, particularly cellulose, was affected, and its crystallinity increased. Additionally, changes in hemicellulose and lignin within the wood contributed to the improvement of its acoustic and mechanical properties. These findings have potential benefits for the production of higher quality musical instruments.

Research on the effect of UV treatment on the acoustic properties of wood is limited, as UV radiation primarily alters the wood's surface properties [65]. However, Gurău et al. [40] found that in 4 mm spruce and maple samples, UV treatment increased the speed of sound and elastic modulus, and decreased density. They suggested that these improvements could be attributed to both UV and the associated heat treatment.

UV exposure is known to change the color of wood surfaces by accelerating the photodegradation of lignin compounds [65]. Lignin chromophores absorb UV light, initiating the formation of free radicals that react with oxygen, resulting in color changes. Severe photodegradation from weathering can also lead to the loosening of surface fibers. UV light penetrates the wood surface to a depth of approximately 2 mm.

However, exposure to UV radiation in accelerated weathering chambers also generates heat, potentially raising the temperature to 40–90 °C. Therefore, the changes observed may be due to heat treatment induced by UV radiation, making it difficult to isolate the specific effects of UV.

A socio-technical study by Carlier et al. [17] surveyed violin makers to document their views and methods for selecting acoustic wood. The results indicated that violin makers believed short-term aging (from a few years to decades) improved the wood's physic-acoustic properties, whereas long-term aging (spanning tens or hundreds of years) influenced aesthetic qualities more significantly than acoustic properties. This was also highlighted later in another research by Obataya et al. [72]. The empirical selection of wood by makers, relying on perceptual criteria (visual, physic-mechanical, and auditory), and correlates with measured acoustic properties. However, a more detailed analysis is necessary to ascertain the precise contribution of each of these criteria. Notably, spruce tonewood possess different visible structural features (e.g., uniformity of growth rings) than conventional softwoods, and these distinctions are attributed to the unique characteristics of these pre-selected woods. The researchers proposed that a model could be created to assess the properties of a complete violin top using cut scraps, potentially refining the wood selection process. In conclusion, the study underscored that both tonewood selection and its age are key determinants of violin quality, but their effects are contingent on several factors that warrant further research for full elucidation.

One of the most comprehensive recent studies on artificial wood aging was conducted by Nefdt and Meincken [66]. This study, which will be discussed a bit more, investigated various wood modification methods to enhance the acoustic properties of tonewood, particularly the acoustic radiation ratio similar to what is happening in natural aging. The primary goal was to increase the radiation ratio by reducing wood density without significantly compromising the dynamic elastic modulus. The methods explored included thermal, fungal, and chemical treatments (using NaOH and Na_2SO_3 solutions as already introduced in previous paragraphs), as well as combinations thereof. Initial experiments were performed on the lower-cost pine species (*Pinus patula*) before successful methods were applied to the higher-quality tonewood, spruce (*Picea abies*). The modification methods were as follows:

- **Thermal**: Samples were heated at temperatures ranging from 160 to 250 °C for durations of 5 min to 8 h. To prevent surface burning at temperatures exceeding 200 °C, samples were wrapped in aluminum foil.
- **Fungal**: Samples were exposed to white rot (*Schizophyllum commune*) and brown rot (*Laetiporus sulphurous*) fungi for periods of 1 to 20 weeks.

- **Chemical**: Samples were immersed in 0.2 M solutions of NaOH, Na_2SO_3, or a combination of both, for durations of 8–168 h.
- **Combined**: Effective methods (white rot, NaOH treatment, and heating at 250 °C for 1 h) were combined.

Sound velocity was measured using a Lucchi meter (introduced in Chap. 4, Fig. 4.8), followed by the calculation of the sound radiation coefficient. Results showed that:

- **Thermal**: Temperatures below 200 °C had no significant effect on radiation and, in some cases, even decreased it. Optimal results were achieved at 250 °C for 1 h, resulting in an increase in radiation of 8% for pine and 5.9% for spruce. This improvement was attributed to a decrease in density while maintaining or slightly increasing the dynamic modulus. The reduction in density was due to the removal of hemicellulose and extractives, accompanied by an increase in cellulose crystallinity.
- **Fungal**: White rot (*S. commune*) demonstrated greater effectiveness than brown rot in reducing density. In pine, white rot exposure for 20 weeks increased radiation by 11%, compared to a 6% increase with brown rot. After 6 weeks, white rot increased radiation by 4%. In spruce, white rot exposure (6 weeks) resulted in only a 0.3% increase in radiation, likely due to the inherent resistance of seasoned spruce wood. Density reduction resulted from the degradation of hemicellulose and lignin without damaging cellulose. Fungal treatment also enhanced cellulose crystallinity, which is similar to what is happening during the aging.
- **Chemical**: NaOH treatment proved more effective than Na_2SO_3, increasing radiation by 2–4% (with a maximum increase of 5.84% after 120 h in 0.2 M NaOH). Na_2SO_3 treatment, alone or in combination with NaOH, yielded less favorable outcomes. In spruce, NaOH treatment led to cracking and a honeycomb structure due to poor solution penetration and uneven drying, rendering it unsuitable for this species. Density reduction was attributed to the removal of hemicellulose and increased cellulose crystallinity; however, prolonged exposure to higher concentrations damaged the cell wall structure.
- **Combined**: The combination of white rot treatment (6 weeks) and thermal treatment (250 °C, 1 h) increased radiation by 13.3% in pine and 5.5% in spruce. Combining NaOH treatment (120 h) and heat treatment in pine increased radiation by 19.8% (the most effective overall), but this method was not viable for spruce due to cracking.

All methods (thermal, fungal, and chemical) successfully increased R by reducing density and maintaining or increasing elastic stiffness. This enhancement was linked to the removal of hemicellulose and extractives, along with increased cellulose crystallinity.

The optimal method for spruce involved combining heat and white rot fungus, while the combination of NaOH and heat proved most effective for pine.

NaOH treatment is unsuitable for seasoned spruce wood due to poor penetration and structural damage.

Their suggestions for future research directions include exploring more effective white-rot fungi for spruce, treating freshly harvested wood, and investigating the reverse sequence of combined treatments (thermal followed by fungal or chemical).

Furthermore, research has shown that while extracting extractives from a particular wood species can deteriorate its acoustic properties, adding those same extractives to a different wood species might improve its acoustic properties [2, 58]. In these studies, extractives from Pernambuco wood (Brazilwood) for example were extracted and introduced to spruce wood in various ways. Spruce wood treated with these chemical additives, essentially Pernambuco extractives, exhibited enhanced acoustic properties suitable for use in soundboards.

5.3.4 The Effects of Varnishes

Varnishes (lacquers or protective coatings) play an important role in the construction and finishing of musical instruments, particularly stringed instruments like violins, guitars, and cellos. While varnish serves several purposes, its effect on an instrument's sonic properties is a complex issue with both positive and negative aspects. This discussion will examine the purposes of varnish, its effects on sound, and whether its use is beneficial or detrimental. Purposes of Varnish are:

- **Protection of wood**: Varnish provides a protective layer on the wood's surface, shielding it from moisture, temperature fluctuations, UV radiation, and physical damage like scratches. This is especially important for tonewoods such as spruce and maple, which are used to make instruments, because moisture can lead to warping or cracking. However, it doesn't always cover the inner wood surface. This long-term protection enhances the instrument's durability and prevents wood decay.
- **Aesthetics**: Varnish enhances the wood's appearance by accentuating its natural color and giving the instrument a polished and attractive look. This is important for both luthiers and musicians from a visual standpoint. Clear or colored varnishes can highlight the wood's grain patterns.
- **Effect on acoustic properties (indirect)**: Varnish can indirectly influence acoustic vibrations by altering the mechanical properties of the wood's surface, such as its hardness or flexibility. Instrument makers may use varnish to fine-tune an instrument's sound. In some cases, varnish helps stabilize acoustic properties over time by minimizing humidity-induced changes in the wood, at least on the coated side.
- **Direct effect on acoustic properties**: Varnish's impact on sound depends on factors like the varnish type, layer thickness, application method, and wood type. This effect can be negative, positive or neutral. Negative Effects are:
  - **Vibration damping**: Thick or heavy varnishes (e.g., some modern polyurethane varnishes) can restrict wood vibrations, potentially reducing the

instrument's volume or warmth of sound. This is especially noticeable on the top of instruments like the Noel, which relies on free vibration.
- **Altered sonic balance**: Excessive or inappropriate varnish application can emphasize or suppress certain frequencies, disrupting the instrument's sonic balance (e.g., between bass and treble).
- **Initial effects**: Immediately after varnishing, the instrument's sound may be slightly muffled or subdued until the varnish fully dries and stabilizes. This effect usually diminishes with time and playing.

Positive effects are:

- **Increased clarity and resonance**: Thin, light varnishes (e.g., traditional oil-based varnishes or shellac) can slightly amplify the wood's surface vibrations, producing a clearer and brighter sound. This is particularly important for instruments like the violin, where the spruce top plate significantly contributes to sound production.
- **Acoustic stabilization**: By reducing moisture penetration into the wood, varnish minimizes acoustic changes caused by environmental conditions, stabilizing the instrument's sound over time.
- **Improvement over time**: Some traditional varnishes (like those used on Stradivari violins) crystallize and become less elastic as they age, potentially enhancing sound quality in the long run.

In many cases, when varnish is applied correctly and thinly, its effect on sound is minimal. Professional luthiers carefully select and apply varnish to preserve the wood's sonic properties. The type of varnish (oil, alcohol, shellac, or resin) and the application method (brush, spray, or pad) can yield different results. For example, oil varnishes are often preferred over harder varnishes due to their greater flexibility.

Is Varnish Good or Bad? It is good if the varnish is carefully chosen and applied in thin, even coats. Traditional varnishes (such as shellac or oil varnishes) used on classical violins typically enhance appearance and protection without significantly compromising sound. Varnish can also contribute to the instrument's sonic stability and longevity. It becomes bad if the varnish is too thick, heavy, or unsuitable, potentially restricting wood vibrations and diminishing sound quality. Modern industrial varnishes (like polyurethane) are sometimes unsuitable for handmade instruments due to their excessive hardness.

The effect of varnish depends on the luthier's skill and the materials used. In high-quality instruments, varnish acts as a complementary element, providing both beauty and protection while minimizing negative impacts on sound.

Violin varnishes from the sixteenth to eighteenth centuries shared common characteristics: the use of natural resins (such as shellac, amber, and pine resin), protection of the wood from moisture and physical damage, and enhancement of visual aesthetics through transparency and warm colors. These varnishes were typically applied in thin layers to preserve the wood's free vibration and positively influence the violin's sonic quality [35].

However, significant differences existed among varnishes from these periods. In the sixteenth and seventeenth centuries, oil varnishes, like those used by Cremona makers such as Amati, were made from a combination of natural resins and linseed oil, resulting in a flexible texture and yellow to amber colors [35]. The eighteenth century represented the peak of violin varnishing. Cremona varnishes from this period, particularly those of Stradivari and Guarneri, were very thin, transparent, and richly colored (red, orange, and brown). These varnishes were believed to impart superior sound quality because they penetrated the wood and minimally affected vibration [35]. In the nineteenth century, the introduction of alcohol (or spirit) varnishes led to decreased sound quality and varnish flexibility due to their rapid drying time and simpler formulations [35].

Modern varnishes blend traditional formulas with new materials, but oil varnishes remain preferred for their similarity to Cremona varnishes. However, the use of synthetic materials in cheaper instruments can compromise sound quality [35].

Other scientific studies also verify that thin varnishes (a few micrometers thick) have a minimal effect on the wood's resonant frequencies. For example, Schelleng [94] demonstrated that thin varnish layers can slightly increase the surface elastic modulus without causing significant damping.

Some research (e.g., [64]) suggests that old varnishes on Stradivari violins might have contained minerals or additives that contributed to improved acoustic properties, although this remains a topic of debate.

Acoustic experiments have shown that heavy varnishes can attenuate high frequencies, sometimes reducing the brilliance of the sound.

A study examined the effect of three common varnishes used for traditional Persian musical instruments (shellac, matte polyester varnish, and Sandarac powder solution) on the acoustic properties of pine wood. The results indicated that shellac had the least negative impact on acoustic parameters such as specific elastic modulus, acoustic coefficient, and sound conversion efficiency, while polyester varnish had a significantly negative impact. Sandarac showed mixed results but has the potential to improve acoustic properties [36].

While a technical discussion of varnish and its application is beyond this article's scope due to its wide and growing variety, this information can serve as a starting point for further study, experimentation, and iteration using other resources.

Essentially, a luthier can identify a suitable varnish through a few checks and tests. Based on these cited researches, clear instructions on varnish for luthiers should emphasize the following.

Traditional varnishes like shellac or oil varnishes based on natural resins (e.g., pine or amber) are often preferred for their flexibility and compatibility with spruce and maple. Applying multiple thin coats with thorough drying between each coat yields the best results.

Professional luthiers often test the instrument's sound after each varnish coat to ensure sound quality is maintained. Using the knowledge of wood acoustic parameters gained in Chap. 4, we suggest that you initially experiment with your desired varnish, application method, and proportions on scrap wood. This preliminary testing will help you optimize these factors and their effect on the wood's acoustic properties prior to applying them to your finished instrument.

Different instruments (e.g., guitars vs. violins) may require different varnishes. For instance, guitars might need harder varnishes for increased impact resistance.

Lämmlein et al. [48, 50] and Lämmlein [49] studied a variety of varnishes and provided a summary of recommendations for luthiers, which is presented here. Varnish is primarily used on musical instruments for protection and aesthetic enhancement. When selected and applied appropriately, it can also aid in sound stability and quality. However, unsuitable or thick varnishes can dampen the sound or create an imbalance. Traditional, thin varnishes are generally favored for stringed instruments, providing a balance between protection and preserving the wood's sonic properties.

Varnishes typically reduce the longitudinal modulus, increase the radial and shear moduli, and significantly increase damping (by up to 150%). Therefore, applying a very thin layer of varnish is critical again to minimize these undesirable changes. Over time, some properties, such as the longitudinal modulus, may return to their original state, but the increased damping tends to persist.

Multilayer varnishes offer enhanced protection against environmental factors but exert a greater influence on the wood's vibrational properties. Consequently, selecting a varnish requires careful consideration of the trade-off between environmental protection and its effect on sound.

This information can assist luthiers and researchers in developing improved varnishes that enhance both the sound quality and durability of musical instruments.

In the literatures we reviewed, we mentioned several varnishes. To broaden our perspective, it would be beneficial to also mention some regional varnishes with notable applications.

- **Urushi (oriental lacquer)**: Urushi, also known as Oriental or Eastern lacquer, is a natural resin derived from the sap of the lacquer tree (*Toxicodendron vernicifluum*) and has been used for centuries in East Asian crafts, arts, and coatings, particularly in Japan, China, and Korea [16, 71, 107]. Renowned for its durability, resistance to water, acids, and scratches, and its deep, glossy finish, Urushi dries into a hard, lasting layer through a process involving moisture and heat. In musical instrument making, Urushi is occasionally employed as a finish for wooden instruments, including Western instruments like guitars and violins, and traditional East Asian instruments like the guqin [16]. This coating protects the wood and enhances the instrument's aesthetic appeal with its beautiful appearance and visual depth.
- **Shellac**: Shellac is a natural resin obtained from the secretions of the lac worm (*Laccifer lacca*) and was widely used in Europe and Asia for coating wood [7]. It is suitable for musical instruments due to its hardness, transparency, and high polishability. In Europe, shellac has traditionally been used to coat the bodies and necks of stringed instruments such as violins, cellos, and guitars [41]. In the construction of classical violins, such as those made by Italian masters like Stradivari, shellac was sometimes used alone or in combination with other resins to create a varnish [29]. The thinness and hardness of the shellac layer enhance the resonance of the wood, producing a brighter, clearer sound. The use of shellac

in violin making is documented in historical sources from the seventeenth and eighteenth centuries in Europe, although exact formulas were often kept secret [7].

- **Oil varnishes**: Oil varnishes are made by combining natural resins (such as pine, damar, or amber) with drying oils (such as linseed oil). This type of varnish was common in Europe, especially during the Renaissance and Baroque periods, for coating musical instruments. Oil varnish was widely used for stringed instruments such as violins, violas, and cellos in the Italian and French schools [87]. It provides flexibility and protection from moisture while allowing the wood to vibrate naturally. Due to its flexibility, oil varnish can produce a warmer and deeper sound. The thickness of the layer and its precise composition (e.g., the ratio of oil to resin) affect sound quality, with thinner varnishes generally better for preserving resonance. Historical sources, such as violin-making texts by Giuseppe Guarneri, refer to the use of oil varnishes in classical instrument construction [87].
- **Alcohol varnishes**: Alcohol varnishes are made by dissolving natural resins (such as sandarac) in alcohol, forming a thin, hard layer that dries quickly [36]. These varnishes were used in Europe during the eighteenth and nineteenth centuries to coat stringed instruments [13]. Alcohol varnish was popular in smaller workshops due to its ease of application and rapid drying. In Persia and the Middle East, pine resins (colophony) were sometimes used to coat traditional instruments such as the tar and setar [45]. In India, damar resin, obtained from the tree family Dipterocarpaceae, was used for instruments like the sitar to protect the wood surface and enhance its beauty [28]. These resins, due to their flexibility and adhesion, help maintain wood resonance, though their sonic effect depends on layer thickness and application method.

Unlike Urushi, which hardens by oxidative polymerization and creates a smooth, resistant surface [71], shellac and oil varnishes are more flexible and may have a different effect on sound [7]. Alcohol varnishes dry faster but can be more brittle [49].

- **Polyurethane and nitrocellulose lacquer**: The varnish commonly used on modern guitars, which creates a clear, glossy, glass-like surface, is typically polyurethane or nitrocellulose lacquer, with polyurethane predominating in mass-produced guitars due to its durability and ease of application [60]. Polyurethane can slightly reduce the natural resonance of wood due to its hardness and thickness, making it less desirable for sound-sensitive acoustic guitars, such as classical guitars. It is also more difficult to repair than traditional varnishes like nitrocellulose if damaged (e.g., cracks or scratches). Nitrocellulose, a more traditional varnish made by combining nitrocellulose with volatile solvents, is thinner and more flexible, allowing better wood vibration, which is critical for the sound quality of acoustic guitars. It remains used in custom and luxury guitars. Due to polyurethane's acoustic limitations, custom instrument makers, such as violin makers or hand-crafted classical guitar makers, prefer traditional materials like shellac or oil varnishes that are more compatible with manual processes and specialized skills.

Information on traditional lacquers and varnishes for musical instruments is limited. To remain focused on this book's goals, we must limit our discussion. Those interested can consult professional instrument makers' practices and conduct acoustic tests on wood before and after applying varnish to determine the most suitable varnish and curing method for their instruments.

5.4 Physical or Mechanical Wood Treatments

Physical–mechanical manipulations, similar to chemical manipulations of wood, are based on changing or maintaining the wood's density and elastic stiffness. This type of treatment and tuning can be applied to the entire piece of wood used in a musical instrument component, or only to a specific area that requires it. Changes in elastic stiffness and the springy constant influence stiffness reactance, while changes in density and mass influence mass reactance. Consequently, the resonant frequency of a mode shape, whose nodes and antinodes are adjusted physically and mechanically, will shift or change its pitch.

First, it's necessary to determine the effects of decreasing density on the physical and acoustic parameters of wood. A decrease in density, while elastic stiffness remains unchanged, results in increased sound radiation, decreased damping capacity, increased acoustic conversion efficiency, and decreased characteristic acoustic impedance.

Reducing density while maintaining stiffness is challenging in most engineering materials. However, it's achievable in wood. For example, it was shown that a proper solvent extraction can reduce wood density without significantly damaging cell walls, leaving elastic stiffness nominally unchanged.

Manipulating stiffness without altering density is also generally difficult. However, increasing the structural apparent stiffness or springiness, is practically possible through certain physical and mechanical changes. One of the oldest traditional methods is adjusting structural materials to create a curve. Let's begin with that.

5.4.1 Curving Wood

Wood is an orthotropic material, exhibiting distinct directional properties not found in isotropic materials. Curving wood by carving produces very different results than elastic bending [83]. As shown in Fig. 5.6 (middle), elastic bending maintains the orientation of the wood fibers parallel to the curve of the beam's surface. Meanwhile, making a pretension wood, raises its elastic constant.

However, carving wood to create an arch (Fig. 5.6, bottom) reduces longitudinal elastic stiffness. This is because the direction of the severed fibers and the direction of the curved beam deviate from the longitudinal direction of growth, moving

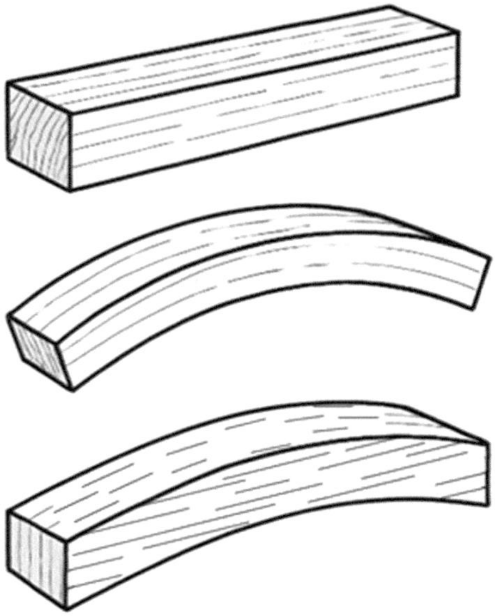

Fig. 5.6 Diagram showing a straight wooden beam (top), one curved by elastic bending (middle), and one curved by carving (bottom), illustrating how each method affects wood fiber orientation and apparent structural elastic modulus. Elastic bending preserves fiber continuity, while carving can disrupt it

towards intermediate directions with a transverse component, which lowers stiffness. This concept is not new in structural engineering, where arched structures are commonly used to enhance the structural apparent stiffness and the mechanical resistance. In musical instruments, these arching methods have been passed down through experience and practice, with their scientific perspective elucidated by Roohnia and Policardi-Antoncich [83]. This technique, supported by a scientific understanding, has been successfully applied by skilled Iranian instrument makers to several Iranian stringed instruments [83].

Elastically arching the soundboard increases the apparent longitudinal stiffness of the wood, virtually increasing the radiation coefficient. However, this process also increases stiffness reactance, causing the vibration modes of the treated wooden plate to shift to higher frequencies, as discussed in Chap. 4.

The effect of arching is better recognized for lower frequency modes and decreases with increasing frequency, as the wavelengths of the flexural vibrations become much shorter than the characteristic radii of curvature [14].

Traditionally, elastic arching of the soundboard in stringed instruments is achieved through the tensile or compressive force of the bridge. In violin family instruments, the elastic arch of the top-plate can also be adjusted by modifying the sound post's height. This additional elastic bending approach requires careful tuning of the wooden soundboards' radiation coefficients and acoustic impedances. Otherwise, the instrument may lose certain harmonics within its frequency range.

Therefore, to maximize the anisotropy ratio (the ratio of the board's longitudinal stiffness to its width), it is suggested that longitudinal curvature be achieved through elastic bending, and transverse curvature through carving. If this ratio is exaggerated, balance can be restored by employing either bending method or a combination of both, to optimize the wood's sound-producing parameters.

The alignment of wood fibers with the wood orthogonal direction is critical. This is exemplified by the oud and Persian setar, where bowls are made from bent ribs despite the irreversible damage to cell structure and loss of stiffness caused by the intense moisture and heat treatments used in the bending process. The resulting sound of these ribbed bowls become preferred (Fig. 5.7, left) because, compared to bowls carved from a single piece of wood, this method results in less fiber damage, and the fiber direction remains aligned with the ribs. This alignment provides greater elastic stiffness in the bowl's back and side components. When constructing a bowl from bent ribs is impractical due to sound preferences or shape constraints, it is recommended, as seen in the Persian tar (Fig. 5.7, right), that the back component's radial surface with longitudinal direction be maintained, and the tangential surfaces be conducted towards the sides.

On some traditional single-bowl musical instruments, such as the Turkmen dotar, the tangential surface is often placed on the back of the bowl due to the influence of tradition and historical precedent.

However, a radial surface is generally preferable for the back of the instrument because, as previously mentioned, the back plays a significant role in transmitting and amplifying sound resonance.

Fig. 5.7 The left panel is the back of an oud (by Luthier Sina Asadi) with a bowl constructed of ribs (tarkeh). The ribs create a segmented, rounded shape. The right panel is a more detailed image, depicting the back of a Persian Tar (by Luthier Hamid Mahmoodi). This view highlights that the radial surface forms the rear of the bowl, while the tangential surfaces form the sides

While the back is important, we do not consider it as sonically sensitive or structurally crucial as the top-plate.

The greater uniformity and strength of a radial surface can contribute to improved sound quality and instrument stability.

Positioning the radial surface on the sides of the bowls carved from a single piece of wood, might offer structural advantages, but it is often considered less aesthetically appealing than a tangential surface.

This is because the decorative patterns of the tangential grain are typically placed on the back and are less visible when the instrument is held.

5.4.2 Brushing or Sandblasting to Adjust the Anisotropic Ratios

The woods used in the soundboards of musical instruments, especially stringed instruments such as violins, Persian setars, and santurs, have an orthotropic structure. This means their mechanical properties differ in different directions: along the fiber, radially, and tangentially. A key characteristic affecting the acoustic performance of an orthotropic wooden soundboard is the velocity ratio, or more precisely, the ratio of stiffness along the fibers to stiffness perpendicular to them (radially or transversely). This is also known as the mechanical anisotropy ratio [81, 101], introduced in Chap. 4. Research indicates that a higher ratio correlates with greater sound radiation coefficient and lower damping.

Brushing the wood's surface along the fiber direction, particularly with hard brush, selectively erodes the softer earlywood (Fig. 5.8). The harder latewood is more resistant and remains intact. This process can influence the board's mechanical properties in several ways:

- **Mass reduction without compromising longitudinal stiffness**: Since the less dense earlywood with a lower modulus is removed while the latewood remains, the longitudinal fiber structure is largely unaffected, and the effective stiffness in that direction changes minimally. However, the mass reduction lowers the apparent density, increasing the radiation ratio (see Chap. 4 for the radiation coefficient relationship).
- **Reduced transverse stiffness due to damage to radial structures**: Brushing, especially if the brush penetrates the inter-ring areas, can damage or sever radial structures like ray cells and transverse fibers. This structural damage reduces the stiffness modulus in the radial or transverse direction.
- **Increased mechanical anisotropy ratio**: The combination of relatively constant longitudinal stiffness and decreased transverse stiffness increases the anisotropy ratio. Therefore, brushing significantly enhances the plate's mechanical anisotropy. This alteration affects the sound radiation and damping capacity of the plate.

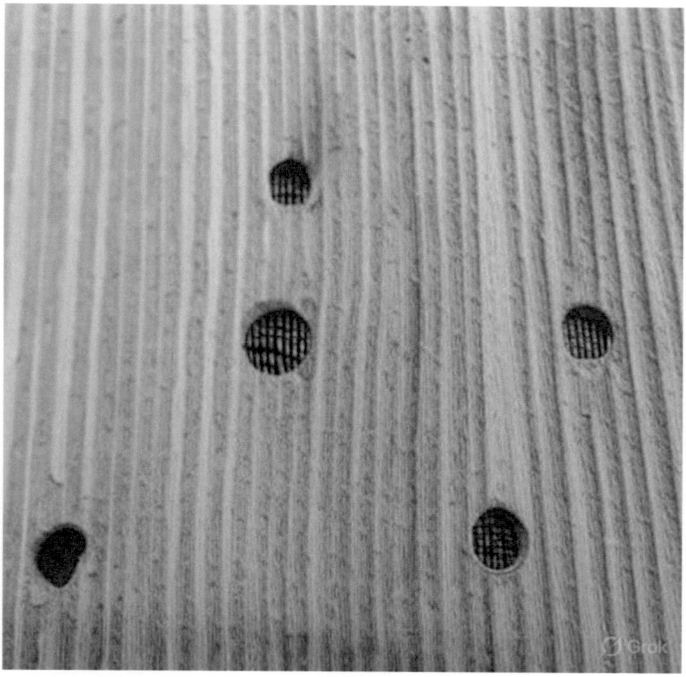

Fig. 5.8 Image of the inner surface of a mulberry soundboard from a Persian setar, demonstrating a traditional brush-finishing technique. The raised texture of the harder latewood is evident, contrasting with the subtly recessed areas where the softer early wood was carefully removed

In essence, brushing a wooden plate's surface along the fibers not only lightens the plate and increases the sound radiation ratio but also increases the mechanical anisotropy ratio by reducing transverse stiffness relative to longitudinal stiffness. This ratio is a critical parameter for optimizing the acoustic behavior of musical instruments and directly influences energy transfer quality, sound clarity, and instrument intensity. Thus, controlled brushing can be a simple yet effective physical method for enhancing wood's acoustic properties.

However, while brushing can yield desirable acoustic properties, it demands careful execution. Maintaining the latewood's integrity is paramount, as these structures primarily govern longitudinal stiffness. Damage or wear to the latewood's surface reduces stiffness and can cause local wave scattering and increased damping due to surface inhomogeneity.

To prevent such damage, the brush should move strictly along the fiber direction; transverse or rotational movements must be avoided.

Brush pressure should be controlled and uniform; excessively hard brushes or deep cuts can also harm latewood.

If more material removal is necessary, it should be done gradually over several passes.

The goal of brushing is selective earlywood removal, preserving the latewood to maintain its structural role. This approach significantly increases the longitudinal-to-transverse stiffness ratio (mechanical anisotropy) without compromising the longitudinal structure.

Potential of sandblasting as an alternative is also remarkable. In certain situations, sandblasting with very fine, light particles might be considered as a brushing alternative. Similar to brushing, this method removes softer wood (earlywood) but requires even stricter control.

- Sand particles must not penetrate or adhere to the wood structure, as residual particles can increase local mass and decrease the sound radiation ratio.
- The applied pressure must not damage the latewood.
- Sandblasting should employ softer, rounded-edge minerals or natural abrasives like walnut shells to minimize damage risk.

While sandblasting offers uniformity and speed, it carries greater risks. For delicate acoustic applications, based on the experience, hand-brushing provides more precise control. For highly sensitive instruments like the setar and santur, brushing is preferable to sandblasting.

5.4.3 Multi-piece Plates: A Homogenization of Density and Stiffness

Bookmatching is a widely used technique in instrument making, particularly for stringed instruments such as the oud, violin, and guitar, for several practical and acoustic reasons. The limited availability of large, high-quality solid wood often necessitates the use of two smaller pieces. Suitable wood for soundboards or back plates, such as spruce, cedar, rosewood, or mulberry, is frequently unavailable in pieces wide enough to cover the entire board of large instruments or, if available, is prohibitively expensive [39]. Additionally, soundboards require homogeneity of narrow grain across their surface to ensure optimal sound vibration transmission, a condition that solid wood does not always satisfy. Inhomogeneities in grain, variations in density, and inconsistencies in elastic stiffness can occur in solid wood, potentially leading to the entanglement of natural frequencies due to mechanical reactance inhomogeneity [103]. Furthermore, the mass of each component can negatively impact sound quality.

Pairing two symmetrical plates not only provides visual symmetry, which is aesthetically pleasing, but also allows builders to use woods with more uniform grain, density, and stiffness characteristics—essential for optimal sound quality—even when these woods are not available in the desired dimensions [34].

This method involves placing two pieces of wood, symmetrically cut from the same block, side by side to create a single panel with a symmetrical design. Bookmatching is significant not only for aesthetic reasons but also for its acoustic and

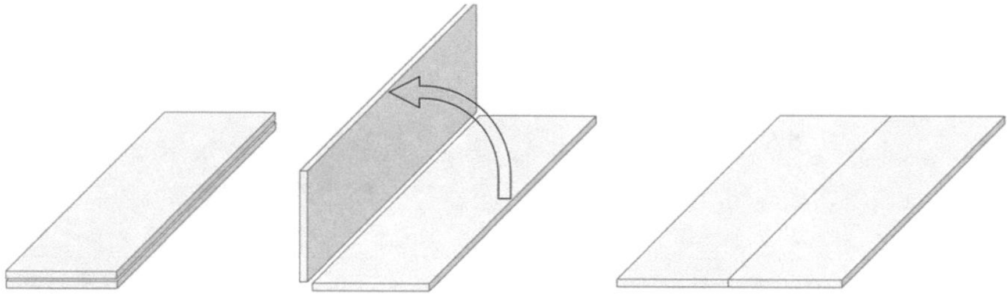

Fig. 5.9 The image displays three steps (labeled **a**, **b**, and **c**) illustrating the book-matching process of two identical pieces of wooden plate (back or top). Step a: Shows a single, rectangular wooden board being cut in half along its thickness. Step b: Depicts the same board that one of the resulting thinner pieces is being flipped open, like turning a page in a book, to lay beside the other piece. Step c: Shows the two thinner pieces of wood now placed side-by-side, with their faces that were originally the inside of the thicker board now facing outwards and joined together to create a symmetrical grain pattern

structural implications. Selecting wood with a thin, even grain, typically from species like spruce or mulberry, is crucial, as regular grain optimizes sound vibration transmission, producing a clear, balanced sound (Fig. 5.9) [39].

From an acoustic perspective, pairing two symmetrical boards can enhance vibration transmission, provided the wood has even grain. Wood grain significantly influences sound wave transmission [103]. Moreover, the symmetry achieved in the soundboard or back plate through bookmatching aids in evenly distributing mechanical stresses across the board, preventing sound distortion and improving the instrument's resonance. However, the joint at the board's center can be a potential weak point, with the risk of cracking or separation if gluing is not performed carefully or if humidity and temperature fluctuate [34]. Structurally, joining two boards requires considerable skill in preparation and gluing. Meticulous gluing and the use of high-quality adhesives designed for musical instruments can minimize the risk of joint separation. Aesthetically, joining two identical boards creates a symmetrical and visually pleasing design, highly valued in professional and traditional instruments. This visual symmetry, particularly in instruments like the oud where the top is exposed, imparts a professional and artistic appearance, enhancing the instrument's visual appeal [39].

Therefore, bookmatching in soundboard and back plate construction balances sound quality, structural integrity, and aesthetics. This technique enables builders to utilize smaller, more accessible pieces of wood while achieving high-quality acoustic and visual results. The success of this method depends on selecting appropriate wood, precise cutting and joining, and controlling environmental conditions to ensure the board's long-term stability and durability.

The decision of which side of the wood piece to cut and open like a book for joining depends on symmetry and access to more regular grain. Assuming the removal of the outer sapwood layers and juvenile wood layers—which lack the stiffness, density, contraction, and swelling properties of mature wood—it is preferable for the symmetry line to align with the outer mature wood layers, specifically those closer to the tree's bark (while avoiding sapwood and juvenile wood).

5.4.4 *Carving and Thinning the Plates and Bracings*

The acoustic behavior of stringed instruments is strongly dependent on the vibrational properties of their soundboards [103]. The precise shaping of these soundboards, often achieved by carving and thinning, is critical for achieving the desired sound qualities. This process requires a deep understanding of modal shapes and how local changes in stiffness and mass affect modal frequencies [34]. As shown in Eq. 4.21 of Chap. 4, altering stiffness or mass can shift modal frequencies. Specifically, the equation indicates that increasing springy constant (stiffness constant) raises the natural frequency, while increasing mass lowers it [23]. However, these changes pertain to simple spring and mass systems and do not uniformly affect plates. This non-uniformity arises because the influence of mass reactance and stiffness reactance varies across the mode shape (e.g., near a node versus an antinode), and the impact of mass and spring stiffness changes depends on the location of the alterations (Table 4.3, Chap. 4). Furthermore, a node in one mode may coincide with an antinode in another, causing the effect of reducing mass or spring stiffness to differ significantly between the two modes [84]. Therefore, the initial step in manipulating a mode's frequency is to understand its shape and the locations of its nodes and antinodes.

- **The Chladni method for identifying mode shapes**: The Chladni Method is a classical and relatively simple technique used to visualize vibration patterns, or "mode shapes," on a surface like an instrument's soundboard. It is named after Ernst Chladni, who developed it in the eighteenth century (Chladni 1787). For plate preparation, the plate (e.g., a wooden plate) is placed horizontally for example on fixed supports boundary condition (or free on soft thin rubber elastic supports resembling free boundary condition). A thin, uniform layer of fine powder (e.g., talcum powder, flour, salt, or dried tea powder) is sprinkled over the surface. The plate is vibrated using a source such as a loudspeaker at various frequencies. The plate vibrates at its resonant frequency modes, which can be already identified by tapping the plate and analyzing the FFT spectrum [85].

When the plate vibrates at a specific frequency, nodes (areas of minimal vibration) accumulate powder, forming distinct lines, while antinodes (areas of maximum vibration) cause the powder to move away, creating patterns known as "Chladni figures" (Fig. 5.10). These patterns reveal the locations of nodes and antinodes, providing insights into the surface's vibration behavior [34].

This method enables instrument makers and researchers to visualize vibration modes, investigate how changes in shape, material, or thickness affect vibration patterns, and gain a deeper understanding of sound production [103]. However, the Chladni method lacks precision for complex objects like rectangular wooden plates

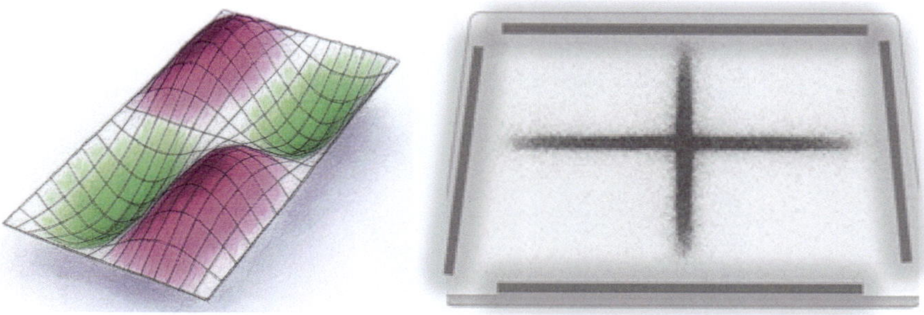

Fig. 5.10 The left panel shows a schematic, three-dimensional representation of a hypothetical rectangular plate (fixed boundary condition) vibrating in its natural mode 2.2. A grid overlays the plate, and the surface is colored to indicate the displacement during vibration: green areas represent downward movement (bellies), while pink areas represent upward movement (bellies). The lines of the grid that remain relatively still represent the nodes of the vibration. The right panel shows a photograph of this plate covered with a fine powder to indicate its chladni pattern. The powder has been agitated by the plate's vibration in its mode 2.2, causing it to be expelled from the areas of maximum displacement (bellies) and accumulate along the nodal lines, which remained relatively stationary. This accumulation forms a distinct " + " pattern on the plate's surface, visually representing for example the nodal lines of the 2.2 vibration mode. Both images are the handarts by author, enhanced using AI

used in musical instruments. We have previously observed that Chladni's results, for instance, for the Persian setar plate under free boundary conditions (Chap. 3, Fig. 3.37), are reliable only as an approximation. Consequently, the patterns of nodes and antinodes, along with their resonant frequencies, may not precisely correspond to reality and might exhibit slight variations. More modern techniques offer greater accuracy.

- **Modal analysis using accelerometers**: Small, lightweight accelerometers are attached to various points on the wooden plate. The plate is then excited with an actuator in certain and known frequencies. The accelerometers measure the plate's vibration response at each location (Fig. 5.11).

  Specialized software processes this data to calculate and display the plate's mode shapes, resonant frequencies, and damping characteristics [13]. This method offers high accuracy, providing detailed quantitative and qualitative information.
- **Interference holography**: This technique uses the interference of light waves to measure minute displacements on the plate's surface (Fig. 5.12). A hologram of the vibrating plate is recorded, and interference patterns reveal the mode shapes [79]. Interference holography is a highly accurate, non-contact method, but it requires complex and expensive equipment [56, 85].
- **Finite element method (FEM)**: FEM is a numerical method that divides the wooden plate into a large number of small elements. By defining the material properties (e.g., density, elastic modulus) and boundary conditions (e.g., how the plate is supported), the equations governing the plate's vibration are solved for

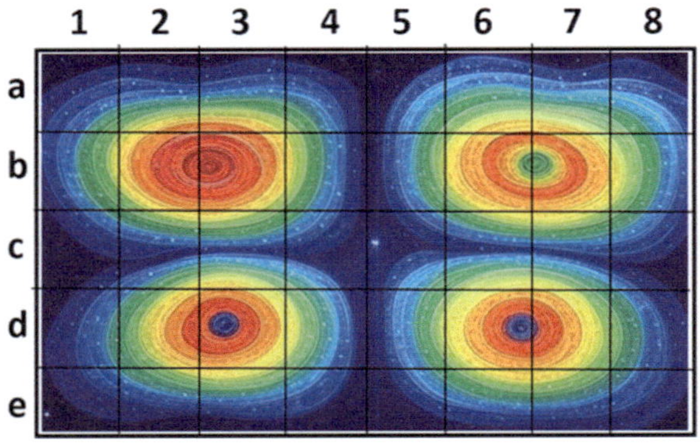

Fig. 5.11 A visual representation of modal analysis using accelerometers on a plate. The image is divided into four quadrants (corresponding to potential mode shapes) overlaid with a grid labeled a-e vertically and 1–8 horizontally, indicating various measurement points where accelerometers might be attached. Each quadrant displays a colorful, contour-like pattern on a dark blue background, representing the vibration amplitude and phase across the plate for a specific mode shape. Concentric colors, ranging from red and yellow at the center to green and blue towards the edges, illustrate varying levels of vibration intensity, with the central regions likely indicating areas of maximum displacement (antinodes). The grid provides a spatial reference for the numerous measurement locations used to generate this modal analysis data, which specialized software would process and do interpulations of the blind points to determine resonant frequencies in addition to these visual mode shapes. A handart by author, enhanced using AI

each element. Performed with computer software, this method enables the prediction of mode shapes and resonant frequencies before the plate is manufactured [13]. FEM is a powerful tool for structural design and optimization but demands specialized knowledge and advanced software.

Using software, the Finite Element Method (FEM) generates images similar to those in Fig. 5.11. However, unlike physical tests that require installing sensors within cell blocks, FEM is a theoretical simulation. This allows for the reduction of cell block dimensions, enabling the acquisition of point-to-point results and higher-resolution images. Nevertheless, theoretical FEM calculations can deviate from real-world conditions due to local inhomogeneities in wood, potentially compromising the accuracy of the results.

- **Laser scanning vibrometry**: A laser is directed at the plate's surface, and changes in the frequency of the reflected light (the Doppler Effect) measure the velocity and displacement of different points. Scanning the laser across the surface generates a complete map of the plate's vibrations and their corresponding mode shapes very similar to what is seen in Fig. 5.11. This method is non-contact, accurate, and fast, but it is relatively costly [84].

Each method has specific advantages and disadvantages. The choice of the most suitable method depends on factors such as the required accuracy, cost constraints, complexity of the plate's geometry, and available resources. Currently, the Chladni method remains more accessible for basic instrument-making workshops, though, as

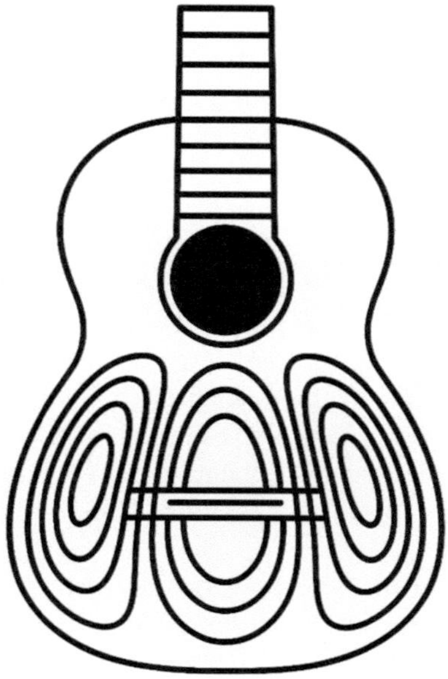

Fig. 5.12 A monochromatic schematic image of a classical guitar's soundboard, displaying a series of dark and light concentric fringe patterns. These interference fringes, created using interference holography, reveal the minute displacements and resulting mode shapes of the soundboard as it vibrates [63]. The patterns are concentrated in distinct areas of the lower bout of the guitar, indicating regions of significant vibration amplitude. The image, demonstrates a highly accurate, non-contact method for visualizing the complex vibrational behavior of a musical instrument soundboard through the interference of light waves. A handart by author, enhanced using AI

experienced, it may lack precision and clarity for identifying some subtle vibration patterns. Combining the Chladni method with modal frequency identification using Fast Fourier Transform (FFT) analysis of tap-tone spectra could provide a more effective approach to detecting and identifying mode shapes. Chapter six will detail a method using a unidirectional microphone moved across different positions of the instrument plate. However, for now, we will continue our discussion using the Chladni method.

As demonstrated by Chladni patterns, soundboards exhibit nodal (zero displacement) and antinodal (maximum displacement) regions for each mode. The effect of carving on frequency shift varies significantly across these regions due to the interaction of stiffness reactance and mass reactance, as shown in the mechanical impedance analysis in Chap. 4. Furthermore, precise control of modal frequency is critical to mitigate "wolf-notes," undesirable howling-like sonic artifacts caused by the exact matching of resonant frequencies among the vibrating string, top-plate, and back-plate. Therefore, carving becomes a vital technique for fine-tuning an instrument's response and preventing these issues [68, 69]. Achieving the desired sound quality necessitates a systematic approach that carefully considers the modal shape, nodal/antinodal locations, and their influence on frequency.

If we refer back to Table 4.3 in Chap. 4, we observe that the stiffness and mass reactance, which determine the natural modal frequencies of a simple vibrating system, change as the dimensions of the vibrating material and the boundary and support conditions of the piece under study are altered.

It is possible to tap-tune any object of any shape and boundary conditions, identify its modal frequencies, and manipulate these frequencies higher or lower through carving and thinning or adding a mass or a stiff body on an appropriate point. However, it's remind to remember that the tuned plate or piece will change after installation in the musical instrument, as the boundary conditions (i.e., the way the plate is connected to the instrument's structure) are altered.

Table 4.3 indicates that as the plate's length decreases, mass reactance decreases, while spring stiffness reactance increases. Consequently, we should expect an overall decrease in the fundamental mode of longitudinal vibration. Reducing the plate's width will also affect the plate's longitudinal vibration and disrupt its intended behavior.

Regarding boundary conditions, transitioning from free to simply supported, and then to a fully fixed support, increases stiffness reactance and decreases mass reactance. This suggests that the plate's natural vibration mode will be raised.

Therefore, we recommend fixing the plate to its final shape, dimensions, and boundary conditions—similar to those it will experience in the instrument—before planning any carving or tapering. This will help stabilize the parameters influencing the modal frequencies in an open-box tuning procedure.

For instance, guitar luthier Giuliano Nicoletti, sometimes leaves the guitar's back plate unattached to evaluate the plate's mode shape and identify the nodes and antinodes of each mode before making any cuts to the front plate [68, 69]. Similarly, luthiers might prefer a specialized frame to temporarily mount guitar plates, simulating their installed conditions on the instrument, then starting to manipulate the thickness or bracings (Fig. 5.13).

Musical instrument makers typically adhere to specific ideal patterns for the front and back plates of their instruments, patterns established by previous makers [14]. For instance, violin makers often follow the circular and contour curves of equal thickness found in Stradivarius or Guarneri violins [14]. Leon [52, 53] introduced several other maps of equal thickness curves for the front and back plates of famous and renowned violins. However, even after carving and achieving these target thickness patterns, differences in the modal frequencies compared to the original instrument can still exist.

As shown in Table 4.3, variations in material properties, specifically the modulus of elasticity and density, influence the interaction between spring stiffness reactance and mass reactance. These variations cause frequency deviations that professional instrument makers attempt to correct by carefully carving and thinning specific areas of the wood. This highlights the importance of measuring the acoustic properties of wood before instrument construction to enable the selection of wood with suitable initial properties. Furthermore, if the wood sample lacks sufficient initial stiffness, excessive thinning is risky because it can drastically reduce the springy constant and

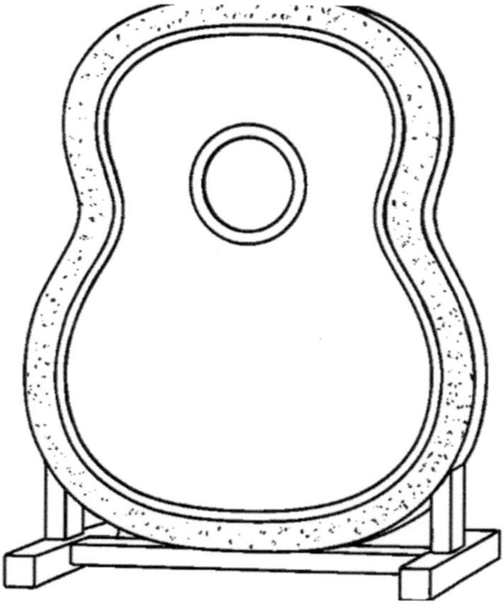

Fig. 5.13 A schematic image shows a specialized frame designed to temporarily mount guitar plates, simulating their installed conditions on the instrument. The frame is a rounded shape, mimicking the shape of a guitar body. The drawing is a simplified representation. A handart by author, enhanced using AI

even lead to the plate breaking or collapsing under the load of the bridge and string tension.

Through carving and thinning (and occasionally attaching additional pieces to adjust mass and stiffness), luthiers try to minimize discrepancies in the tuning of the plate's various modes. It's important to note that carving and thinning the plate reduces both its mass and springy constant. However, as indicated in Table 4.3, moving away from the nodes towards the antinodes (or "belly") of a mode-shape can have varying effects on mass and stiffness reactance.

Changes to mass (either adding or removing it) precisely at the nodal points of a mode have no effect on that specific mode's frequency. As one moves from a node toward an antinode, the influence of mass reactance increases, while the influence of stiffness reactance gradually decreases. Consequently, if a luthier carves and thins the plate near a node, the decrease in stiffness reactance will have a greater effect than the decrease in mass and the mode's frequency will shift towards lower frequencies (bass).

Suppose you have cut a plate according to the desired pattern and achieved the target thickness profile, but one or more frequency modes still require adjustment. Employing the Chladni technique, vibrate the plate at the specific frequency of that mode to identify its nodal and antinodal patterns. To lower the frequency of a mode, carve and thin the plate in the areas near the nodes and away from the antinodes, or add mass to the antinode of that mode. Conversely, to raise the frequency, lighten the antinode by carving and stiffen the areas near the nodes by adding stiff material. If a node is located at the junction of the plate and the instrument's side, you can stiffen its boundary conditions using a more rigid adhesive or other means to increase the springy stiffness in that area.

Braces are primarily employed to reinforce and increase the mass or the stiffness of the large soundboards of instruments [89] like guitars and ouds. These structural elements can also be used to fine-tune the mass, springy constant, and vibrational mode frequencies to serve as a tone-bar, e.g. the bass bar in violin. However, their attachment initially will increase the overall tap-tone frequency of the soundboard [27].

Note that tapering the bracing from its sides, reduces its mass, but due to the brace beam's untouched height (thickness), its bending resistance (and its springy constant) changes minimally (its momentum of inertia depends mainly on the brace's height). So, tapering the brace from the side primarily reduces mass while largely preserving its springy constant. Consequently, when only a mass reduction is desired without significantly altering the spring constant, taper these beams from the side. Conversely, carving the top surface of the bracing reduces both its mass and spring constant (Fig. 5.14). Therefore, when carving from the top surface of the brace, carefully consider the bracing's location relative to the nodes and antinodes of the relevant mode.

It is also important to identify and account for any overlap or interaction between the antinodes and nodes of different modes. This prevents inadvertently altering another mode while adjusting a specific one, or requiring repeated adjustments. To address this issue, brace shape and cutout modifications e.g. scalloped braces (Fig. 5.15) have been developed [26, 27, 68, 69].

Figure 5.16 shows two vibration modes of a Persian setar plate for an example. Both modes share the same vibration nodes along the plate's edge, but the 329 Hz mode (left) has its central antinode overlapping with a nodal line of the 659 Hz mode (right). Therefore, carving around the central nodal line in the right figure will lower the frequency of this mode, but raises the frequency of the left mode. Carving around

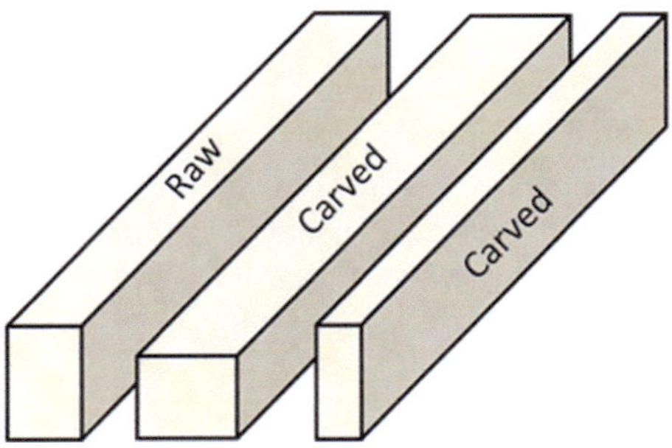

Fig. 5.14 Three rectangular wooden beams are shown side-by-side, illustrating different methods of carving. The leftmost beam is labeled "Raw" and represents the original, uncarved state. The middle beam, labeled "Carved," shows material removed from its top surface, reducing its height. The rightmost beam, also labeled "Carved," shows material removed from its side, reducing its width. Carving from the top (reducing height) reduces both mass and springy constant, while carving from the side (reducing width) results in a greater reduction in mass relative to the reduction in its spring constant. These braces are often tapered non-uniformly, with the carving varying based on their placement and the mode shapes

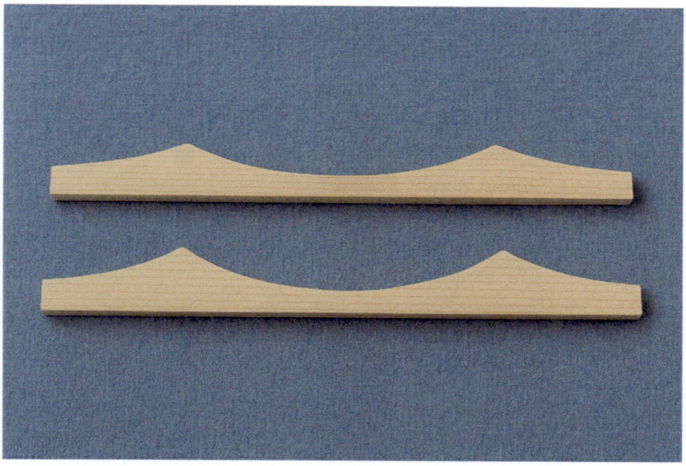

Fig. 5.15 A scalloped wooden brace, for a musical instrument soundboard, is shown. The brace has a curved shape, with carved-out sections (scallops) along its length. These scallops are designed to adjust the brace's mass and springy constant, influencing the instrument's vibration modes. A handart by author, enhanced using AI

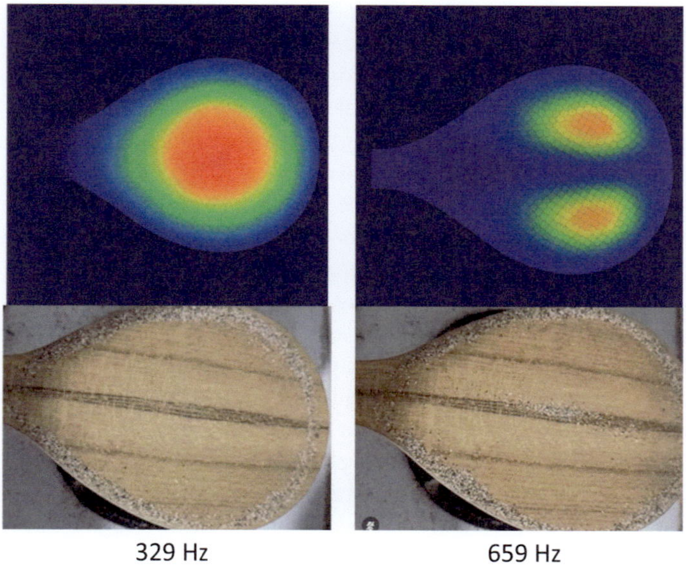

Fig. 5.16 The image shows two vibration modes of a Persian setar plate. The left (329 Hz) and right (659 Hz). The top row are computer aided visualizations, enhanced using AI and the bottom row are Chladni patterns. The 329 Hz mode has a central antinode, while the 659 Hz mode has two antinodes separated by a nodal line

the plate's edge, on the other hand, will lower the frequencies of both the left and right modes.

However, the complexity of mode-shape interference means that we can only effectively tune a limited number of important modes (typically the lower modes in soundboards frequency response range) at any one time using brace carving [26, 27, 68, 69]. This limitation leads to some ambiguity regarding the effectiveness of localized brace carving techniques like scalloped braces [27]. Nevertheless, the relationships between stiffness and mass reactance, as outlined in Table 4.3, provide a theoretical basis for these techniques. The calculation or precise understanding of their effects becomes complex and time-consuming, though ultimately solvable, for a larger number of mode shapes and at higher frequencies, where numerous nodes and antinodes appear across the instrument's surface.

Furthermore, reducing the stiffness constant of a mode shape or increasing the mass will lower its frequency (towards bass frequencies), and it may also decrease its amplitude, depending on the excitation frequency and the mode shape [23]. It is reiterated that the initial plate should be selected from wood with high elastic stiffness to minimize the loss of sound radiation coefficient during the carving and thinning process. This entire process is referred to as "plate tuning".

During the carving process, avoid excessive material removal in any area, as over-carving will compromise the desired sound. Additionally, even if the desired sound is achieved, a plate made from wood that was not initially selected for appropriate stiffness may develop warping (convexity or concavity) or even fail due to its inability to withstand the pressure from the instrument's bridge and strings.

Tuning the modes of the back plates of the instrument is generally less complex. Wood with appropriate initial stiffness and density will typically achieve the desired pitch after being shaped according to the established patterns [14]. Thinning the area around the nodes of a back plate's vibrational mode can enhance sound production from that mode, while keeping the area thick will make it less active in sound production and more effective in reflecting sound within the instrument.

Occasionally, a specific mode frequency of the back plate coinciding with the resonant frequency of the top-plate for a played note or one of its harmonics can produce a wolf-note. Correcting this requires identifying the mode-shape of that frequency in either the back or top-plate and slightly carving the identified area to disrupt the match and eliminate the wolf-note. Otherwise, the back plate of the instrument generally does not require further tuning beyond the established patterns.

Now that the term "wolf-note" has been introduced, it might be assumed that carving and thinning are the only ways to correct it. However, any method that slightly disrupts the precise tuning match between a resonant body mode and the wolf-note frequency can resolve the issue. For example, slightly altering the position of the violin's soundpost can change the tension applied to the back and top-plate, effectively shifting the mode's frequency higher or lower [38].

Therefore, carving and thinning are not the sole methods for shifting frequencies. As demonstrated by the difference between elastic curving and curving with carving, any system capable of altering the stiffness and mass of the tonewood will influence the interaction of mass reactance and stiffness reactance, and consequently shift the frequency [23]. This includes the bridges,its loading on the top-plate varies with string tension, bridge stiffness or flexibility and bridge's height, all of which can affect the top-plate's tuning and the overall tuning of the soundbox's vibration modes.

While we're discussing carving and thinning, we suggest first experimenting with this process and its effects on a piece of scrap wood with a regular shape, such as a disc or rectangle plate, to develop proficiency. In practice, instrument makers shape and adjust soundboards and bracings using these principles, sometimes scientifically and sometimes empirically, to influence the plate's acoustics [27]. By carving parts of the soundboard and braces to reduce mass or altering its thickness to change stiffness, they can precisely control the instrument's resonant frequencies and achieve their desired sound [23, 69]. Therefore, we recommend initially practicing this procedure on a piece of scrap wood to observe how local carving and thinning, or adding mass or stiffness, near or far from a node or antinode, affects the frequency shift of that mode in the plate's tap tone. It's also important to note any interference with other mode shapes during this exercise.

5.4.5 Drilling for Sound Holes

In the sound boxes of musical instruments, a type of acoustic resonance occurs within the air chamber. For example, in the guitar, this resonance is known as the first mode of its tap-tone FFT spectrum and closely resembles Helmholtz resonance [68, 69]. To clarify, Helmholtz resonance happens when air trapped in a cavity exchanges energy with the outside air through a small opening, oscillating at a specific frequency. In guitars and similar musical instruments, the air volume inside the guitar body acts as the cavity and the sound hole functions as the opening.

The sound hole in a guitar plays a critical role in determining the Helmholtz resonance frequency. The mass and dimensions (size and shape) of the air within the hole influence this resonance's frequency [19, 68, 69]. The Helmholtz resonance frequency is directly related to the sound hole's dimensions and inversely related to the volume of the musical instrument's sound box chamber. Specifically, increasing the hole's dimensions shifts the Helmholtz mode's sound toward higher frequencies (treble), and vice versa [67]. Similarly, reducing the resonance chamber's volume shifts this mode's sound toward higher frequencies, and vice versa [33].

However, it's important to note that while advantageous, a sound hole on the resonance plate introduces a discontinuity, disrupting the plate's uniformity and reducing its elastic or stiffness constant. This effect is particularly noticeable in instruments like classical guitars with large sound holes, where the area between the neck and the sound hole becomes less involved in vibration modes and practically ineffective in vibrating [20, 34]. To compensate for the stiffness reduction caused by the sound hole, luthiers, e.g. guitar makers employ techniques such as using specialized bracing patterns or reinforcing the areas around the sound hole. These methods enable the resonance plate to maintain effective vibration transmission. So, the sound hole's effect differs between classical guitars (with larger sound holes) and instruments like folk guitars or violins (with f-holes). In classical guitars, the discontinuity effect is more pronounced due to the larger sound hole size.

To eliminate these defects, the soundhole could be removed from the top and moved to the side, similar to the Turkish bağlama; however, this design is not yet widespread due to aesthetic preferences and tradition.

Conversely, because luthiers perform final carvings of guitar braces after the guitar is assembled, reaching through the sound hole to access the underside of the soundboard and backplate [68], large sound holes remain common on guitar soundboards.

The Helmholtz resonance frequency within the sound box of a guitar-like instrument is often apparent in the tap-tone of the instrument body, amidst the instrument's other body vibrational modes. As will be discussed in Chap. 6 during the evaluation of the completed instrument, positioning a unidirectional microphone close to the sound hole can emphasize the intensity of one or more of these modes.

This technique allows for the identification of the frequency or frequencies associated with the air pump effect caused by the instrument's body vibration. To illustrate this, consider the guitar example provided, where previous research has established the sound-hole frequency [19, 68].

Figure 5.17 depicts a series of schematic diagrams showing a test guitar. The diagrams illustrate the identified frequencies of the body's first mode and the first mode of the Helmholtz resonance and how these frequencies shift when manipulating the hole diameter and plate's surface by adding a light mass to the soundboard's first mode's antinode, inserting a collar into the sound-hole, and finally, completely blocking the sound hole. Using a cardboard strip collar, shaped into a 3–4 cm high cylinder to fit the sound hole, increased the effective length of the air volume within. Because the soundhole diameter and cavity volume remained unchanged, this resulted in a lowering of the Helmholtz resonance frequency [19].

5.4.6 *Bridge Tuning*

In Chap. 4, we discussed how the bridge in stringed instruments plays a fundamental role in converting the strings' horizontal vibration into a vertical force. This force is then transferred to the instrument's top, producing sound. In essence, the bridge functions to convert vibration—that is, to change the strings' horizontal movement into vertical force—and to transfer energy—moving the strings' vibration energy to the top of the instrument [14].

Additionally, the bridge acts as an intermediary, adjusting the impedance match between the strings and the instrument's body, which influences the instrument's timbre. Therefore, a balance of mechanical impedance between the strings and the bridge is critical, and a skilled luthier's tuning focuses on adjusting the bridge's mechanical impedance in relation to the impedance of the strings and the top-plate.

According to the concept of mechanical impedance, manipulating the bridge's stiffness and mass can significantly affect the sound of the instrument [105]. Even

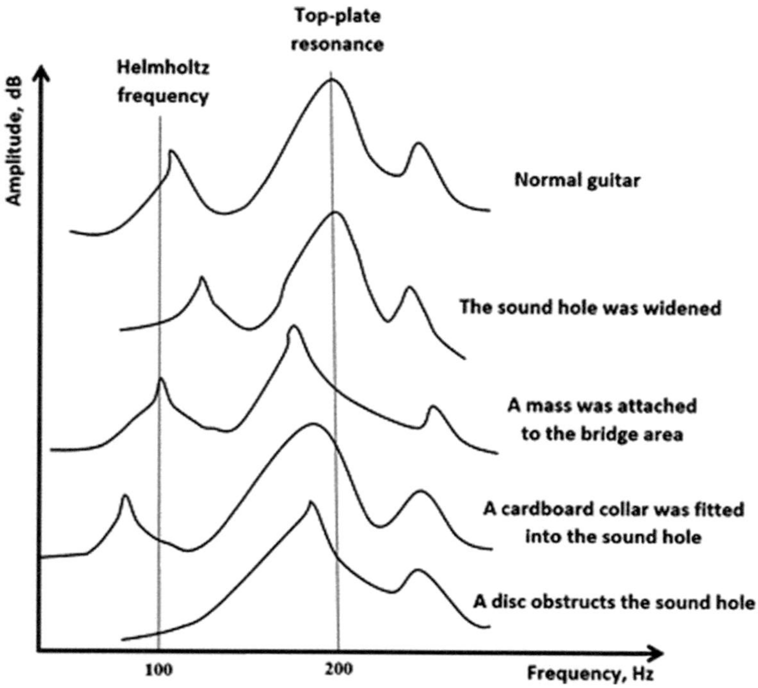

Fig. 5.17 Image shows five overlaid frequency response curves, plotting Amplitude (dB) against Frequency (Hz). Each curve represents a different modification to a test guitar. The top curve shows a "Normal guitar" with peaks indicating the Helmholtz frequency around 100 Hz and the top-plate resonance around 200 Hz. Subsequent curves show the effect of: widening the sound hole (shifting Helmholtz peak to higher frequencies); attaching a mass to the bridge area (primarily lowering the top-plate resonance); fitting a cardboard collar into the sound hole (decreasing the Helmholtz frequency); and obstructing the sound hole with a disc (eliminating the Helmholtz resonance and significantly lowering the top-plate resonance). Vertical lines are drawn at approximately 100 and 200 Hz to highlight the initial Helmholtz and top-plate resonance frequencies of the normal guitar. These schematic diagrams are handarts by author, inspired and redrawn from researches presented by Christensen and Vistisen [19] and Fernandez [33]

slight changes to the mass and stiffness of the bridge affect its resonant frequencies and its admittance ability to the vibrational energy. Instrument makers typically alter the mass, stiffness constant, and flexibility of the bridge by adjusting its thickness and carving grooves.

The mass and stiffness constant of the bridge in stringed instruments are fundamental factors in tuning and shaping the instrument's frequency range. The bridge acts as an acoustic filter, transmitting the strings' vibrations to the instrument's body. During this process, the resonances of the bridge and the body interact, creating unique sonic characteristics.

This interaction is particularly evident in the violin, where it produces a phenomenon called the “bridge hill.” The “bridge hill” is characterized by a concentration of modal frequencies in the body’s FFT tap-tone spectrum, forming a hill-like curve structure.

For example, increasing the bridge’s mass, without altering its stiffness constant, shifts the “bridge hill” resonance to lower frequencies and reduces loudness (Fig. 5.18a). Conversely, increasing the bridge’s stiffness constant, while keeping its mass constant, shifts the “bridge hill” resonance to higher frequencies and again reduces loudness (Fig. 5.18b). Even simultaneously increasing both the stiffness and mass of the bridge without any shift to the “bridge hill” resonance frequency is adjustable (Fig. 5.18c) but reduces loudness [105]. Therefore, achieving optimal bridge tuning by carefully balancing these factors should be a key objective for instrument makers (luthiers).

The flexibility and mass of the bridge are primarily determined by the wood species, its density, and its elastic stiffness, and secondarily by its dimensions and volume. For an example, violin makers often reduce mass by removing wood from the height or the thickness of bridge. They may also reduce both mass and stiffness constant by making cuts, holes and grooves (such as the heart and kidney holes and grooves in a violin family bridge) to decrease the mechanical impedance and increase its admittance, or input acceptance (Fig. 5.20).

However, care is essential. If the mechanical impedance of the bridge becomes too low, falling below that of the top-plate, the transfer of vibrational energy into the soundboard will be impeded. Adding mass to the bridge is similar to applying a violin mute piece (Fig. 5.19), which significantly increases the mass impedance of the bridge, hindering energy transfer to the top-plate and weakening the instrument’s sound [30]. Consequently, adjusting the bridge involves considering its flexibility and mass and carefully monitoring the resulting changes.

The height of bridge also influences the vertical force exerted on the plate, as calculated in Chap. 3, making a pretension soundboard and shifts some modal frequencies of the soundboard. Therefore, reducing the height to lighten the bridge requires careful consideration of this effect coming from string tension vertical components [44].

Since the antinodes of many of the plate’s vibration modes are influenced by the bridge placement, increasing the bridge’s mass causes these resonances to shift toward lower frequencies while simultaneously reducing their sound intensity. Conversely, increasing the bridge’s stiffness and reducing its flexibility shifts the resonance modes toward higher frequencies. Due to the increased mechanical impedance and reduced vibration admittance, the intensity and loudness of the related modes are again diminished, and the sound becomes weaker. Thus, the luthier must find the precise tuning point by carefully adjusting the bridge’s mass and flexibility, which can otherwise compromise the musical instrument’s sound-producing capability. Johnson et al. (1990) specifically investigated whether addig removing wood (or adding a local mass) from different areas of the violin bridge would result in a greater change in mass or in stiffness, and how these changes would affect the shift in resonant frequency. Based on their research, carving the top and bottom edges

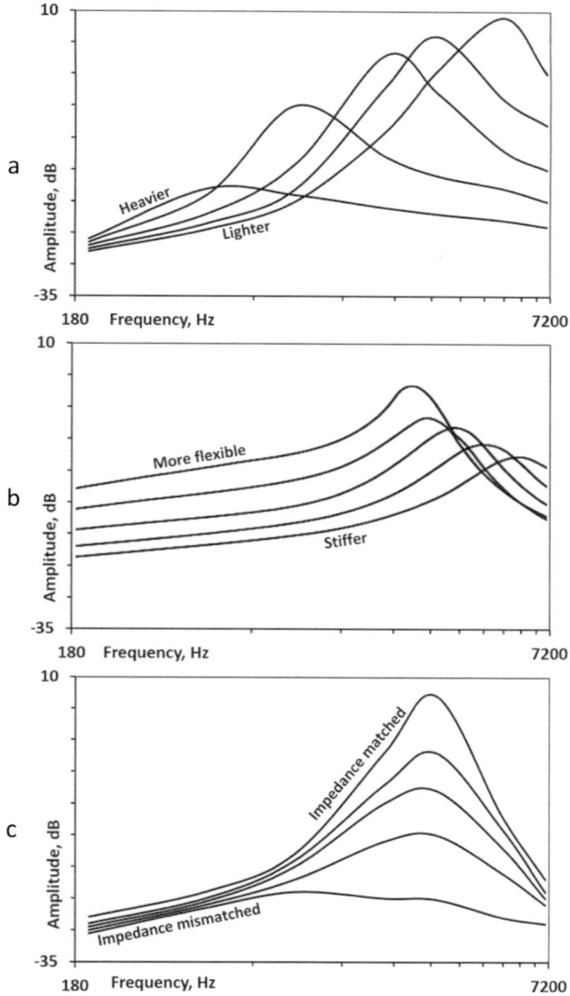

Fig. 5.18 Three graphs (**a**, **b**, and **c**) illustrating the "bridge hill" phenomenon in the tap-tone spectrum of a violin body. Each graph plots Amplitude (dB) on the vertical axis against Frequency (Hz) on the horizontal axis, ranging from 180 to 7200 Hz. Each graph displays five overlaid curves, representing changes to the bridge. **a** Shows that increasing the bridge's mass shifts the "bridge hill" resonance peak to progressively lower frequencies, with a corresponding decrease in loudness. **b** Shows that increasing the bridge's stiffness constant shifts the "bridge hill" resonance peak to progressively higher frequencies, also with a corresponding decrease in amplitude. **c** Shows that simultaneously manipulating both the stiffness and mass of the bridge can maintain the "bridge hill" resonance frequency at approximately the same frequency range, but still results in a decrease in amplitude. This schematic diagrams are inspired, adapted and redrawn from researches presented by Jansson et al. [44] and Woodhouse [105]

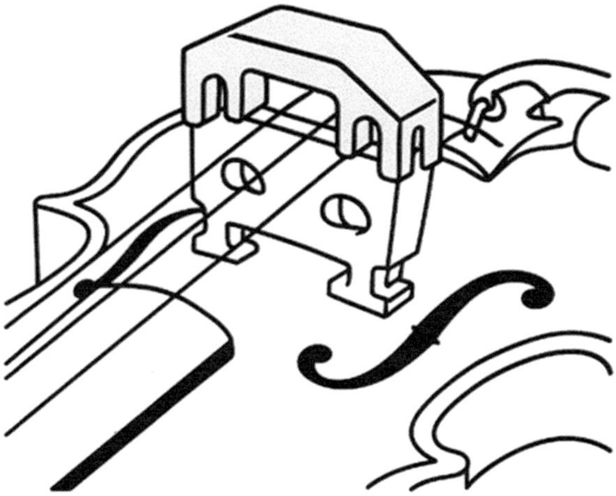

Fig. 5.19 A violin bridge with a mute (or sordine) attached. The mute, a clamp-like piece, is positioned on top of the bridge, adding mass to it. This added mass increases the bridge's mass reactance and reduces the vibration energy transmitted from the strings to the violin's body, creating an impedance mismatch and thus quieting the instrument. The form of the bridge and the mute are exaggerated. A handart by author, enhanced using AI

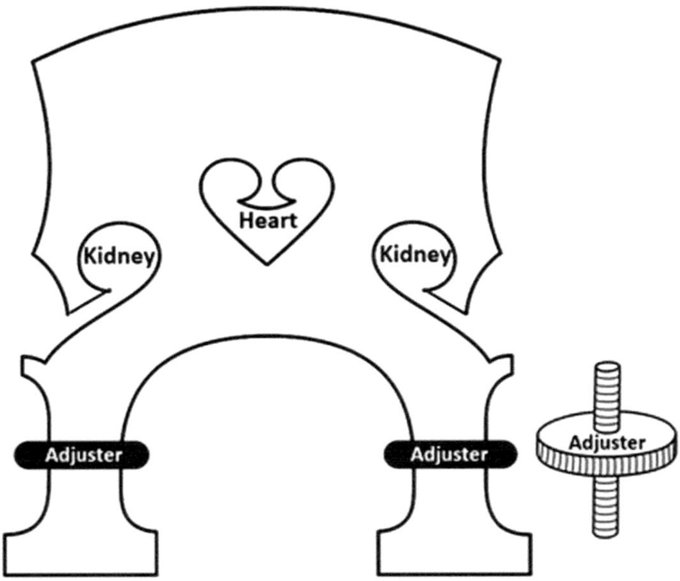

Fig. 5.20 A schematic drawing of a bass violin bridge. The bridge is shown with a heart-shaped hole in the center and kidney-shaped cutouts on either side. The image highlights the presence of adjusters at the base of the bridge's legs, enabling modifications to the bridge height. The adjusters are labeled in the diagram. Bass violin bridges require frequent adjustments due to their size and sensitivity to environmental changes, and that these adjusters represent an innovation in addressing this need. A handart by author, enhanced using AI

between the bridge legs and the kidneys' outer edge, as well as reducing the bridge thickness, results in a greater reduction in the bridge mass and raises the resonance frequency.

However, carving the inner wall of the kidneys increases the bridge flexibility, effectively reducing its stiffness. This shifts the resonance toward lower frequencies.

The bridge placement is generally determined by established patterns, so that the bridge mass and the compressive or rotational force it applies to the soundboard adjust the plate's vibration center of vibration. However, even slight displacement from this placement, although sometimes employed by luthiers, can cause a corresponding shift in the vibration modes' center of gravity and may not produce the desired sound. For example, in the violin, the bridge is held in place by friction according to traditional designs, and its position on the violin's top is between the f-holes. Additionally, one of the bridge's feet rests on the bass bar, and the other is near the soundpost. Violin makers are well aware that altering the shape and material of the bridge can significantly influence the instrument's sound.

Regarding the distance between the bridge legs, research indicates that there is an optimal adjustment and tuning. Increasing this distance beyond the traditional and accepted patterns can initially cause a decrease in the sound's loudness, followed by a subsequent increase. However, further widening the distance between the bridge legs results in a decrease in sound again. This phenomenon is likely related to the interaction of the bridge's contact points with the antinodes and nodes of the vibration modes. For example, in a violin, increasing the distance between the bridge legs from 5 mm to 2 cm reduces the loudness of the bridge's fundamental resonance. However, when the distance reaches 30 mm, the loudness peaks. Further increasing the distance to 40 mm weakens the loudness once more. Concurrently, widening the distance between the violin bridge legs shifts the peak of the bridge's fundamental resonance towards lower frequencies [105]. Consequently, instrument bridge design relies heavily on tradition and experience, leading to slow progress in innovation. Many musicians also favor traditional designs and resist change. Nevertheless, efforts to design and construct new and different bridges for instruments even in the violin family persist. The reasons for pursuing new bridge designs are varied:

- **Improving sound quality**: A primary goal is to enhance the frequency response and produce a richer, more balanced sound.
- **Addressing specific problems**: For example, the bass violin bridge requires frequent adjustments due to its large size and sensitivity to temperature and humidity variations. Devices like bridge height adjusters represent innovations in this area (Fig. 5.20).
- **Utilizing new materials**: Advances in materials science enable the use of novel materials with desirable acoustic impedance properties. Examples include bridges made of cross-layered maple with ebony bases or those with asymmetrical shapes inspired by Art Nouveau and Gothic architecture.

However, evaluating the success of these new designs necessitates careful examination of their mechanical and acoustic properties. For instance, tests have shown

Fig. 5.21 A schematic line drawing shows an asymmetric bridge for a violin, as designed by Schneider [90]. The bridge has a distinctive shape with two large, circular openings within its upper section. A handart by author, enhanced using AI

that some new bridges can increase mechanical admittance compared to traditional ones, resulting in a brighter sound.

While there are ongoing efforts to refine musical instrument bridge design using new materials, varied shapes, and engineering approaches focused on flexibility and mass (Fig. 5.21), widespread acceptance of these innovations among violinists and makers may take time.

The hardness of the bridge edge, which is modified in instruments like guitars by mounting a thin saddle layer, also affects the sound. This modification alters the boundary conditions of the vibrating string support, depending on the string's angle on the bridge and its tension. Although these structures are typically not wooden and are outside this book's scope, the impact of changing bridge coupling edge hardness is a topic worthy of investigation by luthiers.

Like all bridges, the violin bridge is under constant pressure from the strings [14] and can deform or wear out over time. Using more durable wood in the areas subjected to greater stress can extend the bridge's lifespan. This is particularly important for musicians who play their instrument frequently. Furthermore, mounting a small piece of wood with a different hardness, which alters the boundary conditions of string coupling on the bridge, affects the string's vibrational characteristics and the transfer of vibrational energy into the bridge system (Fig. 5.22). This design modification is especially suitable for professional musicians seeking a balanced and high-quality sound for a specific string.

5.4.7 Double-Top Plates

Double-top plates are considered an innovation in the construction of soundboards for stringed instruments, particularly guitars, aimed at increasing volume and enhancing resonance and dynamic response. This design consists of two very thin wood layers with desirable acoustic properties, separated by a core material (honeycomb form). A key challenge in designing these plates is maintaining adequate bending behavior while minimizing overall weight [18, 97].

Fig. 5.22 A close-up shows a luthier's hands carefully working on a violin bridge made of light-colored maple wood. A small, dark triangular piece of ebony, is being precisely mounted onto the top edge of the bridge where the E string would rest. The name "MASOUD GHAFOURI" is engraved on the bridge. This modification, alters the string coupling on the bridge, influencing the string's vibration and the transfer of energy into the instrument. Image courtesy of Luthier Masoud Ghafouri, gifted to be included in this book

In a bending beam or plate, the neutral axis is an imaginary plane where there's no tensile or compressive stress (Fig. 5.23). Bending stresses increase linearly with distance, y, from this axis [9]. In a solid plate of uniform thickness, the neutral axis lies at its midpoint. So, in a double-top structure, composed of two thin wood layers spaced apart, the neutral axis has a reduced effect on the plate's overall bending strength due to the separation of the layers.

In other words, the material near the neutral axis contributes minimally to the plate's flexural strength. Consequently, a lightweight core material can be used in this laminated design to decrease the plate's overall weight without significantly compromising its bending behavior. This approach increases the soundboard's strength-to-weight ratio, which is beneficial for producing a louder sound with a richer tonal spectrum.

Various materials are suggested to serve as the core in double-top plates, including:

- **Balsa wood**: Its very low weight and specific cellular structure can enhance the plate's resonance [97].
- **Lightweight synthetic foams**: Low-density foams with controlled damping properties can reduce weight and fine-tune the plate's frequency response [1].
- **Thin wood veneer**: Sometimes, a very thin layer of another lightweight wood, or even the same wood as the outer layers, is used as the core [37].
- **Kevlar**: A commercial synthetic para-aramid fiber, introduced by Dupont company, is known for its exceptional strength-to-weight ratio. Kevlar has seen

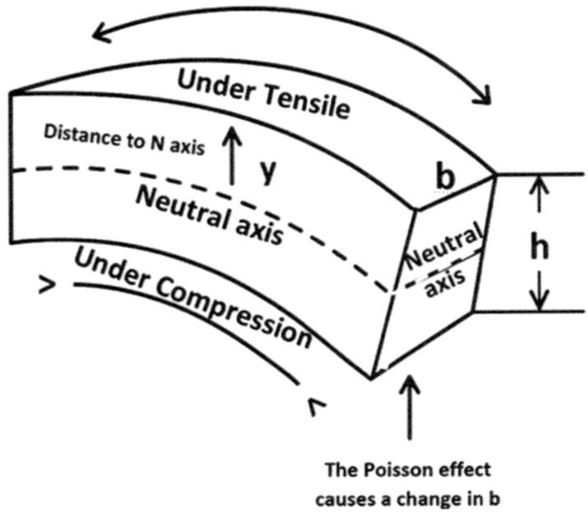

Fig. 5.23 The stress distribution in a bent rectangular beam. A dashed line through the center represents the "Neutral axis," an imaginary plane within the beam experiencing neither tension nor compression during bending. The top surface of the beam, labeled "Under Tensile", experiences tensile stress. The bottom surface, labeled "Under Compression" and experiences compressive stress. The distance 'y' from the neutral axis to a point within the beam is shown. The beam has a width 'b' and a height 'h'. The diagram also indicates that the "Poisson effect causes a change in b," showing a slight narrowing of the width in the tensile region and a slight widening in the compression width and compressing the tensile region width

research use as a core for double-top plates [18] but it hasn't gained widespread popularity compared to other core materials. Its lightweight compared to many materials, Kevlar in solid or fabric form is generally denser than other honeycombs for a given volume, potentially adding more weight to the soundboard.

- **Nomex**: A registered trade mark, also introduced by Dupont company, is a common and efficient core material. Nomex, a fire-resistant meta-aramid material, is increasingly popular as a core material in double-top guitars and other stringed instruments. Nomex offers unique advantages for this application. Nomex's low density helps minimize the top's overall weight. Despite its light weight, Nomex is remarkably stiff. This stiffness, combined with thin wood layers, helps maintain or even improve the top's bending behavior and prevents loss of sound quality. Nomex possesses good mechanical strength, contributing to the double-top structure's durability. So, using Nomex as a core in double-top plates enables a lighter plate while preserving or enhancing stiffness, thereby increasing the instrument's volume and responsiveness (Fig. 5.24) [3, 18].

5.5 Final Words

As we draw the threads of this chapter together, it becomes abundantly clear that the journey from a standing tree to a resonant musical instrument is one steeped in both artistry and scientific understanding. The selection and preparation of tonewood are

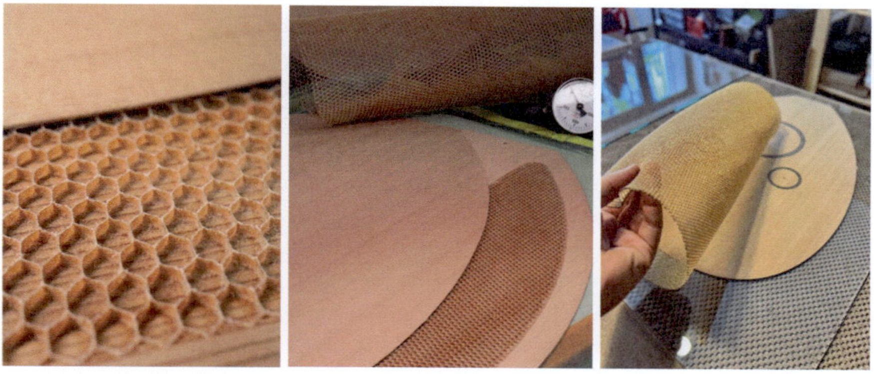

Fig. 5.24 Depicted in this image is a double-top soundboard construction, often found in stringed instruments such as the oud and guitar. The illustration clearly shows its layered composition: two thin outer wood layers encasing a honeycomb-patterned core. This design is intended to achieve a balance in bending behavior and to minimize weight. The photographs feature the Nomex double-top for oud, generously provided by Dr. Emir Değirmenli, Musical Instruments Scientist, for this publication

not merely about choosing a species; it is a nuanced process of identifying and optimizing the inherent acoustic potential within each piece of wood. We have explored the limitations of relying solely on visual inspection and the critical need for a scientific approach, one that embraces the measurement and manipulation of key acoustic parameters.

The inherent variability of wood, a living material shaped by its unique environment and growth, necessitates a move beyond simplistic classifications. Factors such as the specific section of the tree, the methods of drying and storage, and even the prevailing environmental conditions all contribute to the unique sonic signature of a piece of wood. Recognizing this complexity underscores the importance of employing precise acoustic measurements to truly understand and evaluate the suitability of tonewood for musical instrument construction.

This understanding naturally leads us to the important role of wood treatment. As discussed, various techniques exist to modify and enhance the acoustic properties of wood, allowing luthiers to mitigate natural variations and steer the material towards desired tonal characteristics. These treatments, whether focused on altering density, enhancing stiffness, or controlling damping, represent a powerful set of tools in the luthier's arsenal. They enable a level of control over the final sound of an instrument that would be impossible to achieve by simply relying on the raw, untreated material.

However, the realm of wood treatment is broad and diverse. While this chapter has explored various methods, it is important to acknowledge the ongoing dialogue and differing perspectives within the lutherie community. While certain chemical treatments have been explored and implemented, a strong tradition and compelling arguments exist for prioritizing non-chemical methods. Techniques such as careful seasoning, thermal modification, and even specific carving and shaping approaches can significantly influence the acoustic properties of wood without introducing external substances. These methods often align with a philosophy of working with the natural properties of the wood, gently coaxing out its inherent sonic potential rather than fundamentally altering its chemical composition.

The art of tuning, closely intertwined with treatment, further refines the sonic character of the instrument. Through meticulous adjustments to the instrument's components, such as the top and back plates, the soundboard bracing, and the bridge, luthiers can fine-tune the resonant frequencies and overall tonal balance. This process, often guided by both scientific measurements and the trained ear of the craftsman, represents the final stage in optimizing the instrument's voice.

Ultimately, the creation of a truly exceptional musical instrument hinges on a deep understanding of the material at hand. It requires a synthesis of scientific rigor in evaluating wood properties and the artistic intuition to guide treatment and tuning processes. By embracing a scientific approach to tonewood selection and preparation, while thoughtfully considering the spectrum of available treatment methods, luthiers can continue to elevate the craft and create instruments that resonate with both sonic beauty and enduring quality. The journey from forest to concert hall is a testament to the intricate relationship between nature, science, and the enduring art of music.

References

1. N. Abaeian, Finite element design and manufacturing of a nylon-string guitar soundboard from sandwich-structured composites (Publication No. 28249174). Master's thesis, McGill University]. ProQuest Dissertations and Theses Global (2017)
2. E.S. Alves, E.L. Longui, E. Amano, Pernambuco wood (Caesalpinia echinata) used in the manufacture of bows for string instruments. IAWA J. **29**(3), 323–335 (2008). https://doi.org/10.1163/22941932-90000190
3. R. Archee, The Australian school of Lutherie: origins and achievements. Int. J. Human. Soc. Sci. **4**(1), 101–107 (2014)
4. ASTM, *ASTM D1105-21 Standard Test Method for Preparation of Extractive-Free Wood*. ASTM International (2021)
5. I. Barducci, G. Pasqualini, Misura dell'attrito interno e delle costanti elastiche del legno. Nuovo Cim. **5**, 416–446 (1948). https://doi.org/10.1007/BF02780887
6. C. Barlow, P. Edwards, G. Millward, R. Raphael, D. Rubio, Wood treatment used in Cremonese instruments. Nature **332**(6162), 313–313 (1988)
7. C. Barlow, J. Woodhouse, The influence of varnish on the properties of spruce plates. Proc. Inst. Acoust. **12**(2), 765–770 (1990)
8. D.N. Barman, M.A. Haque, M.M. Hossain, S.K. Paul, H.D. Yun, Deconstruction of pine wood (*Pinus sylvestris*) recalcitrant structure using alkali treatment for enhancing enzymatic saccharification evaluated by congo red. Waste Biomass Valor. **11**, 1755–1764 (2020). https://doi.org/10.1007/s12649-018-00547-z
9. F.P. Beer, E.R. Johnston, J.T. DeWolf, D.F. Mazurek, *Mechanics of Materials*, 7th edn. McGraw-Hill Education (2015)
10. J. Berglund, D. Mikkelsen, B.M. Flanagan, S. Dhital, S. Gaunitz, G. Henriksson, M.E. Lindström, G.E. Yakubov, M.J. Gidley, F. Vilaplana, Wood hemicelluloses exert distinct biomechanical contributions to cellulose fibrillar networks. Nat. Commun. **11**, 4692 (2020). https://doi.org/10.1038/s41467-020-18390-z
11. J. Bodig, B.A. Jayne, *Mechanics of Wood and Wood Composites* (Van Nostrand Reinhold, 1982)

12. I. Brémaud, N. Amusant, K. Minato et al., Effect of extractives on vibrational properties of African Padauk (*Pterocarpus soyauxii* Taub.). Wood Sci. Technol. **45**, 461–472 (2011). https://doi.org/10.1007/s00226-010-0337-3
13. V. Bucur, Wood species for musical instruments. In *Acoustics of Wood* (Springer, Berlin, Heidelberg, 2006), pp. 163–191. https://doi.org/10.1007/3-540-30594-7_7
14. V. Bucur, *Handbook of Materials for String Musical Instruments* (Springer, Berlin, 2016). https://doi.org/10.1007/978-3-319-32080-9
15. W. Cai, H.C. Tai, Three millennia of tonewood knowledge in Chinese guqin tradition: science, culture, value, and relevance for Western lutherie. Savart J. **1**, 27 (2018). https://www.savartjournal.org/articles/27/about.html
16. W. Cai, H.C. Tai, The materials chemistry of Chinese guqin zithers—decoding the mysteries of an intangible cultural heritage. ARKIVOC **2023**(i), 1–21 (2023). https://doi.org/10.24820/ark.5550190.p012.008
17. C. Carlier, I. Brémaud, J. Gril, The role of tonewood selection and aging in instrument "quality" as viewed by violin makers, in *2nd Annual Conference COST FP1302 WoodMusICK, Effects of Playing on Early and Modern Musical Instruments* (2015), pp. 24–25
18. Ü. Çiçekçioğlu, A.M. Alaskan, Investigation of kevlar application in baglama soundboard and its effects on sound properties. ARTS Artuklu Sanat Ve Beşeri Bilimler Dergisi **12**, 131–147 (2024). https://doi.org/10.46372/arts.1492472
19. O. Christensen, B.B. Vistisen, Simple model for low-frequency guitar function. J. Acoust. Soc. Am. **68**(3), 758–766 (1980). https://doi.org/10.1121/1.384814
20. I. Curtu, M.D. Stanciu, N. Cretu, I. Rosca, Modal analysis of different types of classical guitar bodies, in *Proceedings of the 10th WSEAS International Conference on Acoustics & Music: Theory & Applications–AMTA09 (ISTP/ISI Proceeding of Thomson Scientific-Institute for Scientific Information)* (2009), pp. 23–25
21. Z. Daeepour, A. Lashgari, M. Roohnia, A. Jahan-Latibari, V. Safdari, Pine wood extraction by fermentation to improve its acoustical efficiency. BioResources **18**(4), 8062–8075 (2023). https://doi.org/10.15376/biores.18.4.8062-8075
22. Z. Daeepour, A. Lashgari, M. Roohnia, A. Jahan Latibari, V. Safdari, Effect of fermentation pretreatment and extraction on the acoustic properties of Ash wood (*Fraxinus excelsior*). Iranian J. Wood Paper Sci. Res. **39**(1), 27–40 (2024). https://doi.org/10.22092/ijwpr.2023.361567.1750
23. E.B. Davis, Designing soundboards with flexural disk models. Proc. Meetings Acoust. **12**(1), 035003 (2011). https://doi.org/10.1121/1.3681324
24. F. Divos, F.S. Kiss, Strength grading of structural lumber by portable lumber grading-effect of knots, in *The Future of Quality Control for Wood & Wood Products* (2010), pp. 4–7
25. M.O. Downey, R.L. Hanlin, Comparison of ethanol and acetone mixtures for extraction of condensed tannin from grape skin. S. Afr. J. Enol. Vitic. **31**(2), 154–159 (2010)
26. P. Dumond, N. Baddour, Effects of using scalloped shape braces on the natural frequencies of a brace-soundboard system. Appl. Acoust. **73**(11), 1168–1173 (2012). https://doi.org/10.1016/j.apacoust.2012.05.015
27. P. Dumond, N. Baddour, Can a brace be used to control the frequencies of a plate? Springerplus **2**, 558 (2013). https://doi.org/10.1186/2193-1801-2-558
28. M.-G. Durier, Contribution of geochemistry and geochronology to the knowledge of the history of the uses of modern musical instruments. Doctoral dissertation, Université Paris-Saclay (2022)
29. J.-P. Echard, L. Bertrand, A. von Bohlen, A.-S. Le Hô, C. Paris, L. Bellot-Gurlet, B. Soulier, A. Lattuati-Derieux, S. Thao, L. Robinet, B. Lavédrine, S. Vaiedelich, The nature of the extraordinary finish of Stradivari's instruments. Angew. Chem. Int. Ed. **49**(1), 197–201 (2010). https://doi.org/10.1002/anie.200905131
30. B. Elie, G. François, D. Bertrand, Analysis of bridge mobility of violins, in *Stockholm Music Acoustics Conference 2013* (Stockholm, Sweden, 2013), pp. 54–59
31. EN 338, *Structural timber—strength classes*. European Committee for Standardization (CEN) (2016)

32. F. Farvardin, M. Roohnia, A. Lashgari, The effect of extractives on acoustical properties of Persian silk wood (*Albizia julibrissin*). Maderas. Cienc. Tecnol. **17**(4) (2015). https://doi.org/10.4067/S0718-221X2015005000065
33. A. Fernandez, *The Relationship Between Resonant Frequency, Sound Hole Diameter, and Body Depth in Acoustic Guitars.* Master's thesis, Purdue University (2023). https://www.proquest.com/openview/b7f446f130b7c93837f3ba33bbd356b2/1?cbl=18750&diss=y&pq-origsite=gscholar
34. N.H. Fletcher, T.D. Rossing, *The Physics of Musical Instruments* (Springer, 1998)
35. G. Fry, *The Varnishes of the Italian Violin-Makers of the Sixteenth, Seventeenth and Eighteenth Centuries and Their Influence on Tone* (Stevens & Sons, San Antonia, TX, 1904)
36. M. Ghaznavi, A. Rostamisani, M. Roohnia, A. Jahanlatibari, A. Yaghmaeipour. Traditional varnishes and acoustical properties of wooden soundboards. Sci. Int. **1**(12), 401–407 (2013). https://doi.org/10.17311/sciintl.2013.401.407
37. S. González, E. Chacra, C. Carreño, C. Espinoza, Wooden mechanical metamaterials: towards tunable wood plates. Mater. Des. **221**, 110952 (2022). https://doi.org/10.1016/j.matdes.2022.110952
38. C. Gough, The violin: Chladni patterns, plates, shells and sounds. Eur. Phys. J. Special Top. **145**, 77–101 (2007). https://doi.org/10.1140/epjst/e2007-00149-0
39. T. Gore, G. Gilet, *Contemporary Acoustic Guitar Design and Build: Volume 2–Build* (Trevor Gore, Terrey Hills, NSW, 2011), pp. 19–22
40. L. Gurău, M.C. Timar, C. Coşereanu, M. Cosnita, M.D. Stanciu, Aging of wood for musical instruments: analysis of changes in color, surface morphology, chemical, and physical-acoustical properties during UV and thermal exposure. Polymers **15**(7), 1794 (2023). https://doi.org/10.3390/polym15071794
41. W.H. Hill, A.F. Hill, F.A. Davis, A.E. Hill, A.E. Hill, S. Beck, American Musicological Society--Music Library Association, *Antonio Stradivari, His Life and Work: 1644–1737.* Courier Corporation, North Chelmsford, MA (1963)
42. D. Holz, Zum Alterungsverhalten des Werkstoffes Holz-Einige Ansichten, Untersuchungen, Ergebnisse (1981). http://pascal-francis.inist.fr/vibad/index.php?action=getRecordDetail&idt=PASCAL8130417029
43. ISO 13910. *Timber Structures—Strength Graded Timber—Test Methods for Structural Properties.* International Organization for Standardization (2014)
44. E.V. Jansson, L. Frydén, G. Mattsson, On tuning of the violin bridge. J Catgut Acoust Soc **1**(6), 11–15 (1990)
45. E. Karami, Effet de traitements thermiques modérés et de revêtement sur les propriétés vibratoires des bois d'Epicéa et de Mûrier. Doctoral dissertation, Université Montpellier ⟨tel-01816988⟩ (2016)
46. S. Kim, K. Kim, G. Jun, W. Hwang, Wood-nanotechnology-based membrane for the efficient purification of oil-in-water emulsions. ACS Nano **14**(12), 17233–17240 (2020). https://doi.org/10.1021/acsnano.0c07206
47. M. Kohantorabi, M.A. Hossein, M. Shahverdi, M. Roohnia, Vibration based NDT methods to verify wood drying efficiency. Drvna Ind. **66**(3), 221–228 (2015). https://doi.org/10.5552/drind.2015.1352
48. S.L. Lämmlein, D. Mannes, B. van Damme, T.K. Bader, P. Niemz, G. Oebel, Influence of varnishing on the vibro-mechanical properties of wood used for violins. J. Mater. Sci. **54**(11), 8063–8095 (2019). https://doi.org/10.1007/s10853-019-03440-9
49. S.L. Lämmlein, Violin varnishes and tonewood: relationships of vibro-mechanical and moisture sorption properties. Doctoral dissertation, ETH Zurich (2020)
50. S.L. Lämmlein, B. Van Damme, D. Mannes, F.W.M.R. Schwarze, I. Burgert, Violin varnish induced changes in the vibro-mechanical properties of spruce and maple wood. Holzforschung **74**(8), 765–776 (2020). https://doi.org/10.1515/hf-2019-0182
51. M. Lehr, M. Miltner, A. Friedl, Removal of wood extractives as pulp (pre-) treatment: a technological review. SN Appl. Sci. **3**, 886 (2021). https://doi.org/10.1007/s42452-021-04873-1

52. J.S. Loen, Thickness graduation mapping: methods and goals. CASJ **4**(4 Series II), 5–6 (2001)
53. J. Loen, *Thickness Graduation Maps: Classic Violins, Violas, and Cellos*. Jeffrey Loen (2004)
54. J. Li, C. Chen, J.Y. Zhu, A.J. Ragauskas, L. Hu, In situ wood delignification toward sustainable applications. Acc. Mater. Res. **2**(8), 606–620 (2021). https://doi.org/10.1021/accountsmr.1c00075
55. R. Malvermi, M. Albano, S. Gonzalez et al., The impact of alkaline treatments on elasticity in spruce tonewood. Sci. Rep. **12**, 13335 (2022). https://doi.org/10.1038/s41598-022-17596-z
56. M.V. Mathews, J. Kohut, Electronic simulation of violin resonances. J. Acoust. Soc. Am. **53**(6), 1620–1626 (1973). https://doi.org/10.1121/1.1913511
57. M. Matsunaga, K. Minato, Physical and mechanical properties required for violin bow materials II: comparison of the processing properties and durability between pernambuco and substitutable wood species. J. Wood Sci. **44**(2), 142–146 (1998). https://doi.org/10.1007/BF00526260
58. M. Matsunaga, K. Minato, F. Nakatsubo, Vibrational property changes of spruce wood by impregnation with water-soluble extractives of pernambuco (*Guilandina echinata* Spreng.). J. Wood Sci. **45**(6), 470–474 (1999). https://doi.org/10.1007/BF00538955
59. J.C. Maturana, C. Arroyave, A. Hurtado et al., Effect of alkaline delignification process on the anatomical characteristics and some mechanical properties of tropical woods of different densities. Res. Square (2022). https://doi.org/10.21203/rs.3.rs-2188263/v1
60. L. McKeen, The effect of sterilization on plastics and elastomers, 3rd edn, in *Plastics Design Library*. Elsevier (2012). https://doi.org/10.1016/C2011-0-06772-8
61. Y. Miao, R. Li, X. Qian, M. Jiang, Y. Zhang, B. Fei, Effect of extraction on the acoustic vibrational properties of *Picea jezoensis* var. microsperma (Lindl.) W.C.Cheng & L.K.Fu. Ann. For. Sci. **78**(24) (2021). https://doi.org/10.1007/s13595-021-01048-1
62. E. Meier, The Wood Database (2025). https://www.wood-database.com
63. N.-E. Molin, Optical methods for acoustics and vibration measurements, in *Springer Handbook of Acoustics* by ed. T.D. Rossing (Springer, New York, NY, 2014), pp. 847–884. https://doi.org/10.1007/978-1-4939-0755-7_27
64. J. Nagyvary, R.N. Guillemette, C.H. Spiegelman, Mineral preservatives in the wood of Stradivari and Guarneri. PLoS ONE **4**(1), e4245 (2009). https://doi.org/10.1371/journal.pone.0004245
65. K.C. Nefdt, Improvement of the acoustic properties of tonewood. Doctoral dissertation, Stellenbosch University (2024)
66. K. Nefdt, M. Meincken, Modifying the radiation ratio of tonewoods through wood degradation. Wood Mat. Sci. Eng. **20**(1), 1–8 (2024). https://doi.org/10.1080/17480272.2024.2376858
67. H.T. Nia, A.D. Jain, Y. Liu, M.-R. Alam, R. Barnas, N.C. Makris, The evolution of air resonance power efficiency in the violin and its ancestors. Proc. R. Soc. A Math. Phys. Eng. Sci. **471**(2173), 20140905 (2015). https://doi.org/10.1098/rspa.2014.0905
68. G. Nicoletti, Measuring and tuning the performances of the acoustic guitar (2022). https://www.iuliusguitars.com/wp-content/uploads/2025/03/Measuring_Eng_060222.pdf
69. G. Nicoletti, Mastering the sound of the acoustic guitar. Julius Guitars. Printed in Poland by Amazon Fulfillment Poland Sp. z. o. o. Wrocław. ISBN 979-10-415-1253-9 (2024)
70. E. Obataya, M. Norimoto, Acoustic properties of a reed (*Arundo donax* L.) used for the vibrating plate of a clarinet. J. Acoust. Soc. Am. **106**(2), 1106–1110 (1999). https://doi.org/10.1121/1.427118
71. E. Obataya, Y. Ohno, M. Norimoto, B. Tomita, Effects of oriental lacquer (urushi) coating on the vibrational properties of wood used for the soundboards of musical instruments. Acoust. Sci. Technol. **22**(1), 27–34 (2001). https://doi.org/10.1250/ast.22.27
72. E. Obataya, N. Zeniya, K. Endo-Ujiie, Effects of seasoning on the vibrational properties of wood for the soundboards of string instruments. J. Acoust. Soc. Am. **147**(2), 998–1005 (2020). https://doi.org/10.1121/10.0000723
73. S. Omrani Ghahjaverstani, A. Lashgari, M. Roohnia, Nondestructive evaluation of flexural strength comparing to static bending strength in wood from Pine and Beech. Iran. J. Wood Paper Sci. Res. **34**(2), 157–166 (2019). https://doi.org/10.22092/ijwpr.2019.125019.1519

74. T. Ono, M. Norimoto, Study on Young's modulus and internal friction of wood in relation to the evaluation of wood for musical instruments. Jpn. J. Appl. Phys. **22**(4R), 611 (1983). https://doi.org/10.1143/JJAP.22.611
75. E.C. Peck, Shrinkage of wood (Report No. R1650). U.S. Department of Agriculture, Forest Service, Forest Products Laboratory (1947)
76. R.C. Pettersen, The chemical composition of wood, in *The Chemistry of Solid Wood* by ed. R.M. Rowell. Advances in Chemistry Series 207 (1984), pp. 57–126. https://doi.org/10.1021/ba-1984-0207.ch002
77. Y. Qi, Z. Zhou, R. Xu, Y. Dong, Z. Zhang, M. Liu, Effect of NaOH pretreatment on permeability and surface properties of three wood species. ACS Omega **8**(43), 40362–40374 (2023). https://doi.org/10.1021/acsomega.3c04745
78. J.E. Reeb, Wood and moisture relationships. Oregon State University Extension Catalog (1995). https://extension.oregonstate.edu/catalog/pub/em-8600-wood-moisture-relationships
79. B. Richardson, Mode studies of plucked stringed instruments: application of holographic interferometry, in *Proceedings of the Second Vienna Talk*, ed. R. Bader (University of Music and Performing Arts Vienna, Austria, 2010)
80. M. Roohnia, S.-F. Hashemi-dizaji, L. Brancheriau, A. Tajdini, A.-H. Hemmasi, N. Manouchehri, Effect of soaking process in water on the acoustical quality of wood for traditional musical instruments. BioResources **6**(2), 2055–2065 (2011)
81. M. Roohnia, A. Tajdini, N. Manouchehri, Assessing wood in sounding boards considering the ratio of acoustical anisotropy. NDT and E Int. **44**(1), 13–20 (2011). https://doi.org/10.1016/j.ndteint.2010.09.001
82. M. Roohnia, M. Kohantorabi, A. Tajdini, Maple wood extraction for a better acoustical performance. Eur. J. Wood Prod. **73**, 139–142 (2015). https://doi.org/10.1007/s00107-014-0871-z
83. M. Roohnia, F. Policardi, Tonewood treatment ideas for musical instruments soundboards. Zbornik mednarodne Elektrotehniške in računalniške konference (2022), pp. 313–316
84. T.D. Rossing, N.H. Fletcher, *Principles of Vibration and Sound* (Springer Science and Business Media, Berlin, 2004)
85. T.D. Rossing (ed.), *The Science of String Instruments*. Springer New York, NY (2010). https://doi.org/10.1007/978-1-4419-7110-4
86. F. Ruffinatto, A. Crivellaro, *Atlas of Macroscopic Wood Identification: With a Special Focus on Timbers Used in Europe and CITES-Listed Species* (Springer Nature, Berlin, 2019)
87. S.F. Sacconi, *Secretos de Stradivari: Catálogo de objetos de recuerdo del Museo Stradivari, Ala Ponzone de Cremona*. Biblioteca del Congreso Cremona (1972)
88. H. Sakagami, J. Matsumura, K. Oda, In situ visualization of hardwood microcracks occurring during drying. J. Wood Sci. **55**(5), 323–328 (2009). https://doi.org/10.1007/s10086-009-1041-5
89. S. Šali, J. Kopac, Positioning of braces on a guitar soundboard, in *Proceedings of SEM, IMAC-XX: A Conference on Structural Dynamics, Los Angeles, California*, vol. 1 (2002), pp. 709–715
90. J. Schneider, *Free New Design Violin Bridge Sounds like a Strad*. violinist.com discussion forum (2010). https://www.violinist.com/discussion/archive/17185/
91. A. SeGolpayegani, I. Brémaud, J. Gril et al., Effect of extractions on dynamic mechanical properties of white mulberry (*Morus alba*). J. Wood Sci. **58**, 153–162 (2012). https://doi.org/10.1007/s10086-011-1225-7
92. J. Shi, Y. Lu, Y. Zhang et al., Effect of thermal treatment with water, H_2SO_4 and NaOH aqueous solution on color, cell wall and chemical structure of poplar wood. Sci. Rep. **8**, 17735 (2018). https://doi.org/10.1038/s41598-018-36086-9
93. N. Sobue, S. Okayasu, Effects of continuous vibration on dynamic viscoelasticity of wood. Mokuzai Gakkaishi **38**(4), 364–369 (1992). https://doi.org/10.2472/jsms.41.164
94. J.C. Schelleng, Wood for violins. Catgut Acoust. Soc. Newslett. **37**, 8–19 (1982)
95. M. Stanciu, I. Curtu, Researches concerning the acoustic properties of tone wood used on musical instruments structures, in *Proceedings of the 11th International Research/Expert*

Conference "Trends in the Development of Machinery and Associated Technology" (2007), pp. 807–810
96. M.D. Stanciu, H.D. Teodorescu, S. Vlase, M. Mihalcica, M. Cosniță, A. Savin, Multiscale assessment of artificial aging treatment of polysaccharides from tonewood species. Int. J. Biol. Macromol. **274**(Part 1), 133310 (2024). https://doi.org/10.1016/j.ijbiomac.2024.133310
97. S. Stenzel, From a guitar maker's notes: aplead for the traditional construction of classical guitar soundboards (2004). https://stenzel-guitars.de/resources/pdf/trad.vs.modern.pdf
98. C.K. Su, S.Y. Chen, J.H. Chung, G.C. Li, B. Brandmair, T. Huthwelker et al., Materials engineering of violin soundboards by Stradivari and Guarneri. Angew. Chem. **133**(35), 19293–19303 (2021). https://doi.org/10.1002/ange.202105252
99. H.-C. Tai, G.-C. Li, S.-J. Huang, C.-R. Jhu, J.-H. Chung, B.Y. Wang, C.-S. Hsu, B. Brandmair, D.-T. Chung, H.-M. Chen, C.C.J. Chan, Chemical distinctions between Stradivari's maple and modern tonewood. Proc. Natl. Acad. Sci. **114**(1), 27–32 (2017). https://doi.org/10.1073/pnas.1611253114
100. TAPPI Test Method. TAPPI T 204 cm-17, (2017). Technical Association of the Pulp and Paper Industry. Solvent Extractives of Wood and Pulp.
101. G. Tsoumis, *Science and Technology of Wood: Structure, Properties, Utilization*, vol. 115 (Van Nostrand Reinhold, New York, NY, 1991)
102. I. Wachter, T. Štefko, M. Rolinec, Optimization of two-step alkali process of lignin removal from basswood. Res. Pap. Facul. Mater. Sci. Technol. Slovak Univ. Technol. **27**(44), 153–161. https://doi.org/10.2478/rput-2019-0016
103. U.G.K. Wegst, Wood for sound. Am. J. Bot. **93**, 1439–1448 (2006). https://doi.org/10.3732/ajb.93.10.1439
104. WoodBin. Wood Shrinkage Table (2025). http://woodbin.com/ref/wood-shrinkage-table
105. J. Woodhouse, On the "bridge hill" of the violin. Acta Acust. United Acust. **91**(1), 155–165 (2005). https://www.researchgate.net/publication/228632416
106. E. Xu, D. Wang, L. Lin, Chemical structure and mechanical properties of wood cell walls treated with acid and alkali solution. Forests **11**(1), 87 (2020). https://doi.org/10.3390/f11010087
107. K. Yatsui, H. Asaoka, Y. Yoshida, T. Yatsui, Change of sound characteristics of yidaki (didgeridoo) coated with thick films of lacquer (Wajima-nuri). Jpn. J. Appl. Phys. **59**(7), 077002 (2020). https://doi.org/10.35848/1347-4065/ab9277
108. S. Zamaninasab, A. Lashgari, M. Roohnia, A. Jahan-Latibari, A. Tajdini, Fermentation pretreatment and extraction's effect on the acoustic properties of walnut wood (*Juglans regia*). BioResources **18**(3), 5085–5095 (2023). https://doi.org/10.15376/biores.18.3.5085-5095
109. S. Zamaninasab, A. Lashgari, M. Roohnia, A. Jahan-Latibari, A. Tajdini, Effect of removing extractives on the acoustic properties of cypress wood (*Cupressus arizonica*). J. Renew. Nat. Resour. Res. **9**(1), 1–14 (2024). https://doi.org/10.71916/jrnr.2024.10304

Chapter 6
Final Wooden Musical Instrument: Acoustic Tests

Abstract This chapter discusses the acoustic testing of wood in finished wooden musical instruments, emphasizing the multifaceted nature of what constitutes a good sound. The chapter highlights that achieving a desirable sound involves a combination of resonance, clarity, and other factors that go beyond personal taste. While advanced technologies discussed in earlier chapters might not be universally available, this chapter argues for the necessity of accessible acoustic evaluation facilities for all luthiers. The chapter outlines the key acoustic parameters that define the sound of wood in a stringed instrument, including resonance, richness, clarity, and definition. Resonance refers to the instrument's ability to sustain sound, while richness relates to the complexity of the sound's frequency response. Clarity and definition pertain to how distinct each note is and the balance and flexibility of the instrument's sound. The chapter also discusses the influence of traditional designs, player preferences, and the importance of removing unwanted resonances from the instrument's sound. It emphasizes that a good instrument results from the careful integration of design, materials, construction, and an understanding of acoustic physics. Furthermore, the chapter details the use of software and instrumentation in sound analysis, highlighting the importance of tools like FFT (Fast Fourier Transform) and unidirectional microphones. These tools enable luthiers to measure and analyze the acoustic properties of instruments with adequate precision. Finally, the chapter provides methodologies for analyzing the frequency response of instruments, including tap tests for the tonewood mounted as the top plate and back of the bowl of an instrument. These tests help luthiers identify and address acoustic issues in their instruments.

6.1 What Is a Good Instrument Sound?

Of course, understanding the characteristics of a good sound in a stringed instrument is a delicate matter, and there are many different aspects involved [17]. Let's take a look at these dimensions together.

The desirable sound of a stringed instrument appears to be something of a general consensus, moving beyond purely personal taste towards a more comparable ideal.

M. Roohnia, *Properties and Treatment of Tonewood for String Instrument Construction*, Springer Series in Materials Science 355, https://doi.org/10.1007/978-3-032-00824-4_6

Therefore, to identify the sound-producing aspects of a musical instrument and adapt it to our preferences, we measure its physical acoustic parameters. These characteristics may include the following.

6.1.1 Resonance and Richness

The sound of the instrument should be resonant, warm, and deep. This means that the sound continues to vibrate for a while after the note is played (small attenuation and strong sustain) and produces various harmonics (a rich frequency response) that add to its richness.

- **Sustain**: The note's amplitude should be sustained for a reasonable length of time. This allows the player to create more continuous musical phrases.
- **Responsiveness**: The instrument should be sensitive to playing and respond quickly and accurately to changes in dynamics (loudness) and bowing. This allows the player to convey more emotion and subtlety in the performance.

6.1.2 Clarity and Definition

Along with resonance, the sound should be clear and distinct. Each note should be heard well and not mixed up with others, especially when playing complex pieces or multi-part chords.

- **Balance**: All the strings of the instrument should be in balance with each other in terms of volume and sound quality [10]. No string should sound noticeably weaker or stronger than the others. Also, the balance between the different frequencies (highs, mids, lows) is important.
- **Flexibility**: A good instrument should be able to produce a wide range of sounds and tonal colors. This includes the ability to produce soft and delicate sounds to strong and powerful sounds [10].
- **Intonation**: The instrument should be made in such a way that the notes are accurate and in tune throughout the entire length of the fretboard and on all strings [10].

6.1.3 Other Factors

- **Old patterns and tradition**: Taking inspiration from old patterns of famous makers (such as Stradivarius in the violin) can be an excellent starting point, as these patterns have stood the test of time. However, mere imitation is not enough, and modern makers can also achieve extraordinary sounds through innovation and a deep understanding of the physics of the instrument. Traditional prejudices can be limiting if they prevent progress and experimentation.

- **The taste of the player**: The taste of the player plays an important role in the choice of the final instrument [17]. One player may prefer a warmer and softer sound, while another likes a brighter and more powerful sound. However, this taste usually falls within the framework of the general characteristics of a "good" instrument.
- **Physical noise removal**: Trying to eliminate unwanted noise (such as bow clicks or extra vibrations) is certainly important. A good instrument should produce a clean, focused sound. However, some of the subtle, natural sounds that come from the vibrations of the wood and strings can add character and appeal to the sound. The goal is to eliminate annoying noises, not to eliminate all of the instrument's unique sonic characteristics.
- **General agreement**: Despite the diversity of tastes, there is a general agreement on the essential characteristics of a good instrument. This agreement is usually based on the quality, balance, responsiveness, and musical potential of the instrument.

It seems, then, that a good stringed instrument is an artistic combination of careful design, quality materials, skillful construction, and a deep understanding of the physics of sound. It should be able to produce a rich, clear, balanced, and responsive sound that allows the player to fully express her/his musical feelings and ideas. While the player's taste plays a role in the final choice, the fundamental characteristics of sound quality usually go beyond personal preference.

For any taste, physics, tradition, or other scale, there are parameters in the instrument that need to be evaluated and addressed. We will introduce these for the purposes of this book, from the perspective of wood acoustic sound.

Now that we already became familiar with sound analysis techniques through observing the FFT spectrum, the first step is to identify the sound ranges, their loudness, and attenuation characteristics for the specific instrument we are crafting. This will enable us to adjust the instrument's sound to suit various sonic preferences using the treatment and tuning methods discussed in Chap. 5.

For many musical instruments, this sound analysis has already been conducted and is available in existing literature. You can utilize these resources to tune and adjust the resonance and timbre of your instruments accordingly. For example, the characteristics of a good violin have been specifically defined as follows [14, 15]:

- **Large response at low frequencies**: Good violins exhibit a strong response in the low-frequency range, below 1000 Hz. Subjectively, this translates to the fundamental tones and lower harmonics of each played note being pleasant, expressive, and carrying well.
- **Small amplitudes at high frequencies**: Often, small amplitudes at high frequencies (above approximately 3000 Hz) contribute to a harmonious softness and a delicate, pure response, described by musicians as a "soft and pure sound."

- **Minimum amplitude around 1500 Hz**: Reduced amplitude in the frequency range around 1500 Hz helps to avoid a nasal tonal quality.
- **Emphasis in the 2000–3000 Hz region**: A prominent loudness in the frequency region between 2000 and 3000 Hz imparts a desirable bright and strong quality to the sound, a characteristic often referred to by musicians. As previously discussed, this frequency range is influenced by the bridge characteristics and is known as the Bridge Hill frequency range [22].

So, we should be able to identify specific physical characteristics within the frequency response of our specialized instrument, analogous to those of the violin, that are associated with good sound quality.

After introducing the capabilities of the available software and tools, we will discuss how to identify these characteristics and provide specialized exercises to facilitate this learning.

6.2 Software and Instrumentation

6.2.1 Importance

In the construction and research of stringed instruments, a thorough understanding of the factors influencing the final instrument's sound quality has always been paramount. While traditional knowledge and the experience of master builders form the bedrock of this craft, recent advancements in measurement tools and sound analysis software have enabled a more accurate and quantitative examination of instrument and component sound characteristics. These developments are playing an increasingly significant role not only in scientific research but also in the experimental workshops of instrument makers.

Historically, the assessment of instrument sound quality relied primarily on human hearing and qualitative judgments. Although the trained ear of an expert can discern subtle sonic differences, this method lacks the objectivity and repeatability essential for detailed research and the optimization of manufacturing processes. Traditional electronic and electrical equipment for sound analysis was often bulky, expensive, and required controlled laboratory environments with complete silence. These limitations hindered the accessibility of accurate sound analysis tools for both makers and researchers.

The advent of computers and specialized sound analysis software has revolutionized this field. Today, luthiers and researchers can conduct a wide array of sound measurements and analyses within their workshop environment using relatively inexpensive microphones and powerful software installed on personal computers. This capability has not only significantly reduced costs but has also simplified and accelerated the evaluation process.

One important application of these tools is the frequency spectrum analysis (Fast Fourier Transform—FFT) of the sound produced by various instrument components, illustrated in previous chapters. By analyzing the tone of the whole body, soundboard, back, soundholes, and strings during vibration, valuable information regarding natural resonances, sound propagation, and the frequency response of each component can be obtained. This data assists builders in understanding the impact of design changes, wood type, thickness, and dimensions on the instrument's final sound.

Beyond analyzing the main components, these tools also allow for the examination of the effect of various attachments, such as the bridge, nut, pegs and pegbox. The stiffness and density of these parts, and their attachment to the instrument body, significantly influence vibration transmission, reflection or radiation and consequently, the instrument's timbre and volume. Through controlled experiments and the analysis of the frequency spectrum resulting from variations in these variables, optimal settings can be achieved to produce the desired sound.

Modern sound analysis software applications offer advanced features for controlling and reducing environmental noise during measurements. While workshop conditions may not match ideal laboratory settings, signal processing algorithms and averaging techniques can substantially mitigate the impact of unwanted noise, providing reliable results. This is particularly important in workshop environments where achieving complete silence is challenging.

However, it is necessary to interpret the results of these measurements and analyses in conjunction with the manufacturer's knowledge and experience. The numerical data and graphs represent only a part of the overall picture; a comprehensive understanding of an instrument's sound quality necessitates integrating this quantitative data with qualitative and auditory assessments.

Therefore, the use of evaluation tools and software for stringed instruments represents a significant step towards a more scientific approach in manufacturing and research within this field. These tools empower manufacturers and researchers to examine the factors affecting sound more accurately, optimize design and construction processes, and gain a deeper understanding of the relationship between an instrument's physical characteristics and its sonic qualities. Although the accuracy of these methods may not always reach the standards of dedicated laboratory research, the results provide a powerful resource for experimental designs and the continuous improvement of musical instrument quality in instrument-making workshops.

6.2.2 *Sound Analysis with Data Export-Capable Software*

For accurate analysis and sound engineering using FFT software, software that provides more than just a graph is necessary, as shown in Chap. 3, Fig. 3.31. Micro-data from the graph is essential because a luthier's evaluations go beyond what ordinary software displays. Suitable software should:

- Depending on the computer sound card capabilities, capture continuously generated sound and divide it into a number of discrete data points, called the sample rate (sampling frequency). A higher sample rate provides more sound data points for analysis. Common sound cards sample at 44,100 sound points per second, which is usually adequate. These points should be arranged in a three-dimensional coordinate system where:

- **X**: Represents the time from the origin to each sound point, measured in seconds (s), milliseconds (ms), or microseconds (μs).
- **Y**: Represents the sound's intensity at that point, measured in decibels (dB, logarithmic scale). The SI unit of intensity, which includes sound intensity, is the watt per square meter (W/m^2).
- **Z**: Each sound point also has a frequency, represented on the Z-axis, and measured in hertz (Hz).

The graph's appearance changes with the viewing angle:

Viewing the graph with frequency as the primary focus yields the Fourier spectrum, which displays sound peaks corresponding to vibration modes (Chap. 3, Fig. 3.31). The loudest peaks indicate the resonance frequencies of those modes.

Viewing the graph with time as the primary focus reveals the sinusoidal oscillation of the sound, as described earlier. Instead of a continuous curve, this graph is a collection of closely spaced data points that, due to their density, appear continuous.

- Export this coordinate data to a spreadsheet for further analysis (Fig. 6.1).

Software with these capabilities seems to be well-suited for evaluating the sound of musical instruments.

6.2.3 *Input Devices: Unidirectional Microphone*

A unidirectional microphone is a type of microphone primarily sensitive to sounds originating from a specific direction, typically the front [20]. These microphones significantly attenuate ambient sounds and those arriving from the sides or behind (Fig. 6.2).

Unidirectional microphones are ideal for isolating a specific sound source and minimizing unwanted ambient noise. Consequently, they serve as a low-cost alternative for capturing point-to-point vibration of musical instruments when expensive laser or other advanced equipment is unavailable. This is particularly important for introducing affordable scientific methods to luthiers. Obviously, well-funded research laboratories will utilize laser and similar equipment that offers higher resolution sound observation points for capturing vibration data.

Common applications for these microphones include recording in noisy environments like interviews or non-acoustic rooms, and capturing the sound of an instrument alongside other ambient sounds. A unidirectional microphone can record a speaker's voice or any point on the instrument while ignoring or reducing surrounding disturbances.

It's important to note that eliminating or filtering noise would be unsuitable for these studies. Filtering a point on the instrument body that produces noise would

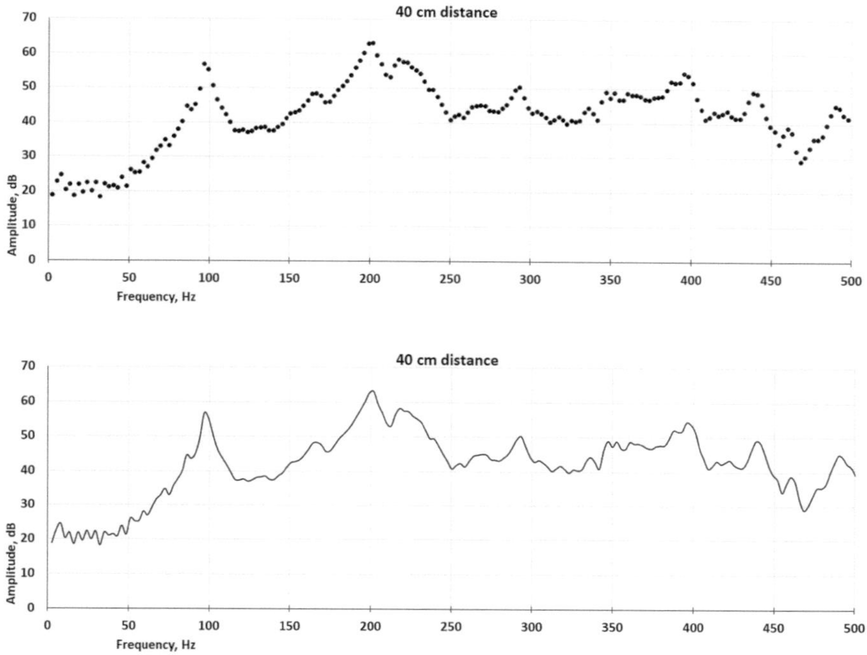

Fig. 6.1 Two graphs display the Fast Fourier Transform (FFT) spectrum of an acoustic guitar soundboard, evaluated using Audacity® software but exported the data to Microsoft Excel. In both graphs, the horizontal axis represents Frequency (Hz), ranging from 0 to 500 Hz, and the vertical axis represents Amplitude (dB), ranging from 0 to 70 dB. The microphone was positioned 40 cm away from the guitar during the recording. The top graph presents the data points as individual markers, generated from sampling sound points. The bottom graph displays the same data as a continuous line graph. Both graphs show similar trends, with peaks indicating resonant frequencies and varying amplitude levels across the frequency spectrum of the guitar's soundboard

mask that sound. Therefore, while disturbing environmental noise should not enter our computational data, any inappropriate sound emanating from a point on the instrument body under study must be identified. Consequently, sound filtering is prohibited in these studies. A completely silent study environment is preferred, but in non-acoustic workshops, these unidirectional microphones can still provide good results.

6.3 Frequency Response

It was said that every solid body has specific natural frequencies, called "vibrational modes." These frequencies can be identified by applying a free impact to the body and using spectral analysis software, such as an FFT Analyzer. These frequencies characterize the material's dynamic behavior in response to mechanical excitation.

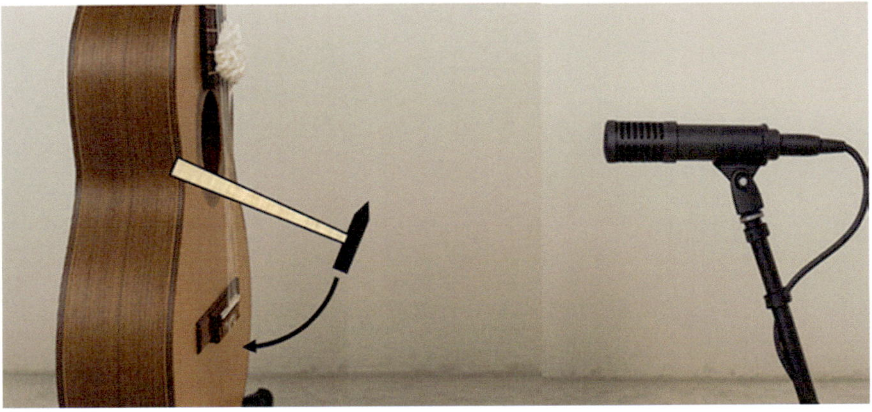

Fig. 6.2 The image shows a setup for analyzing soundboard tap-tones using a light, soft hammer. A unidirectional microphone is positioned on a stand to the right of the frame. On the left, the side of an acoustic guitar rests on a stand, with its soundhole and muffled strings (with cotton). The distance between the microphone and the guitar indicates an experimental arrangement designed to record the guitar's sound, where the unidirectional microphone would primarily capture audio directly from the instrument's front

However, the soundboard's behavior in the final instrument differs from its behavior in the laboratory due to several factors. These factors include the type of supports (boundary condition), the material and properties of the adhesive used (elastic, plastic, or glassy), the soundboard's geometry (length, width, thickness), and the forces applied by the instrument's structure and attachments. Therefore, acoustic studies and analyses should be conducted on the soundboard under conditions that closely simulate its final installation in the instrument to achieve more accurate and reliable results. The previous chapter explained the construction of frames and molds to simulate these conditions.

Now, in this chapter, as we enter the stage of acoustic testing of the completed instrument, we can apply the knowledge gained from the previous stages. This stage allows us to identify the instrument's acoustic strengths and weaknesses and, if possible, correct them. Instruments that can be disassembled and reassembled allow for the removal, correction, and reinstallation of defective or improvable components.

A precise understanding of the effect of the instrument's structure on the soundboard's frequency distribution, combined with techniques like local carving or adding mass at specific accessible points (through the soundhole or other access points), allows for many acoustic adjustments even after the instrument is fully assembled. However, it is still recommended that each component be checked and adjusted before final assembly, under conditions similar to those of the actual installation.

Furthermore, some fundamental acoustic properties, such as sound propagation velocity, sound radiation coefficient, damping capacity, elastic modulus, material density, and acoustic characteristic impedance, are independent of the specimen's geometric shape [19]. Therefore, these characteristics must be carefully measured and adjusted before the instrument's final manufacture and assembly, as they significantly influence the instrument's final sound quality.

6.3.1 *Top Plate Tap Test*

Alright, now we're at an important point in understanding sound production in a stringed instrument. To address potential sound issues at specific frequencies, we need to know both the frequency spectrum of the soundboard's tap tone and the vibration pattern (mode shape) associated with each frequency. Specifically, we need to identify the locations of the vibrational peaks (antinodes) and the areas of minimal vibration (nodes) for each mode on the soundboard [4]. This knowledge, combined with Table 4.3 from Chap. 4, will guide us in strategically manipulating mass and stiffness to fine-tune the instrument's sound.

While sophisticated equipment exists for identifying mode shapes, we'll focus on achieving good results using the readily available tools in most luthiery workshops.

Consider Fig. 6.3 as an example, using a guitar soundboard for illustration, although the principles apply to identifying mode shapes (antinodes and nodes) on all stringed instruments. Here's a practical approach:

- **Soundboard grid**: Divide the guitar soundboard into a grid of squares and number each one. The size of these squares depends on the microphone's ability to isolate sound from a small area. A more focusable unidirectional microphone enables the use of smaller analysis squares.
- **Gentle tap recording**: Perform a gentle tap with a light hammer (as shown in Fig. 6.2) and record the overall tap tone sound from a distance (about half a meter). Store this recording. Attaching a small rubber piece, such as a rubber eraser, to the striking tip of the light hammer can yield better results by muffling unwanted and distracting vibrations from the hammer itself.
- **Sound mapping**: Now, position the unidirectional microphone very close (about half a centimeter) to the center of each numbered square. At each square, tap that specific point with the light hammer and record the resulting vibration sound.
- **FFT analysis**: Once you have collected the tap tone recordings for all the hypothetical squares, begin your Fast Fourier Transform (FFT) analysis.

When the microphone is positioned at a distance during the initial tap, it captures the combined vibration modes from all parts of the soundboard, particularly the bellies, reaching your ear. The resulting Fourier spectrum displays these as peaks, representing the various vibration modes.

However, when the microphone is brought close to a specific square and that point is tapped, it will primarily pick up the vibration sound of that particular square. Consequently, the Fourier diagram for that recording will show a higher amplitude (stronger peak) for the frequencies whose bellies correspond to the location of

Fig. 6.3 A classical guitar, positioned vertically on a black stand against a neutral background, is overlaid with a grid pattern. The grid consists of horizontal lines numbered 1 through 9 from bottom to top and vertical lines labeled a through e from left to right, creating a matrix of squares across the guitar's soundboard. A white crumpled cotton is placed centered on c7, under the strings to muffle their unwanted resonance. The image illustrates a methodology for mapping the vibration bellies of the guitar's soundboard modes by tapping individual squares on the grid and analyzing the resulting Fourier diagrams, where a stronger peak in the frequency spectrum would indicate a belly corresponding to the tapped location

that square on the soundboard. By repeating this process for each square, you can successfully map the approximate locations of the mode shapes' bellies.

If our grid includes squares over the soundhole(s) (guitars typically have one, while ouds, Persian setars, violins, etc., may have two or more), we can also perceive and record the Helmholtz air resonance sound associated with those openings and manipulate it if necessary.

Alternatively, to obtain a tap tone, we can strike the soundboard at a fixed point (for example, on the bridge, which often coincides with the belly for many modes) [5] and move the microphone across each square. However, experience suggests that striking the soundboard near the recording point often yields a clearer result. Advanced commercial products are available that simultaneously taps and records, which simplifies analysis (Fig. 6.4).

One more important point to mention regarding tapping is that we may have seen many luthiers and researchers on social media performing this advanced tapping

Fig. 6.4 A line drawing depicts a hand holding a specialized tapping hammer against the internal bracing structure of a guitar top plate mounted in a frame. The hammer has a handle and a head with a simple LCD displaying "440 Hz," suggesting a specific excitation frequency. The frame, secured with small bolts around its perimeter, holds the guitar plate. This setup illustrates an advanced commercial product used for modal analysis, where the hammer simultaneously taps the plate and records the vibration response, facilitating easier analysis of the instrument's acoustic properties. The rubber eraser, typically attached to such hammers to dampen unwanted resonance, is implied by the cushioned appearance of the hammerhead. A handart by author, enhanced using AI. *Note* In this schematic view, the braces and the soundhole are drawn arbitrarily and do not represent any advantageous or disadvantageous pattern

analysis. Some first fix the entire musical instrument in a rigid frame, while others, as demonstrated in this tutorial, test the instrument in a free and relaxed position or on a simple support. Therefore, it is important to emphasize that sound analysis tests should be conducted under conditions most similar to the final, real-world playing condition of the instrument. This suggests that testing in a free state (like with the cello and Persian kamancheh) or, at most, while held in the player's knees, hands or against their shoulder (like with the Persian and Azeri tar, guitar, and similar instruments) would provide more realistic results. Consequently, we do not recommend performing sound analysis under unrealistic or far-from-reality conditions.

On the other hand, comparisons of similar data in musical instrument research have shown no considerable difference in the identified mode frequencies between testing in a free state and testing with the entire instrument unrealistically fixed. While the mode shapes might exhibit some virtual deformation or displacement, the frequencies remain largely consistent [12].

Practical Exercise 6.1

Analyze the sound of a tap on the bridge of a classical guitar from a unidirectional mic at a distance of 40 cm. Determine which of the resulting frequencies corresponds to the Helmholtz resonance of the guitar's body.

Solution

To solve this exercise, we will record the soundboard and compare it with the sound of the c6-center point (from Fig. 6.3), displaying both sounds on the same diagram (Fig. 6.5).

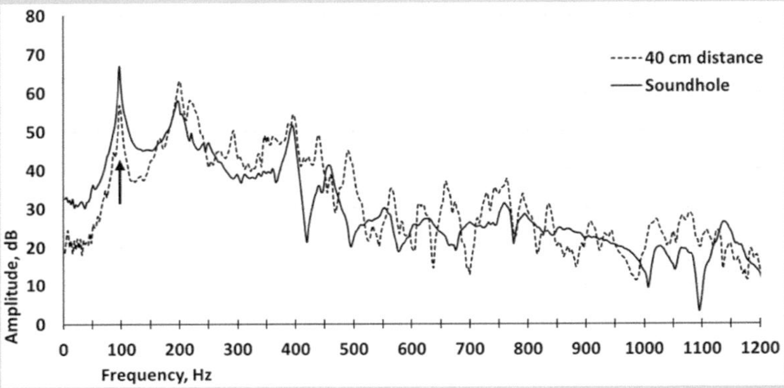

Fig. 6.5 Based on the visual analysis of the provided acoustic response curves, the frequency around 100 Hz (98 Hz) is related to the Helmholtz resonance of the air within the guitar's body. The significant amplitude increase at this frequency, relative to other vibrational modes, suggests it corresponds to the Helmholtz resonance

Similar to Exercise 6.1, other recording points can also indicate the likely location of the antinode of a particular vibrational mode by showing an increase in its loudness. This helps identify where to adjust stiffness or mass to manipulate that mode if necessary.

6.3.2 *Back Bowl Tap Test*

Similar to the soundboard, other parts of a musical instrument, such as the neck, fingerboard, nut, peg box, or the back of the resonance box, can be divided into a grid of small squares. By comparing the tap tone sound of each square with the overall sound of the body tap tone, luthiers can analyze and identify the source of unwanted frequencies within the instrument's entire acoustic system. The loudness of the tap tone in each square helps pinpoint the location of a frequency, allowing for targeted adjustments.

Since the soundboard is primarily responsible for sound generation, the back of the instrument, or the resonance bowl, is the next most acoustically significant part. We will therefore focus our discussion here. It's important to note that the degree to which the back walls of the resonance bowl actively contribute to or passively influence the sound varies depending on musical tastes and styles.

Continuing the soundboard test methodology, we now examine the back of the classical guitar (Fig. 6.6). This component is typically constructed from stiff, dense wood to achieve high characteristic acoustical impedance and effectively reflect sound waves. Ideally, the bowl itself is made of wood with high acoustic impedance, minimizing sound wave entry and maximizing reflection back towards the soundboard, thus optimizing the instrument's projection (Fig. 4.24 Chap. 4).

Similar to Exercise 6.2, other recording points can also indicate the likely location of the antinode of a particular vibrational mode by showing an increase in its loudness. This helps identify where to adjust stiffness or mass to manipulate that mode if necessary.

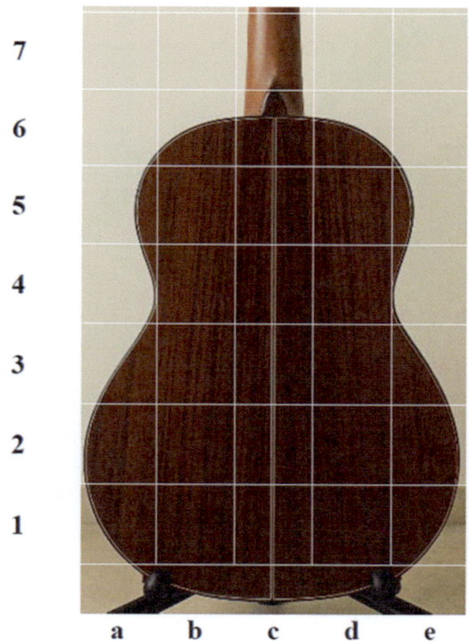

Fig. 6.6 Shows the back of a classical guitar, overlaid with a grid. The grid consists of horizontal lines numbered 1 through 7 from bottom to top and vertical lines labeled a through e from left to right, creating a matrix of squares across the guitar's back. This gridding system is used to systematically record the vibration response of different spots on the guitar back, by tapping each square and measuring the resulting sound or movement. The guitar is positioned vertically on a black stand against a neutral background. This method allows for a detailed analysis of the vibrational characteristics of the instrument's back

Practical Exercise 6.2

Analyze the sound of taps on the back of a classical guitar, specifically at the C2 square (shown in Fig. 6.6), using a unidirectional microphone positioned 5 mm away. Determine which of the resulting frequencies corresponds to the back's resonance activity, and compare it to that of the soundboard.

Solution

To solve this exercise, we will record the soundboard and compare it with the sound of the c2-center point (from Fig. 6.6), displaying both sounds on the same diagram (Fig. 6.7).

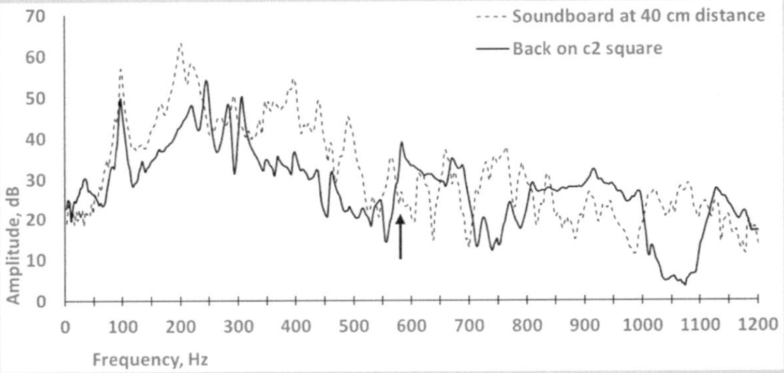

Fig. 6.7 A line graph comparing the frequency response (amplitude in dB vs. frequency in Hz) of a guitar soundboard recorded at 40 cm distance (dashed line) and the back of the guitar at the c2 square (located at the lower center of the back; Fig. 6.6) recorded at 5 mm distance (solid line). An arrow points to a peak around 584 Hz on the response curves

Based on the provided data, the frequency of approximately 584 Hz corresponds to the back's resonance activity at the c2 square. While the back generally exhibits less resonance activation compared to the soundboard (as evidenced by the weaker amplitudes across most frequencies), the 584 Hz frequency shows a noticeable response, suggesting a specific vibration mode of the back at this central location that even appears to enhance the soundboard's response at that frequency.

It's important to note that this analysis is based on a comparison of recordings taken with different microphone distances. However, based on the information provided, the 584 Hz peak on the back's frequency response is the most prominent indicator of its resonance activity at the c2 square, and it appears to be more pronounced than the soundboard's response at that specific frequency in the compared recordings.

6.3.3 Frequency Response in a Completed Instrument

A luthier can identify problems and their locations within a musical instrument by tapping its components and analyzing the resulting sounds. In contrast, a player evaluates an instrument's sound based on their personal taste and listening preferences, ultimately deciding whether to accept or reject it.

This highlights that beyond the information gained from tapping, a good musical instrument should possess a broad frequency response that accurately reproduces all the harmonics of the notes it can play. While some degree of variation in harmonic strength is expected—some frequencies may be weaker or stronger than others—a well-made instrument avoids complete muting any harmonics.

For example, the desired sound profile of the violin specifies which frequency ranges should have what effect on the instrument's overall sound quality [15].

Therefore, a detailed analysis of every note playable on a musical instrument, using FFT, is necessary to identify weakened or muted harmonics, overly strong harmonics, and any extraneous, undesirable frequencies that deviate from the expected harmonic vibrations of the strings. Following this analysis, the luthier must determine where on the instrument's body to make adjustments to correct these issues.

For instance, Table 6.1 provides a listing of the notes available on a classical guitar, ranging from the lowest note (open sixth string, E2) up to the note produced at the 20th fret of that 1st string (C6). When undertaking an analysis of a given instrument, it is recommended to initially construct such a table, incorporating several harmonics for each note.

Practical Exercise 6.3

Analyze the sound of the open 6th string of a classical guitar (E2, 82.4 Hz) recorded with a unidirectional microphone at a distance of 40 cm. Determine which of the resulting frequencies correspond to unwanted resonance.

Solution

To analyze the unwanted resonances of the open 6th string (nominally E2, 82.4 Hz), we recorded its sound at a distance of 40 cm and compared its frequency response to that of the soundboard (recorded at the same distance) and the back plate (recorded at 5 mm, information carried over from the previous exercise). For this measurement, the 6th string was tuned slightly higher to 83 Hz to prevent calculations resulting in decimal values (Fig. 6.8).

Table 6.1 Presents the scale notes and their corresponding fundamental frequencies (Hz) for a guitar fretboard in standard tuning (E2, A2, D3, G3, B3, E4)

1st string	2nd string	3rd string	4th string	5th string	6th string	Note	Frequency (Hz)
					Open	E_2	82.4
					1st	F_2	87.3
					2nd	$F\sharp_2/G\flat_2$	92.5
					3rd	G_2	98
					4th	$G\sharp_2/A\flat_2$	103.8
				Open	5th	A_2	110
				1st	6th	$A\sharp_2/B\flat_2$	116.5
				2nd	7th	B_2	123.5
				3rd	8th	C_3	130.8
				4th	9th	$C\sharp_3/D\flat_3$	138.6
			Open	5th	10th	D_3	146.8
			1st	6th	11th	$D\sharp_3/E\flat_3$	155.6
			2nd	7th	12th	E_3	164.8
			3rd	8th	13th	F_3	174.6
			4th	9th	14th	$F\sharp_3/G\flat_3$	185
		Open	5th	10th	15th	G_3	196
		1st	6th	11th	16th	$G\sharp_3/A\flat_3$	207.7
		2nd	7th	12th	17th	A_3	220
		3rd	8th	13th	18th	$A\sharp_3/B\flat_3$	233.1
	Open	4th	9th	14th	19th	B_3	246.9
	1st	5th	10th	15th	20th	C_4	261.6
	2nd	6th	11th	16th		$C\sharp_4/D\flat_4$	277.2
	3rd	7th	12th	17th		D_4	293.7
	4th	8th	13th	18th		$D\sharp_4/E\flat_4$	311.1
Open	5th	9th	14th	19th		E_4	329.6
1st	6th	10th	15th	20th		F_4	349.2
2nd	7th	11th	16th			$F\sharp_4/G\flat_4$	370
3rd	8th	12th	17th			G_4	392
4th	9th	13th	18th			$G\sharp_4/A\flat_4$	415.3
5th	10th	14th	19th			A_4	440
6th	11th	15th	20th			$A\sharp_4/B\flat_4$	466.2
7th	12th	16th				B_4	493.9
8th	13th	17th				C_5	523.3
9th	14th	18th				$C\sharp_5/D\flat_5$	554.4
10th	15th	19th				D_5	587.3

(continued)

Table 6.1 (continued)

1st string	2nd string	3rd string	4th string	5th string	6th string	Note	Frequency (Hz)
11th	16th	20th				D♯5/E♭5	622.3
12th	17th					E5	659.3
13th	18th					F5	698.5
14th	19th					F♯5/G♭5	740
15th	20th					G5	784
16th						G♯5/A♭5	830.6
17th						A5	880
18th						A♯5/B♭5	932.3
19th						B5	987.8
20th						C6	1046.5

The table is organized with columns for each of the six strings (1st to 6th), indicating the fret number (from open to the 20th fret). Each cell within the string columns shows the fret number played. Adjacent columns display the resulting musical note and its fundamental frequency in Hz. The frequencies range from 82.4 Hz (open 6th string, E2) to 1046.5 Hz (20th fret of the 1st string, C6). The table illustrates the chromatic scale across the guitar's range. Creating such a table, including overtones, when analyzing a specific instrument is recommended

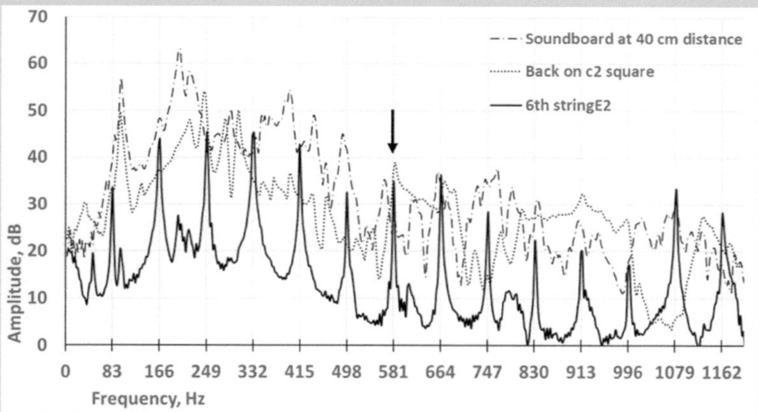

Fig. 6.8 Line graph displaying the frequency spectrum of the open 6th string (≈ 83 Hz). The graph emphasizes the presence of unwanted resonances, including the Helmholtz resonance at approximately 97 Hz and potential soundboard/back resonances around 199 and 220 Hz. The enhanced 7th harmonic due to back plate resonance at 581 Hz is also pointed by an arrow

- **Harmonic series observation**: The overtones of the 83 Hz fundamental frequency generally appear at integer multiples of 83 Hz, indicating a relatively clean harmonic series. However, the amplitude relationships between these harmonics warrant closer examination for potential influences from body resonances.

- **Fundamental and first harmonic**: Interestingly, the first harmonic (83 Hz) exhibits a lower amplitude than subsequent harmonics. It is hypothesized that the soundhole resonance might be influencing the fundamental frequency's projection. Shifting the Helmholtz resonance slightly lower e.g., by adding a collar or reducing the soundhole diameter [2, 6, 16] could potentially enhance the fundamental's amplitude and create a better balance within the harmonic series.
- **Helmholtz resonance (97 Hz)**: A distinct peak at approximately 97 Hz, located between the fundamental (83 Hz) and the second harmonic (166 Hz), strongly suggests the presence of the Helmholtz air resonance of the soundhole. This frequency does not align harmonically with the open 6th string and represents an unwanted vibration mode. Lowering the Helmholtz frequency, as suggested, could potentially mitigate this unwanted resonance in the 6th string's spectrum. It's crucial to be mindful of the "wolf note" phenomenon if the Helmholtz resonance is shifted too close to the string's fundamental or other strong body resonances.
- **Soundboard and back resonances (199 and 220 Hz)**: The presence of non-harmonic peaks at approximately 199 and 220 Hz, coinciding with previously identified resonance modes of the soundboard and back respectively, indicates an undesirable coupling. These body resonances are being excited by the 6th string's vibration. To address these, introducing localized mass near the antinodes of these vibration modes on the soundboard or back could lower their frequencies (Fig. 5.17, Chap. 5). Conversely, increasing stiffness near the nodes could raise their frequencies, potentially moving them away from the stronger harmonics of the 6th string. Given that the amplitudes of these discordant resonances appear weaker than the surrounding second and third harmonics of the string, their audibility might be minimal, but addressing them could improve the overall tonal clarity.
- **Higher order harmonics**: Beyond the third harmonic, no significant, isolated unwanted frequencies with substantial amplitudes are readily apparent. The observed peaks generally align with the higher harmonics of the 83 Hz fundamental, albeit with varying amplitudes likely influenced by complex body resonances.
- **Back plate resonance ınfluence (581 Hz)**: The significant increase in the amplitude of the seventh harmonic (around 581 Hz) is a direct consequence of its near-perfect alignment with the back plate's identified resonance frequency of approximately 583 Hz. This sympathetic vibration of the back plate reinforces this specific harmonic, potentially altering the overall timbre of the note. Even though this strong emphasis on a particular harmonic isn't

necessarily 'unwanted' in a dissonant way, it's a characteristic of the instrument's construction that a luthier might consider when aiming for a specific tonal quality.

In summary, the primary unwanted resonances identified for the open 6th string are the Helmholtz resonance at 97 Hz and the excitation of the soundboard and back resonances at approximately 199 and 220 Hz. The strong reinforcement of the seventh harmonic due to the back plate's resonance at 581 Hz is also a significant observation regarding the instrument's tonal characteristics.

6.4 Sound Color Testing

The ADSR envelope of the modes, beside the frequency spectrum and their amplitude relationships between the harmonics, plays an important role in shaping the timbre and color of a sound. This envelope, an acronym for Attack–Decay–Sustain–Release, defines how the amplitude of a sound evolves over time, significantly influencing its perceived characteristics. Remember that each stage of the ADSR envelope contributes distinctly to the timbre (Fig. 6.9).

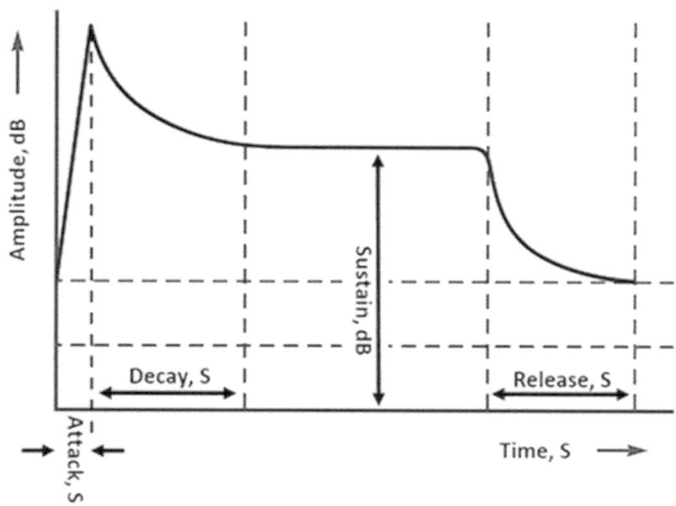

Fig. 6.9 A graph illustrates the ADSR (Attack, Decay, Sustain and Release) envelope, specifically adapted for bowed string instruments and similar sustained-tone instruments [11]. The horizontal axis represents Time in seconds (S), and the vertical axis represents Amplitude in decibels (dB). The graph shows how the amplitude of a sound evolves over time

- **Attack**: This stage determines the speed at which the sound's amplitude rises from silence to its peak. A rapid attack creates a percussive, sharp sound, as heard in pianos or percussion instruments. Conversely, a slow attack produces a softer, gradual onset, characteristic of violins or wind instruments.
- **Decay**: Following the peak, the amplitude decreases during the decay stage. The rate and extent of this decrease affect the sound's perceived dynamism and energy.
  In real-world scenarios, sound attenuation within the decay region exhibits a curve analogous to that seen in the logarithmic representation of damping capacity. Logarithmic interpretation allows for clear observation of these regions, removing the influence of sound attenuation rate variations in loud sounds, from the onset of the vibration wave to its decay.
- **Sustain**: This stage represents a relatively constant amplitude level held until the note is released. The sustain level determines the sound's volume and steadiness.
- **Release**: Finally, upon note termination, the amplitude fades to silence during the release stage. The duration of this stage defines the length of the sound's fade-out, influencing the perceived ending of the note.
  However, in practical examples, in the release region, the vibration sound diminishes to such a degree that its progression becomes indistinguishable unless the experiment is performed in an entirely silent environment, devoid of all forms of ambient noise.

The combination of these four stages allows for a wide range of timbral variations. For instance, subtractive synthesizers utilize the ADSR envelope to control filter cutoff frequency and main gain, both of which significantly affect the sound's timbre [21].

Furthermore, the ADSR envelope is applicable not only to the fundamental frequency but also to the harmonics and overtones of a sound. The fundamental frequency's ADSR envelope shapes the overall amplitude contour of the note [3], while individual ADSR envelopes for harmonics and overtones can modulate their prominence over time, contributing to the sound's complexity and richness. Adjusting the ADSR envelope of individual harmonics alters the amplitude of each harmonic over time. For instance, if the second harmonic has a quicker attack than the fundamental, the note's initial sound will emphasize the second harmonic.

The fundamental harmonic, which defines the note's pitch, significantly influences the overall timbre of the note. Therefore, the ADSR envelope of the fundamental harmonic is crucial in shaping the note's contour and is particularly important. Parameters of the fundamental harmonic's ADSR envelope, such as attack and decay, greatly affect how we perceive the sound. A rapid attack, for example, creates a sharp, percussive sound, whereas a gradual attack produces a gentler, smoother sound.

While harmonics are integer multiples of the fundamental frequency, overtones encompass all frequency components present in the sound, including both harmonic and non-harmonic frequencies. The ADSR envelope can control the amplitude of each overtone over time (Fig. 6.10).

According to Hu and Liu [8], the changes of short-term feature in a musical note can reflect the timbre of musical instrument.

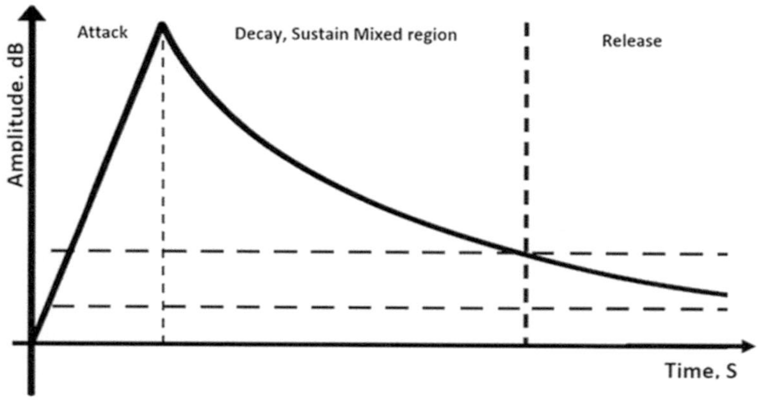

Fig. 6.10 A graph illustrates the ADSR (Attack, Decay, Sustain and Release) envelope, specifically adapted for plucked string instruments [11]. The horizontal axis represents Time in seconds (S), and the vertical axis represents Amplitude in decibels (dB). The graph shows how the decay and sustain phases mix together and the amplitude of a sound evolves over time

Employing a logarithmic scale in ADSR analysis offers several advantages, especially when analyzing plucked string instruments like the guitar or hammered dulcimer.

In these instruments, sound is produced by plucking or striking strings, resulting in a rapid attack followed by a gradual decay. Unlike instruments capable of sustaining a steady sound, plucked strings inherently have a limited sustain. A logarithmic representation similar to what was introduced for damping capacity in Chap. 4, can aid in a more detailed analysis of this decay, comparable to the damping capacities of wood untilized for resonator making.

On the other hand, the decay and sustain stages may not be linear in plucked instruments (Fig. 6.10). In louder sounds, the decibel reduction is initially faster, gradually slowing down. A logarithmic interpretation can represent this non-uniform decay more uniformly, facilitating sound analysis and comparison.

Meanwhile, human hearing perceives sound intensity logarithmically. Therefore, a logarithmic representation of the ADSR envelope may provide a more accurate correlation between the sound's acoustic properties and our perception of it [7].

The logarithmic approach is frequently (not always) employed when analyzing the ADSR (Attack, Decay, Sustain, Release) envelope in sound synthesis. While Liang et al. [11] does not explicitly discuss this logarithmic interpretation of the ADSR envelope, logarithms were used in audio signal processing, including energy envelope calculations, of which ADSR is a component. Similarly, Da Fontoura Costa [3] mentions the common practice of representing musical note frequencies on a logarithmic scale, Genussov and Cohen [7] utilize a logarithmic scale for frequency representation, a concept indirectly related to ADSR. This representation of frequencies using a logarithmic scale suggests that analyzing sound properties, such as the amplitude changes described by the ADSR envelope, logarithmically may better align with human sound perception. Hu and Liu [8] specifically state that a close relationship exists between the ADSR envelope and the logarithmic frequency scale, noting that ADSR properties are calculated based on the zero-beam frequency scale, which they emphasize is strictly logarithmic.

Although Charef and Ifeachor [1] do not directly address the logarithmic interpretation of ADSR, the existence of ADSR envelope generators indicates that the ADSR concept is considered within sound synthesis systems. Logarithmic scales are prevalent in many applications, particularly in audio signal processing, for analyzing and representing sound properties like amplitude. Marozeau and Cheveigné [13] utilized logarithmic interpretation for spectral shaping of stimuli, demonstrating the presence of both ADSR concepts and logarithmic frequency representation in their research. Furthermore, the internal friction index, representing the damping capacity of wood and related to the decay and sustain of wood vibration, is conventionally interpreted logarithmically [19]. Therefore, creating a diagram of the ADSR envelope that incorporates methodologies and relationships for measuring logarithmic decrement and sound attenuation could provide a new avenue for comparative research on both raw wood and finished instruments.

The intensity with which a plucked stringed instrument, such as a guitar, is played significantly affects the sound's initial amplitude and, consequently, the shape of its ADSR envelope. This makes it challenging to accurately compare and analyze the sound's decay, which in these instruments includes the sustain region [11]. A louder sound will exhibit a different decay pattern than a softer one.

So, to address this, it is proposed to analyze the decay, incorporating the sustain region, using a logarithmic approach focused on measuring the logarithmic rate of amplitude decay (or sustain). Instead of measuring the linear decrease in amplitude over time, this method measures the relative amplitude decay, similar to the logarithmic decay calculation for damping capacity discussed in Chap. 4.

By focusing on the relative rate of amplitude decay, the influence of variations in initial amplitude due to playing intensity (hard or soft) is reduced. In essence, the logarithmic attenuation rate can be similar for loud and quiet sounds, even with differing initial amplitudes. This method may offer a more accurate understanding of an instrument's inherent decay characteristics, independent of playing style variations.

While not yet common, using logarithmic interpretation in sound decay analysis is a promising approach, especially for stringed instruments like guitars where initial intensity varies considerably. This method can help extract more precise and meaningful information from ADSR data, leading to a better understanding of the instruments' acoustic properties.

We examined the ADSR (Attack, Decay, Sustain and Release) envelope in stringed instruments in detail to observe the time-domain characteristics of these processes and to distinguish the harmonics within a complex sound. However, many researchers apply ADSR analysis to whole composite sounds without considering individual harmonics (Fig. 6.11), using it primarily for comparative purposes [3, 9, 11].

In Chap. 4, within the discussion of damping, we noted that damping capacity is sometimes expressed using the quality factor, Q, defined as the inverse of tanδ. We

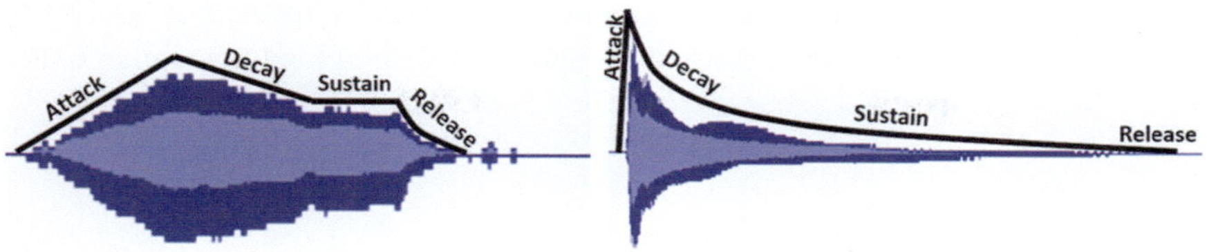

Fig. 6.11 Two graphs compare the ADSR envelopes of a violin (left, bowed) and a guitar (right, plucked) playing an open G3 string (196 Hz) in a full composite sound without eliminating none of the harmonic frequencies. The violin shows a gradual decay and longer sustain (depends on bowing length), while the guitar exhibits a rapid decay mixed with a relatively short sustain, illustrating the different sound characteristics produced by bowing versus plucking. The schematic representations by Liang et al. [11] are similar to these real sound envelopes

suggest that, in cases such as plucked string instruments where decay and sustain are intertwined, the parameter Q (quality factor or sound vibration duration) be used instead of decay and sustain. Furthermore, research has shown that distinguishing between decay and sustain regions, or even identifying the release phase, can be challenging due to offset time ambiguity [11]. Therefore, we recommend calculating the quality factor in these situations (Fig. 6.12).

$$\frac{1}{Q} = \frac{1}{n\pi} Ln \frac{A_i}{A_{i+n}} \tag{6.1}$$

where, Q is the quality factor, n is the number of oscillation peaks after the A_i, A_i or partcularly A_1 and A_{i+n} are the peak height of the sine wave of vibration where a region starts and ends.

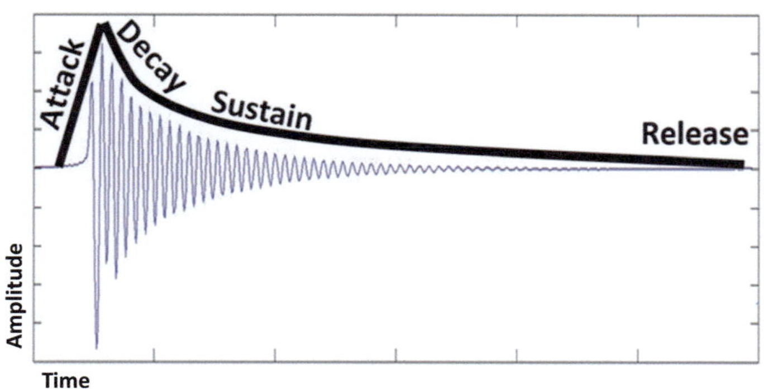

Fig. 6.12 The ADSR envelopes of a guitar playing an open G3 string (196 Hz) in its fundamental monochrome harmonic frequency. Images from NDT-Lab® software [18]

This equation closely resembles Eq. (4.13), which illustrates an overall linear decrease or increase in amplitudes. However, sophisticated software applications may utilize Eq. (4.12) to detect amplitude changes at each individual point and then display these changes as a series of connected points [18, 19].

6.5 Further Exercises

- Determine the changes in frequency and acoustic response of stringed instruments resulting from repositioning an added mass on different parts (top, back, sidewall, pegbox, etc.).
- Determine the changes in frequency and acoustic response of stringed instruments caused by manipulating bridge mass and stiffness, and by altering string force.
- If sound velocity evaluator devices are accessible, measure and compare sound velocity in the wood of completed instrument components.

Any modification you can implement that alters the mass reactance or stiffness reactance of the wood—whether across the entire piece, within specific mode shapes, or at the boundaries where it contacts adjacent components—can positively or negatively shift the frequencies of its associated modes. Experiment with these modifications in the instrument you are building, meticulously recording the results in your personal database. Through this systematic approach, you will progressively develop an extensive understanding of tonewoods for musical instruments.

Then, it is necessary to measure the stiffness, density, and other acoustic parameters of any wood intended for musical instrument construction. Utilizing wood with appropriate properties for each component is essential to avoid complications during treatment and tuning. Given the rapid advancements in knowledge and technology, it is highly recommended to allocate budget for and employ the latest technologies to evaluate the acoustic properties of wood—before, during, and after the instrument-making process, culminating in the final, tuned instrument.

6.6 Final Words

This chapter has taken us through the intricate world of acoustic testing for tonewood in finished wooden stringed musical instruments. In this journey, we've explored the complexities of defining good sound, the important acoustic parameters that shape an instrument's sound, and the increasing role of technology in sound analysis.

It's clear that the creation of a truly exceptional musical instrument is a delicate balance of art and science. While the experience and intuition of master builders remain fundamental, the ability to measure and analyze sound with precision offers a powerful tool for both understanding and refining instrument acoustics.

We've established that good sound is not merely a matter of personal taste. While individual preferences certainly play a role, there are underlying acoustic principles that contribute to an instrument's quality and playability. These principles include:

- Resonance and richness.
- Clarity and definition.

These acoustic characteristics, along with factors like responsiveness and balance, contribute to an instrument's ability to convey musical expression and emotion.

This chapter emphasizes the growing importance of technology in this field. Sound analysis software and advanced measurement tools provide luthiers and researchers with objective data, allowing for more informed decisions in instrument design and construction. This technology complements traditional craftsmanship, enabling a deeper understanding of the complex interplay between an instrument's physical properties and its sonic output.

Furthermore, the chapter highlights the significance of frequency response analysis in evaluating instrument sound. By examining the distribution of sound energy across different frequencies, luthiers can identify areas for improvement and fine-tune the instrument's tonal characteristics.

In closing, this chapter, the final chapter of this book underscores that the pursuit of exceptional instrument sound is an ongoing journey of discovery. It requires a blend of artistic sensitivity, scientific rigor, and a deep appreciation for the materials and craftsmanship that bring music to life.

References

1. B.H. Charef, E. Ifeachor, Artificial intelligence based modeling of musical instruments, in *Proceedings of the 2nd COST G-6 Workshop on Digital Audio Effects (DAFx99)* (1999), pp. 9–11
2. O. Christensen, B.B. Vistisen, Simple model for low-frequency guitar function. J. Acoust. Soc. Am. **68**(3), 758–766 (1980). https://doi.org/10.1121/1.384814
3. L. Da Fontoura Costa, On sound synthesis: single note, single frequency (CDT-42), in *On Sound Synthesis I: Single Note, Single Frequency (CDT-42)* (São Carlos Institute of Physics—DFCM/USP, 2020)
4. E. Değirmenli, Türk Müziği Çalgılarından 'Ud'da Ses Oluşumunun İncelenmesi ve Telli Çalgıların Ses Karakteri Açısından Tasarımlarının Belirlenmesine Dair Yöntem Önerisi (Doktora tezi). Gazi Üniversitesi Güzel Sanatlar Enstitüsü Türk Müziği Anabilim Dalı (2018)
5. E. Değirmenli, Titreşim ve ses yayınım analizleri kullanılarak telli çalgılardaki tınısal niteliklerin incelenmesi: Ud örneği. İdil Sanat ve Dil Dergisi **12**(104), 525–534 (2023)
6. A. Fernandez, *The Relationship Between Resonant Frequency, Sound Hole Diameter, and Body Depth in Acoustic Guitars*. Master's thesis, Purdue University (2023). https://www.proquest.com/openview/b7f446f130b7c93837f3ba33bbd356b2/1?cbl=18750&diss=y&pq-origsite=gscholar
7. M. Genussov, I. Cohen, Multiple fundamental frequency estimation based on sparse representations in a structured dictionary. Digital Signal Process. **23**(1), 390–400 (2013). https://doi.org/10.1016/j.dsp.2012.08.012

8. Y. Hu, G. Liu, Dynamic characteristics of musical note for musical instrument classification, in *2011 IEEE International Conference on Signal Processing, Communications and Computing (ICSPCC)*, Xi'an, China (2011), pp. 1–6. https://doi.org/10.1109/ICSPCC.2011.6061734
9. F. Leccese, G. Salvadori, G. Bernardini, P. Bernardini, The bowed string instruments: acoustic characterization of unique pieces from the Italian lutherie. IOP Conf. Ser. Mater. Sci. Eng. **364**(1), 012022 (2018). https://doi.org/10.1088/1757-899X/364/1/012022
10. D.J. Levitin, S. McAdams, R.L. Adams, Control parameters for musical instruments: a foundation for new mappings of gesture to sound. Organised Sound **7**(2), 171–189 (2002). https://doi.org/10.1017/S135577180200208X
11. C.Y. Liang, L. Su, Y.H. Yang, H.M. Lin, Musical offset detection of pitched instruments: the case of violin, in ISMIR (2015), pp. 281–287
12. H. Mansour, G. Scavone, V. Freour, A comparison of vibration analysis techniques applied to the Persian setar. Acoustics 2012, Nantes, France ⟨hal-00810675⟩ (2012)
13. J. Marozeau, A. de Cheveigné, The effect of fundamental frequency on the brightness dimension of timbre. J. Acoust. Soc. Am. **121**(1), 383–387 (2007). https://doi.org/10.1121/1.2384910
14. M. Meincken, G. Roux, T. Niesler, An African violin—the feasibility of using indigenous wood from southern Africa as tonewood. South African J. Sci. **117**(11/12) (2021). https://doi.org/10.17159/sajs.2021/11175
15. H. Meinel, Regarding the sound quality of violins and a scientific basis for violin construction. J. Acoust. Soc. Am. **29**(7), 817–822 (1957). https://doi.org/10.1121/1.1909064
16. G. Nicoletti, Measuring and tuning the performances of the acoustic guitar (2022). https://www.iuliusguitars.com/wp-content/uploads/2025/03/Measuring_Eng_060222.pdf.
17. M. Rodger, P. Stapleton, M. van Walstijn, M. Ortiz, L.S. Pardue, What makes a good musical instrument? A matter of processes, ecologies and specificities, in *Proceedings of the International Conference on New Interfaces for Musical Expression* (2020), pp. 405–410. https://doi.org/10.5281/zenodo.4813438
18. M. Roohnia, NDT-LAB; System to evaluate the mechanical properties of wood. Iranian Patent, IR-Patent No. 44032/22–08–1386 (2007)
19. M. Roohnia, Wood: vibration and acoustic properties, in *Reference Module in Materials Science and Materials Engineering* (Elsevier, Oxford, 2019), pp. 1–13. https://doi.org/10.1016/B978-0-12-803581-8.01996-2
20. J.R. Sank, Microphones. J. Audio Eng. Soc. **33**(7/8), 514–547 (1985)
21. C. Vahidi, G. Fazekas, C. Saitis, A. Palladini, Timbre space representation of a subtractive synthesizer. Timbre (2020). https://doi.org/10.48550/arXiv.2009.11706
22. J. Woodhouse, On the "bridge hill" of the violin. Acta Acust. Acust. **91**(1), 155–165 (2005)

MIX
Papier aus verantwortungsvollen Quellen
Paper from responsible sources
FSC® C105338

If you have any concerns about our products,
you can contact us on
ProductSafety@springernature.com

In case Publisher is established outside the EU,
the EU authorized representative is:
Springer Nature Customer Service Center GmbH
Europaplatz 3, 69115 Heidelberg, Germany

Printed by Libri Plureos GmbH
in Hamburg, Germany